NONNATIVE OYSTERS IN THE CHESAPEAKE BAY

Committee on Nonnative Oysters in the Chesapeake Bay

Ocean Studies Board

Division on Earth and Life Studies

NATIONAL RESEARCH COUNCIL
OF THE NATIONAL ACADEMIES

THE NATIONAL ACADEMIES PRESS
Washington, D.C.
www.nap.edu

THE NATIONAL ACADEMIES PRESS 500 Fifth Street, N.W. Washington, DC 20001

NOTICE: The project that is the subject of this report was approved by the Governing Board of the National Research Council, whose members are drawn from the councils of the National Academy of Sciences, the National Academy of Engineering, and the Institute of Medicine. The members of the committee responsible for this report were chosen for their special competences and with regard for appropriate balance.

This study was supported by a contract between the National Academies and the following entities: Contract No. 50-DGNA-1-90024 from the U.S. Environmental Protection Agency, Contract No. 50-DGNA-1-90024 from the National Oceanic and Atmospheric Administration, Cooperative Agreement No. 50181-2-J030 from the U.S. Fish and Wildlife Service, Grant from the National Fish and Wildlife Foundation, Grant from the Maryland Department of Natural Resources, Purchase Order No. D227024-4 from the Virginia Sea Grant, Grant No. NA070Z0136-01 from the Virginia Department of Environmental Quality, Grant No. SA7293808L.SA7528018F from the Maryland Sea Grant, and Purchase Order No. D227024-4 from the Connecticut Sea Grant. Any opinions, findings, conclusions, or recommendations expressed in this publication are those of the authors and do not necessarily reflect the views of the organizations or agencies that provided support for the study.

This study was funded in part by a contract from the National Oceanic and Atmospheric Administration. The views expressed herein are those of the author(s) and do not necessarily reflect the views of NOAA or any of its sub agencies.

Library of Congress Catalog Number: 2003117095

International Standard Book Number: 0-309-09052-0 (Book)
International Standard Book Number: 0-309-52696-5 (PDF)

Additional copies of this report are available from the National Academies Press, 500 Fifth Street, N.W., Lockbox 285, Washington, DC 20055; (800) 624-6242 or (202) 334-3313 (in the Washington metropolitan area); Internet, http://www.nap.edu

Printed in the United States of America

THE NATIONAL ACADEMIES

Advisers to the Nation on Science, Engineering, and Medicine

The **National Academy of Sciences** is a private, nonprofit, self-perpetuating society of distinguished scholars engaged in scientific and engineering research, dedicated to the furtherance of science and technology and to their use for the general welfare. Upon the authority of the charter granted to it by the Congress in 1863, the Academy has a mandate that requires it to advise the federal government on scientific and technical matters. Dr. Bruce M. Alberts is president of the National Academy of Sciences.

The **National Academy of Engineering** was established in 1964, under the charter of the National Academy of Sciences, as a parallel organization of outstanding engineers. It is autonomous in its administration and in the selection of its members, sharing with the National Academy of Sciences the responsibility for advising the federal government. The National Academy of Engineering also sponsors engineering programs aimed at meeting national needs, encourages education and research, and recognizes the superior achievements of engineers. Dr. Wm. A. Wulf is president of the National Academy of Engineering.

The **Institute of Medicine** was established in 1970 by the National Academy of Sciences to secure the services of eminent members of appropriate professions in the examination of policy matters pertaining to the health of the public. The Institute acts under the responsibility given to the National Academy of Sciences by its congressional charter to be an adviser to the federal government and, upon its own initiative, to identify issues of medical care, research, and education. Dr. Harvey V. Fineberg is president of the Institute of Medicine.

The **National Research Council** was organized by the National Academy of Sciences in 1916 to associate the broad community of science and technology with the Academy's purposes of furthering knowledge and advising the federal government. Functioning in accordance with general policies determined by the Academy, the Council has become the principal operating agency of both the National Academy of Sciences and the National Academy of Engineering in providing services to the government, the public, and the scientific and engineering communities. The Council is administered jointly by both Academies and the Institute of Medicine. Dr. Bruce M. Alberts and Dr. Wm. A. Wulf are chair and vice chair, respectively, of the National Research Council.

www.national-academies.org

COMMITTEE ON NONNATIVE OYSTERS IN THE CHESAPEAKE BAY

JAMES ANDERSON (*Cochair*), University of Rhode Island, Kingston
DENNIS HEDGECOCK (*Cochair*), University of California, Davis
MARK BERRIGAN, Florida Department of Agriculture and Consumer Services, Division of Aquaculture, Tallahassee
KEITH CRIDDLE, Utah State University, Logan
WILLIAM DEWEY, Taylor Shellfish Company, Inc., Shelton, Washington
SUSAN FORD, Rutgers University, Port Norris, New Jersey
PHILIPPE GOULLETQUER, IFREMER, Paris, France
RICHARD HILDRETH, University of Oregon, Eugene
MICHAEL PAOLISSO, University of Maryland, College Park
NANCY TARGETT, University of Delaware, Lewes
ROBERT WHITLATCH, University of Connecticut, Groton

Staff

SUSAN ROBERTS, Study Director
KIM WADDELL, Senior Program Officer
SUSAN PARK, Christine Mirzayan Science and Technology Policy Intern
DENISE GREENE, Senior Project Assistant
SARAH CAPOTE, Project Assistant

The work of this committee was overseen by the Ocean Studies Board.

Acknowledgments

The committee would like to acknowledge the Scientific and Technical Advisory Committee of the Chesapeake Bay Program (STAC) for co-sponsoring the Workshop on Nonnative Oysters in the Chesapeake Bay, held on October 7-8, 2002, in Fredericksburg, VA (see Appendix H), especially the efforts of Kevin Sellner (executive secretary) and Melissa Bugg (staff). This report was greatly enhanced by the participants at the first meeting and workshop held as part of this study. The committee would like to acknowledge the efforts of those who gave presentations at the meetings. These talks helped set the stage for fruitful discussions in the closed sessions that followed.

Stan Allen, Virginia Institute of Marine Science
Jim Andreasen, U.S. Environmental Protection Agency
Lowell Bahner, National Oceanic and Atmospheric Administration
Gene Burreson, Virginia Institute of Marine Science
Frank Dawson, Maryland Department of Natural Resources
Mike Fritz, U.S. Environmental Protection Agency
Pat Gaffney, University of Delaware
Bill Goldsborough, Chesapeake Bay Foundation
Eric Hallerman, Virginia Polytechnic Institute and State University
Carl Hershner, Virginia Institute of Marine Science
Eileen Hofmann, Old Dominion University
Bob Hume, U.S. Army Corps of Engineers
Chris Judy, Maryland Department of Natural Resources

Tom Kellum, W. E. Kellum Seafood
Victor Kennedy, University of Maryland
Fred Kern, National Oceanic and Atmospheric Administration
Doug Lipton, University of Maryland
Mark Luckenbach, Virginia Institute of Marine Science
Roger Mann, Virginia Institute of Marine Science
Laura McKay, Virginia Department of Environmental Quality
Frederika Moser, Maryland Sea Grant
Roger Newell, University of Maryland
Ken Paynter, University of Maryland
Karen Rivara, East Coast Shellfish Growers Association
Larry Simns, Maryland Watermen's Association
Julie Thompson, U.S. Fish and Wildlife Service
Casey Todd, Metompkin Bay Oyster
Jack Travelstead, Virginia Marine Resources Commission
George Washington, Virginia Watermen's Association

The committee is also grateful to a number of people who provided important material for this report: **Stan Allen, Atlantic States Marine Fisheries Commission, Katherine Holt, Lisa Kline, Charles Peterson, and the Chesapeake Bay Program Scientific and Technical Advisory Committee office.**

This report has been reviewed in draft form by individuals chosen for their diverse perspectives and technical expertise, in accordance with procedures approved by the National Research Council's Report Review Committee. The purpose of this independent review is to provide candid and critical comments that will assist the institution in making its published report as sound as possible and to ensure that the report meets institutional standards for objectivity, evidence, and responsiveness to the study charge. The review comments and draft manuscript remain confidential to protect the integrity of the deliberative process. We wish to thank the following individuals for their participation in the review of this report:

Michael Beck, The Nature Conservancy, Santa Cruz, California
Neil Bourne, Department of Fisheries and Oceans, Nanaimo, Canada
Eugene Burreson, Virginia Institute of Marine Science, Gloucester Point
Edwin Grosholz, University of California, Davis
Eric Hallerman, Virginia Polytechnic Institute and State University
Walter Keithly, Jr., Louisiana State University, Baton Rouge
Chris Langdon, Oregon State University, Newport

Bonnie McCay, Rutgers University, New Brunswick, New Jersey
Charles Peterson, University of North Carolina, Morehead City
Edella Schlager, University of Arizona, Tucson
John Supan, Louisiana State University Sea Grant, Baton Rouge

Although the reviewers listed above provided many constructive comments and suggestions, they were not asked to endorse the conclusions or recommendations nor did they see the final draft of the report before its release. The review of this report was overseen by **Judith McDowell**, Woods Hole Oceanographic Institution, appointed by the Division on Earth and Life Studies and **May Berenbaum**, Univerisity of Illinois, appointed by the Report Review Committee, who were responsible for making certain that an independent examination of this report was carried out in accordance with institutional procedures and that all review comments were carefully considered. Responsibility for the final content of this report rests entirely with the authoring committee and the institution.

Contents

*Appendixes D through G are found on the CD-Rom attached to the inside back cover.

NONNATIVE OYSTERS IN THE CHESAPEAKE BAY

1

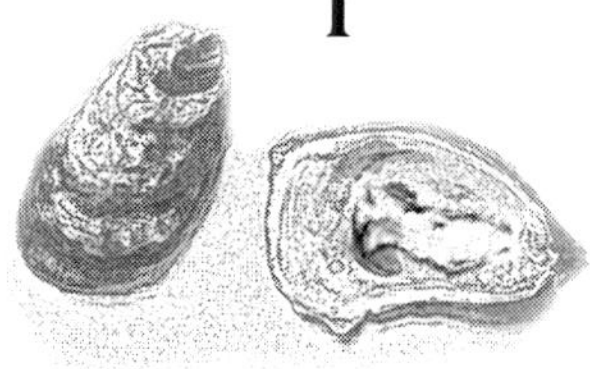

Executive Summary

Are nonnative oysters a potential solution or problem? This simplistic question frames the extremes of opinion over a complex and controversial issue that has embroiled the Chesapeake Bay region ever since the state of Virginia proposed introducing a nonnative oyster from Asia to revive the oyster industry. *Crassostrea ariakensis,* commonly known as the Suminoe oyster, is native to regions in China and Japan but is mostly unfamiliar to oyster growers and consumers in North America. Relatively little is known about the Suminoe oyster, and this has made it difficult for scientists and resource managers to decide whether this oyster has the potential to help or to hurt conditions in the Chesapeake Bay, either for the watermen or for the bay ecosystem. Hence, opinions range from the hope that the Suminoe oyster will revive a threatened industry and restore some of the filtering capacity of the original oyster population to the fear that it will be an invader that outgrows the commercial demand for oysters and upsets the ecology of the Chesapeake Bay.

EXPECTATIONS AND PERSPECTIVE

Declines in the quality and quantity of natural resources in the Chesapeake Bay have taken place over at least the past 150 years. Today, we stand at an unprecedented crossroads, facing a conscious decision to introduce or not to introduce a nonnative oyster in the hope that this action will improve prospects for both the fisheries and the environment. It is unrealistic to expect that this long-term degradation in the Chesapeake Bay could be reversed in 10 years or less through any single management action. The degradation of water quality, increased sedimentation, loss of habitats such as sea grass beds and oyster reefs, and changes in the abundances of many plant and animal species not only serve as indicators of change but also act as barriers to restoration. Improvements in the natural resources of the bay will require a multifaceted approach and sustained commitment from communities throughout the watershed toward the goal of reestablishing some of the ecological functions that have been valued for generations.

Loss of the Native Oyster

The indigenous oyster *C. virginica* has been depleted to less than 1% of its original abundance in the Chesapeake Bay through a combination of heavy fishing pressure during the 19th and 20th centuries and the recent high mortalities due to the spread of two parasites, *Haplosporidium nelsoni* and *Perkinsus marinus,* that cause the diseases MSX and Dermo, respectively. Additionally, nutrient and toxic pollutants, increased sedimentation, and loss of shell bed habitat have made the environment in the bay less conducive to recovery of the oyster population. This combination of factors has threatened the survival of the oyster industry in Virginia and Maryland. As a possible means to address this problem, Virginia has been exploring the option of introducing a nonnative Asian oyster (*C. ariakensis*) into the state's coastal waters, including the Chesapeake Bay.

Opinions on the likely risks and benefits of introducing a nonnative oyster differ among the states and federal agencies that participate in regional agreements through the Chesapeake Bay Program (CBP). Because of the high stakes associated with the decision to introduce a nonnative species, the Chesapeake Bay Commission (a tristate commission that serves as the legislative arm of the CBP) requested that the National Research Council undertake a study of the pros and cons of introducing *C. ariakensis* either as an infertile triploid (triploid oysters contain three rather than the normal two sets of chromosomes and cannot reproduce normally) for use in aquaculture or as a reproductive diploid that could either augment or supplant the diseased populations of the native oyster (see Box 1.1). In the past, introductions of nonnative species were not subjected to this level of scrutiny, but rising awareness of the potential ecological and economic problems associated with invasive nonnative species has made resource managers more cautious. Thus, this study presents a landmark opportunity to identify concerns that should be addressed by decision makers when the introduction of a nonnative marine species is under consideration.

BOX 1.1
Statement of Task

This study will examine the ecological and socioeconomic risks and benefits of open-water aquaculture or direct introduction of the nonnative oyster *Crassostrea ariakensis* in the Chesapeake Bay. The committee will address how *C. ariakensis* might affect the ecology of the bay, including effects on native species, water quality, habitat, and the spread of human and oyster diseases. Possible effects on recovery of the native oyster *C. virginica* will be considered. The potential range and effects of the introduced oyster will be explored, both within the bay and in neighboring coastal areas. The study will investigate the adequacy of existing regulatory and institutional frameworks to monitor and oversee these activities.

The committee will assess whether the breadth and quality of existing research, on oysters and other introduced species, are sufficient to support risk assessments of three management options: (1) no use of nonnative oysters, (2) open-water aquaculture of triploid oysters, and (3) introduction of reproductive diploid oysters. Where current knowledge is inadequate, the committee will recommend additional research priorities.

How Might *C. ariakensis* Affect the Ecology of the Chesapeake Bay?

The potential effects of the Suminoe oyster on the ecology of the Chesapeake Bay may be evaluated based on past experiences with introduced species, both deliberate and accidental, and on a comparison of the biological characteristics of the nonnative oyster with the native oyster.

The Pacific oyster *C. gigas* is the most frequently introduced oyster species. *C. gigas* is native to Japan, China, and Korea but is now the principal oyster harvested worldwide, having been introduced to all continents except Antarctica. Most of the problems associated with previous introductions of nonnative oysters arose from the cointroduction of other marine pest species. Shipments of Pacific oysters contained oyster parasites, pathogens, and predators as well as aquatic plants used as packing material. The risk of introducing a harmful "hitchhiking" species can be greatly reduced through application of protocols adopted by the International Council for the Exploration of the Sea (ICES). The ICES protocols require that imported aquatic organisms be maintained in quarantined facilities as brood stock. The first generation progeny of the brood stock may be transplanted into the environment if they appear to be free of parasites and disease agents.

CONCLUSION: Strict application of the ICES protocols reduces the risk of cointroduction of undesirable organisms, including most pathogens and parasites. Oversight of the importation and deployment of the new species and prevention of a rogue introduction (an unsanctioned,

illegal, direct release of reproductive nonnative oysters) will be required to prevent release of "hitchhiking" species.

There remains some risk that a nonnative oyster by itself will cause serious ecological problems. The Pacific oyster *C. gigas* has not been an invasive species in the United States, Canada, and Europe, but in New South Wales, Australia, and northern New Zealand, it spread rapidly and depressed or eliminated active fisheries based on the indigenous rock oysters. The mixed outcomes observed with *C. gigas* introductions suggest that it is difficult to predict whether or not a species will be invasive.

Another way to assess the potential behavior of a nonnative species in a new habitat is to compare its biological characteristics with similar endemic species. The Suminoe oyster has a range of environmental tolerances comparable to the Eastern oyster *C. virginica* and based on small-scale field trials it grows rapidly in the estuarine waters of the Chesapeake. The most notable characteristic of *C. ariakensis* is its resistance to the two diseases that currently plague the native oyster. Not much is known about the disease susceptibility of the Suminoe oyster in its home range, although there have been reports of disease problems in China. In France, *C. ariakensis* was found to be unsuitable because it is susceptible to a disease caused by *Bonamia ostrea* that also infects the European flat oyster.

CONCLUSION: Based on the limited data available, it appears that *C. ariakensis* **has environmental tolerances that make it well suited for growth and reproduction in the Chesapeake Bay and in other similar estuarine habitats on the Atlantic and Gulf coasts. It is likely to compete with the native oyster, although differences in environmental tolerances might result in these two species occupying different habitats if** *C. ariakensis* **becomes established in the bay.**

Research Recommendations:

- Develop a better understanding of *C. ariakensis* biology in the Chesapeake Bay under various temperature and salinity regimes, particularly its growth rate, reproductive cycle, larval behavior, and settlement patterns in different hydrodynamic regimes; size-specific, postsettlement mortality rates; and susceptibility to native parasites, pathogens, and predators.
- Determine the ecological interactions of *C. ariakensis* and *C. virginica* at all life stages, including interspecific competition and reef-building capacity.
- Determine the genetic and phenotypic diversity of different geographic populations of *C. ariakensis* and other closely related Asian

species of the genus *Crassostrea* and the extent to which they might respond differently to the Chesapeake Bay environment.

- Develop an integrated approach to understanding the responses of native and nonnative oysters to environmental change and multiple stressors, including naturally occurring or introduced diseases, climate change, land use, nutrient loading, sedimentation, pollutants, and the interactions of these factors.
- Develop a model of oyster larval dispersion based on a detailed circulation model for the Chesapeake Bay and incorporating information about differences in the larval behavior or physiology of native and nonnative oysters to predict their dispersal patterns.

What Are the Potential Economic and Social Impacts of a Nonnative Oyster?

As recently as 1980 the Chesapeake Bay accounted for roughly 50% of the U.S. oyster harvest. The harvest dropped by 55% from 1991 to 2001, and the real price of Chesapeake oysters declined by 24% over the same period. The combined effect of reduced landings and reduced price resulted in a roughly 90% decrease in the value of Chesapeake oyster landings from 1980 to 2001.

With this severe decline in oyster harvests, the industry's interest in the nonnative Suminoe oyster has intensified, even if Suminoe production is limited to contained aquaculture. Because production costs are higher, containerized aquaculture is unlikely to replace wild harvest as a significant source of oysters for the shucking houses. Most intensively cultured oysters are targeted for the higher-value half-shell market. Significant price declines might occur if growth in the oyster supply occurs very rapidly, either as a result of recovery of the native oyster or through establishment of a large reproductive population of Suminoe oysters.

Oystering constitutes an important part of the cultural heritage of watermen communities in both Virginia and Maryland. Even at the much-reduced harvest levels of recent years, oysters continue to provide small amounts of much-needed income during periods when watermen have no other fishing income. As winter approaches, watermen shift to the oyster harvest and thereby decrease pressure on the fall harvest of blue crabs. Virginia and Maryland differ in the structure of their oyster fisheries; during the 1990s, more than 96% of the oyster harvest in Maryland came from public beds, while over 60% of Virginia's harvest came from private leased beds. This places Virginia in a better position to take advantage of the introduction of a new oyster that at least initially will require hatchery production of triploid spat and containerized aquaculture production methods.

CONCLUSION: Although communities in both states value the oyster fishery, policy differences regarding the relative distribution of public grounds and privately leased submerged oyster beds in Maryland and Virginia will have a significant effect on the ease with which the industry in each state adapts to dependence on hatchery production.

Research Recommendations:

- **Examine Public versus Leased Oyster Beds**. The mix of public- and leased-bottom oyster fisheries has been evolving. A better understanding of the institutional differences, their consequences, and their possible evolution will be critical for predicting the outcome of the management options, the ability of managers to oversee and control production practices, and the potential for Maryland and Virginia oystermen to compete with producers in other regions.
- **Establish an Ongoing Economic and Sociocultural Data Collection Program**. The ability of social scientists to predict or evaluate the outcomes of potential management actions is impaired by the lack of both baseline and postimplementation data necessary to evaluate the effects of management action separately from the effects of unrelated changes in the fishery. This research program could be organized through Sea Grant or through a multistate entity but should be designed and budgeted for a full 5- to 10- year period.
- **Examine Economic Feasibility of Alternative Production Systems**. Estimates should be developed for the profitability of public grounds versus private leased-bottom fisheries in Virginia and Maryland for different production modes. Triploid nonnative aquaculture estimates should account for the costs of antipredator strategies and biocontainment safeguards.
- **Develop Models of Community and Household Impacts of Alternative Production Systems**. The impacts of shifting traditional fisheries into aquaculture-based production will depend on the economic consequences and effects on local sociocultural norms. Data for building these models should be collected from both comparable case studies of traditional fisheries shifting into aquaculture-based production and structured interviews with watermen and coastal community members to solicit their perceptions of the likely effects of the various management options.

Adequacy of Regulatory and Institutional Structures

The regulatory framework that addresses the deliberate introduction of nonnative species into marine waters presents a patchwork of state, regional, federal, and international legislation and directives that leave significant gaps in the ability to monitor and oversee the use of exotic organisms. In the Chesapeake Bay, nonnative introductions are covered by the 1993 Policy for the Introduction of Non-Indigenous Aquatic Species through the region's CBP, but the recommendations made by the program are not legally binding. At the federal level there is limited regulation of nonnative introductions under the Lacey Act, which prohibits the importation of certain species found to be injurious. Species not named on this "black list" are regulated under authority delegated to the states. The U.S. Army Corps of Engineers has permitting authority over aquaculture operations that use in-water structures or fill. Under this authority, the Corps reviews the entire proposal for compliance with other federal statutes such as the Clean Water Act and the Endangered Species Act.

CONCLUSION: The existing regulatory and institutional framework is not adequate for monitoring or overseeing the interjurisdictional aspects of open-water aquaculture or direct introduction of *C. ariakensis*. **There is no federal legislation that gives specific criteria for regulating the introduction of a nonnative marine species. States may set their own criteria, but when an introduction is likely to affect neighboring states, there is no statutory mechanism for resolving differences among the interests of affected states.**

Research Recommendation: Although the CBP is well positioned to air the concerns of the participating state and federal agencies, its decisions are non-binding. The program should be evaluated as a potential venue for interjurisdictional decision making. Also, the review should identify institutional changes that would be required for the CBP, or other regional body, to assume binding authority over management decisions that will potentially affect coastal areas of more than one state.

MANAGEMENT OPTIONS

The committee was asked to evaluate whether the breadth and quality of existing research are sufficient to support risk assessments of three management options: (1) no use of nonnative oysters, (2) open-water aquaculture of triploid oysters, and (3) introduction of reproductive diploid oysters. The risks and benefits associated with the three management options are discussed individually below. The major concern and greatest uncertainty relates to the likelihood that *C. ariakensis* will become an invasive nuisance species and potentially threaten the ecological integrity of the Chesapeake Bay and adjacent waters along the Atlantic coastline or in the Gulf of Mexico.

Option 1: Prohibit Introduction of Nonnative Oysters

Under the first management option, introduction of any nonnative oyster would be prohibited even if the oysters were reproductively sterile. This option precludes risks associated with the introduction of a nonnative species. Another benefit is preservation of the cultural value associated with native species and natural habitats. The main risk identified with pursuing this option would be a continued failure of native oyster restoration efforts with continued decline of the oyster fishery and erosion of the traditional economies and cultures of Chesapeake Bay watermen. Under this option there are additional risks that would arise if the native oyster population failed to rebound, including an erosion of confidence in the ability of managers to address resource problems and continued loss of the ecological functions associated with healthy oyster beds, such as habitat and filtering capacity.

An additional risk under this option could arise if frustration with the slow pace of restoration leads to the importation and illegal release of Suminoe oysters. Suminoe oysters could be imported for seafood markets, but if they were released into open waters they could carry oyster pathogens or harbor other undesirable marine species. Introduction of an oyster pathogen or nonnative pest species is generally irreversible. The threat of a rogue introduction could be reduced by identification and vigilant monitoring of likely routes of importation and critical points where interdiction might be achieved. Review and enforcement of regulations against rogue introductions by the responsible state agencies would help avoid a situation in which the states were faced with developing burdensome programs for eradication of nonnative oysters and associated organisms. Public education on the high risks associated with a rogue introduction could be used to increase awareness and vigilance.

CONCLUSION: The long-term risk of an outright prohibition on the use of nonnative oysters (either for controlled aquaculture or for deliberate release into open waters) depends on the potential success of restoration programs for the native Eastern oyster.

Research Recommendations:

- Development of an integrated science-based approach to restoration of the native oyster. This would include a selective breeding program focused on building genetically diverse, native oyster stocks less susceptible to Dermo and MSX diseases.

- Determination of the causes of recruitment success or failure for *C. virginica* and evaluation of the success of oyster sanctuaries.
- Determination of the genetic and physiological bases for disease susceptibility, tolerance, and resistance.
- Assessment of the native oyster restoration program for probable level of success and the near- and intermediate-term economic consequences posed for watermen. This should include expected net benefits and variance of expected net benefits for harvesting and processing sectors in Maryland and Virginia with probable levels of recovery for each year of the program.

Option 2: Open-Water Aquaculture of Triploid Oysters

Aquaculture of sterile triploid *C. ariakensis* in controlled settings, as proposed by the Virginia Seafood Council, may result in the establishment of a diploid, self-reproducing population in the Chesapeake Bay because the process of generating mated triploids is not 100% effective and some triploid oysters may become reproductive as they age. The risk of establishment of a reproductive population of Suminoe oysters will increase with an increase in the scale of the commercial aquaculture operations. Many of the risks associated with the use of triploid nonnatives for aquaculture can be identified, quantified, and managed, as elaborated in Chapter 9 of this report.

Aside from the hazard of establishing a self-reproducing population of a nonnative oyster, potential short- and long-term negative impacts of triploid aquaculture include continued or accelerated declines in the traditional oyster fishery, economic exclusion of some harvesters due to the high investment costs required for converting to aquaculture production, potential introduction of pathogens not excluded by adherence to ICES protocols, and conflicts with the cultural value placed on conservation of native species. Managers could face a considerable burden for monitoring aquaculture operations and surveying the bay to detect stray nonnative oysters. Expenses would increase if reproductive oysters were found and it became necessary to eradicate them.

Some of the short- and long-term benefits of this option include regulatory and management control over most aspects of the use of nonnative oysters; improved viability of oyster aquaculture; increased employment in the aquaculture sector; and possibly reduced harvest pressure on sanctuaries established for restoration of the native oyster. A major benefit of the controlled use of triploid *C. ariakensis* would be increased potential for research relevant to assessing the risk of introducing a reproductive population in the Chesapeake Bay. For example, the likelihood of ecological harm or benefit of widespread triploid-based aquaculture or direct introduction could be more accurately assessed if basic information were available on the season of reproduction (triploids, though sterile, still go through an annual reproductive cycle) and the susceptibility of *C. ariakensis* to native pathogens and parasites.

One potential short-term benefit might be the community's perception of progress with respect to resource management, especially if this perception were to reduce the risk of a rogue introduction. This option also buys time for recovery of the native oyster, with either a reduction in the severity of oyster diseases because of more favorable climate conditions or through breakthroughs in the restoration of the native oyster, such as the development of disease-resistant populations. Revival of the native oyster would likely reduce the pressure for nonnative introduction.

CONCLUSION: Contained aquaculture of triploid *C. ariakensis* provides an opportunity to research the potential effects of extensive triploid-based aquaculture or introduction of reproductive nonnative oysters on the ecology of the bay and offers some additional economic opportunities for the oyster industry and the watermen.

Research Recommendations: To fully assess the risks associated with the larger-scale and longer-term aquaculture of triploid *C. ariakensis* requires research in the areas listed below.

- Determination of the susceptibility of *C. ariakensis* to the parasite *B. ostreae* through challenge experiments and comparison of the DNA sequence of the *Bonamia*-like organism associated with a *C. ariakensis* mortality event that occurred in France (archived material from IFREMER, La Tremblade) with known *B. ostreae* sequences.
- Development of sufficient data on the fidelity, stability, and fertility of mated triploid *C. ariakensis* to permit estimates of means and variances in parameters such as the proportions of triploids, diploids, or mosaics in lots of mated triploid seed.
- Determination of the proportion of triploid individuals undergoing gametogenesis, the fecundity of triploids, the types and proportions of gametes produced by triploids, and the fertility of these gametes.
- Determination of the maximum proportion of reproductive oysters that can be raised in containers of triploids without successful spawning (i.e., fertilization) under various conditions of water flow.

Option 3: Introduction of Reproductive Diploid Oysters

This management option has strong support in some sectors because of fear that the native oyster will never recover and the belief that introduction of a nonnative oyster that is resistant to disease is the only option for sustaining the traditional fishery and lifestyle of the watermen in the Chesapeake Bay. Underlying this support is the assumption that a purposeful introduction will yield a large population of Suminoe oysters after a few years with little or no adverse effects on the native oyster or other species.

Potential short- and long-term risks of introducing reproductive *C. ariakensis* into the Chesapeake Bay include the introduction of a new disease (greatly reduced but not eliminated if ICES protocols are followed); competition with *C. virginica* or fouling of boats, marinas, and other marine structures; dispersal of nonnative oysters outside the bay where competitive displacement of robust native oyster populations might occur; low market demand for nonnative oysters; susceptibility to endemic pathogens, parasites, or fouling organisms or to lower consumer acceptance; abandonment of attempts to restore native oyster; and conflicts with the conservation ethic for maintaining native species.

The potential benefits of a deliberate introduction of reproductive nonnative oysters, if successful, would be similar to those expected from recovery of the native oyster population. Some of these benefits may only be realized over the long term and only if the Suminoe oyster withstands environmental stresses of low water quality, limited habitat, and high sedimentation. With a deliberate introduction, the likelihood of a rogue introduction should be reduced. A successful introduction could improve the profitability of the traditional oyster fishery. Establishment of a dense population of nonnative oysters could improve water clarity in shallow embayments due to the oysters' filtering activity. Also, Suminoe oysters may form additional reef structures that provide habitat for fish and other invertebrates. The potential ecological impacts of a reproductive Suminoe oyster population will depend on as yet uncharacterized attributes of this species (e.g., temperature and salinity preferences for spawning and larval development, reef-forming capacity, susceptibility to predators and parasites, substrate preferences for larval settlement) that will affect the size of the population in any given area and hence the magnitude of the ecological effects.

CONCLUSION: It is not possible to predict if a controlled introduction of reproductive *C. ariakensis* will improve, further degrade, or have no impact on either the oyster fishery or the ecology of the Chesapeake Bay.

Research Recommendation: Further research for assessing the risks of establishing a reproductive population of Suminoe oysters in the Chesapeake Bay should include:

- determination of the oyster's capacity to survive and reproduce;
- analysis of the Suminoe's reef-building capacity;
- investigation of competitive interactions with the native Eastern oyster; and
- assessment of the marketability of naturally spawned, nonnative oysters harvested with tongs, rakes, or dredges or taken by divers.

Choosing Among the Management Alternatives

Development of a quantitative risk assessment model would require a great deal of additional research over a long period of time. Because of the dire circumstances faced by the oyster industry, resource managers are under pressure to make a decision about whether or not to proceed with the use of the nonnative oyster despite uncertainty in the type and magnitude of the potential risks. This is a particularly difficult decision due to the uncertainty of all options and the perceptions on all sides that a decision either way will have lasting and serious consequences.

Option 3, introduction of diploid, reproductively competent, nonnative oysters may or may not increase the abundance of oysters in the Chesapeake Bay or have a detrimental impact on the ecology of the bay and adjacent waters. This option would be essentially irreversible and would be ill advised given current knowledge. Option 1 is ecologically reversible, since the nonnative oyster could always be introduced at a later time. However, the economic decline of watermen and fishery-dependent communities may become irreversible if oyster abundance remains extremely low. Under Option 1 the threat of a rogue introduction must be addressed because of the high risk of introducing other potentially harmful species or disease-causing organisms to the bay and mid-Atlantic region. Rogue introductions also threaten the region's fisheries because of the risk of introducing a new predator or disease.

Option 2, aquaculture of triploid, nonnative oysters is unlikely to solve the fishery crisis, but it is reversible, at least in its early stages, and offers more opportunity for adapting management to changing circumstances. Over the long term, the risk of establishment of a nonnative oyster population increases due to the risk of diploid production from triploid stocks. Adoption of triploid *C. ariakensis* aquaculture may be perceived as progress in reversing the decline of the fishery, possibly reducing the incentive to pursue a rogue introduction. Option 2 has already received considerable scrutiny by the CBP and its member states and federal agencies. Limited field trials have been completed in Virginia and North Carolina and larger trials are in advanced planning stages. The risks of proceeding with triploid aquaculture in a responsible manner, using best management practices, are low relative to some of the risks posed under the other management options. Strict standards and protocols are required to reduce risks and enhance benefits of this course of action.

CONCLUSIONS:

- **Option 2 should be considered a short-term or interim action that provides an opportunity for researchers to obtain critical biological and ecological information on the nonnative oyster required for risk assessment. This option also allows for more management flexibility in the future depending on the status of the native oyster and the success of restoration efforts.**
- **Stringent regulations will be necessary to ensure that aquaculture of triploid *C. ariakensis* does not result in the establishment of a self-reproducing population in the Chesapeake Bay region.**

Recommendations for Establishing Standards for Nonnative Oyster Aquaculture

Before the commencement of open-water aquaculture (or pilot-scale field trials) of triploid nonnative oysters, the committee recommends developing a protocol to minimize and monitor the unintentional release of reproductive *C. ariakensis,* similar to the Hazard Analysis Critical Control Point protocol currently used to ensure seafood safety. This protocol should establish:

- acceptable limits for a variety of biological parameters to prevent release of reproductive nonnative oysters from the culture system;
- disease and quarantine certification of brood stock;
- confinement and accounting of nonnatives at all life stages;
- fidelity of triploid induction and the stability and sterility of triploids; and
- parameters of growth, survival, reproductive maturation, and fecundity of cultivated triploids.

Monitoring systems for ensuring these limits at each control point should be established. The protocol for controlled introduction of triploid nonnative oysters should identify corrective actions when monitoring reveals that a critical limit has been exceeded. Parties responsible for corrective actions and for record keeping should be identified. An independent verification of the effectiveness of the protocol and means of assessing failures should be established. For instance, genetic identification of hatchery stock could be used to track the source of any nonnatives discovered outside containment.

2

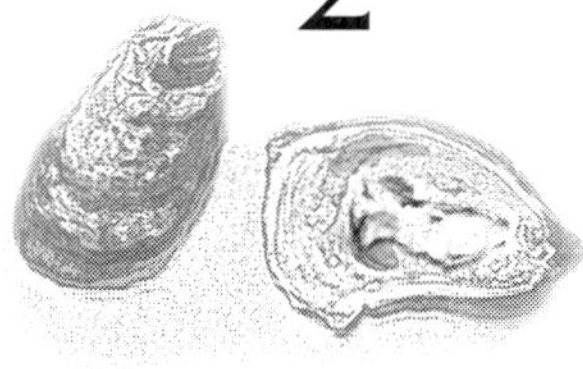

Introduction and Overview

HISTORY OF OYSTERS IN THE CHESAPEAKE BAY

The Eastern oyster, *Crassostrea virginica*,[1] is native to coastal waters from the St. Lawrence River in Canada to the Atlantic coast of Argentina (Carriker and Gaffney, 1996). It was an important dietary component for Native Americans in the Chesapeake region and has supported a major fishery in the Chesapeake since colonial times (Lunz, 1938; Stenzel, 1971; Ham and Irvine, 1975; Waselkov, 1982; Kent, 1988; Mackenzie and Burrell, 1997). This oyster has also been an important component of the bay ecosystem. At peak abundance, sometime prior to 1870 (Wharton, 1957), Newell (1988) estimates that oysters could filter the water of the bay every 3.3 days. In addition to filtering algae and particulates from the water column, the Eastern oyster forms three-dimensional reefs that provide habitat for other species in the bay (Wells, 1961; Bahr and Lanier, 1981; Dame and Patten, 1981; Mann et al., 1991; Coen and Luckenbach, 2000; Lenihan et al., 2001; Newell et al., 2002).

In the late 1880s the Chesapeake Bay was the greatest oyster-producing region in the world, with an oyster harvest twice that of the rest of the (non-U.S.) world. The oyster fishery in the bay represented 39% of the U.S. oyster harvest, 17% of all U.S. fisheries, and employed 20% of all Americans who worked in the fishing industry (Kennedy and Breisch, 1983). Oyster landings peaked in the latter part of the 19th century and

[1]Scientific names for species mentioned in the text are given in Appendix K.

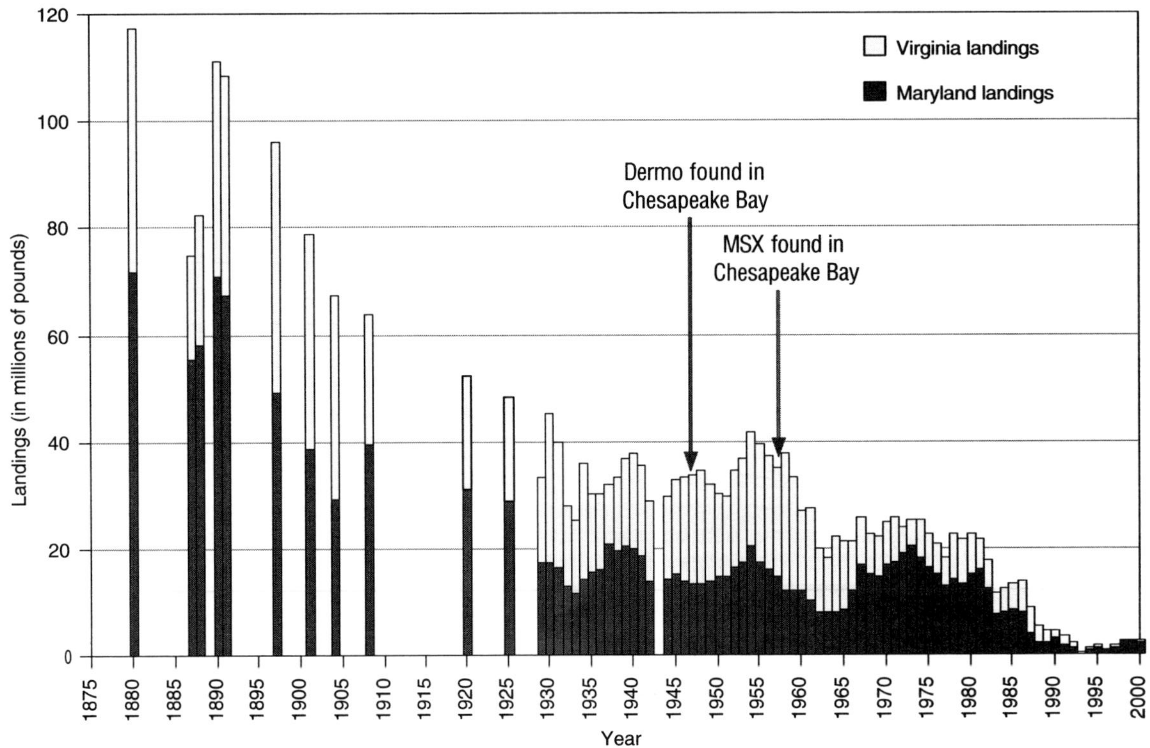

FIGURE 2.1 History of commercial oyster landings in the Chesapeake Bay.
SOURCES: Data from Chesapeake Bay Program, *http://www.chesapeakebay.net/data/historicaldb/livingresourcesmain.htm;* National Marine Fisheries Service, *http://www.st.nmfs.gov/st1/commercial/landings/annual_landings.html*

have declined steadily since then (see Figure 2.1). While the average density of oysters in the bay in 1991 was estimated to be 4% of the 1884 levels (Rothschild et al., 1994), oyster landings hit historically low levels in 2002. The decline in oyster abundance has been attributed to many factors, including fishing pressure, reduced water quality, habitat destruction, diseases, and interaction among factors.

Over the years, fishing pressure increased on the bay's oyster population as the oyster-harvesting gear used by fishermen became progressively more elaborate and efficient, beginning initially with rakes and hand tongs and progressing to dredges (legalized in 1865) and hydraulic patent tongs. Changes in fishing gear improved oyster landings (e.g., 8 to 25 bushels/day for a fisherman using a hand tong versus 30 to 100 bushels/day for a fisherman using patent tongs; Johnson, 1988) but negatively impacted oyster reefs. Dredging reduces the integrity of oyster reef habitat because large clumps of oysters are broken off (Winslow, 1881; Hargis and Haven, 1988, 1999; Rothschild et al., 1994; Lenihan and Peterson, 1998; Coen and Luckenbach, 2000).

Water quality in the Chesapeake Bay has also declined dramatically since colonial times as a result of increased sediment and nutrient loads due to population growth, land-use changes in the Chesapeake Bay watershed, and pollution (Rothschild et al., 1994; Boynton et al., 1995; Seagle et al., 1999; Smodlaka et al., 1999). This is perhaps best documented in the paleological record from sediment cores taken across the Chesapeake Bay near the mouth of the Choptank River (Cooper and Brush, 1991, 1993). For the bay, enhanced sedimentation and eutrophication have translated into increased phytoplankton biomass, reduced water transparency, loss of submerged aquatic vegetation, and expansion of hypoxic conditions in the stratified waters of the estuary during the summer (Harding et al., in Malone, 1999).

By 1959 Eastern oyster populations in the Chesapeake Bay had already been in decline when the disease MSX, caused by the protozoan parasite *Haplosporidium nelsoni*, spread to oyster populations in the lower Chesapeake from the Delaware Bay (Virginia Institute of Marine Science (VIMS),[2] 1998; Burreson et al., 2000). By the 1970s, MSX dramatically reduced oyster densities in Virginia's high salinity oyster grounds. Oysters populating less saline waters (less than 15 ppt) were generally spared the effects of the disease. Another protozoan parasite, *Perkinsus marinus*, which causes the disease Dermo, had been identified in Chesapeake Bay oysters in 1949 but did not become a major problem until the mid-1980s (VIMS, 1998; Ford and Tripp, 1996). Like MSX, Dermo caused high oyster

[2]Acronyms cited in the text are listed in Appendix C.

mortality, particularly at salinities above 12 to 15 ppt. Unlike MSX, Dermo also killed oysters at lower salinities (less than 9 ppt), although under these conditions the disease takes 2 to 3 years to develop.

In assessing the historical changes to the oyster populations in the Chesapeake Bay, it is clear that over the past 110 years the population decline has been the result of multiple interacting factors. During the latter part of the 20th century, diseases had a devastating impact on oyster populations in the Chesapeake Bay, particularly where salinities were greater than 15 ppt; however, they acted on an oyster population that was already compromised by poor water quality, fishing, and habitat loss. Thus, a combination of interacting factors occurring over the past 150 years has threatened the survival of oyster fisheries in both Virginia and Maryland (Ulanowicz and Tuttle, 1992; Rothschild et al., 1994; Newell et al., 2000, 2002; Jackson et al., 2001; US ASMFC, 2002).

There are significant differences in the nature of the Chesapeake Bay oyster fishery as it is pursued in Maryland versus Virginia. In Maryland most of the watermen fish public bottom (96%), while in Virginia more than 60% of the fishing occurs on private beds. Hand tongs, patent tongs, and hand (divers) are used to collect most of the catch in Maryland (Chesapeake Bay Program, 2002) while in Virginia most of the oysters are taken with patent tongs, oyster dredges, and by hand (divers) (Virginia Marine Resources Commission [VMRC], 2003a). Virginia does more processing and marketing than Maryland, with about two-thirds of the oysters processed coming from outside the area (J. Wesson, VMRC, Newport News, personal communication 2003).

Restoration of the Eastern oyster is at the center of the Chesapeake 2000 Agreement, signed by Maryland, Virginia, Pennsylvania, the District of Columbia, the Chesapeake Bay Commission, and the U.S. Environmental Protection Agency. The agreement calls for a 10-fold increase in native oysters in the bay by 2010 relative to the 1994 baseline. This is to be achieved through continued improvements to bay water quality, reduced oyster fishing pressure in selected areas, shell deposition to rebuild the reef structure, and continued development of disease-resistant oyster strains. Past restoration efforts, dating back to the 1920s, were directed toward maintaining the oyster fishery. Recent programs have included establishment of more than 30 oyster reef sanctuaries and production of disease-free oyster seed for stocking both sanctuary and harvest areas (Luckenbach, 2001).

At the same time, a debate has arisen over whether the nonnative Asian or Suminoe oyster *C. ariakensis* should be introduced into the Chesapeake Bay for commercial harvest. In 1995 the Virginia General Assembly, through House Joint Resolution 450, directed VIMS to begin research on nonnative oyster species for possible use in the Chesapeake Bay. VIMS used sterile (triploid) oysters to prevent any unintended introduction, initially working with the Pacific oyster, *C. gigas,* and then *C. ariakensis* (Calvo et al., 1999; Hallerman et al., 2001). Results of the *C. gigas* trials were disappointing; at low- and medium-salinity locations, the oysters had high mortality rates and slow growth rates. However, controlled experiments in Virginia waters with the Asian oyster suggested that it might grow faster and be more disease resistant than the native, Eastern oyster, *C. virginica* (Calvo et al., 2001). The Asian oyster was also well received in marketing trials undertaken by the Virginia Seafood Council (VSC; also see a more recent study by Grabowski et al., 2003). Based on these promising results, the VSC requested state approval to use chemically derived triploid *C. ariakensis* in a field trial in 2002 with 40 participants from the Virginia oyster aquaculture industry.

The proposal, which was withdrawn after criticism from the Chesapeake Bay Program's ad hoc review panel, was the stimulus for the request for this National Research Council (NRC) study. A second VSC proposal recommended the introduction of genetically derived sterile *C. ariakensis* for use in a 2003 economic field trial to test whether *C. ariakensis* aquaculture would provide an economic alternative to the diseased native oyster (Virginia General Assembly, 2002) and potentially contribute to restoration of the ecosystem function historically provided by the extensive beds of native oysters. Because of concerns not addressed in the second VSC proposal, this NRC committee submitted interim comments in the form of a brief letter to the Virginia Marine Resources Commission (VMRC) outlining some of the risks that should be addressed in the field trial (Appendix B). The VMRC approved the VSC proposal with modifications to address many of these concerns. Field trials using triploid *C. ariakensis* are also underway in North Carolina.

Little is known about the interactions of *C. ariakensis* with the native oyster or the ecological consequences of introducing this oyster to the Chesapeake Bay (Zhou and Allen, 2003). Nonnative species introduced by accident over the years have already altered the bay's ecological balance, and several previous experiences with exotic species invasions (e.g., mute swans, nutria, kudzu, zebra mussels, and gypsy moths) argue against unregulated introduction of species that have the potential to spread throughout ecosystems (see Carlton, 2001; Moser, 2002, and references therein). Not all introductions have caused ecological problems and economic losses. Still, many resource managers, environmentalists, and bay scientists fear that introduction of a nonnative species could upset the bay's ecological balance. Maryland officials, who have no apparent legal authority over these experiments in Virginia waters, initially objected to open-water testing of *C. ariakensis*. It is also unclear which, if any, federal agencies have jurisdiction to review further experiments or introductions of *C. ariakensis* (e.g., the North Carolina study, which was originally funded by the state legislature, has proceeded with no federal involvement).

VIMS research to date has shown that aquaculture of triploid *C. ariakensis* offers promise for economic development in Virginia and the region (Calvo et al., 2001). Using current technologies and production methods, large-scale use of triploid *C. ariakensis* would entail the possibility of introducing reproductive (diploid) nonnative oysters over the long term through reversion, production errors, or both. The introduction of diploid *C. ariakensis* into the Atlantic coastal waters of the United States is a resource management decision with potentially far-reaching consequences. These consequences could extend well beyond Virginia if the oysters are shipped to other states or the larvae are spread by natural processes, and the risks and merits of this species may vary in different regions. As a result, there has been a call for more research on the ecological, genetic, and disease potential of *C. ariakensis* because of the possibility of introducing reproductively capable populations over the long term, even with current triploid technology and production methods.

The goal of this study was to examine the ecological and socioeconomic risks and benefits of open-water aquaculture or direct introduction of the nonnative oyster, *C. ariakensis*, in the Chesapeake Bay (see Box 2.1). The committee prepared this report after convening two public meetings (Appendix H). At the first meeting the study's sponsors (see Box 2.2) were invited to explain their goals for the study and to describe what they thought were the most important issues for the committee to consider.

BOX 2.1
Statement of Task

This study will examine the ecological and socioeconomic risks and benefits of open-water aquaculture or direct introduction of the nonnative oyster *Crassostrea ariakensis* in the Chesapeake Bay. The committee will address how *C. ariakensis* might affect the ecology of the bay, including effects on native species, water quality, habitat, and the spread of human and oyster diseases. Possible effects on recovery of the native oyster *C. virginica* will be considered. The potential range and effects of the introduced oyster will be explored, both within the bay and in neighboring coastal areas. The study will investigate the adequacy of existing regulatory and institutional frameworks to monitor and oversee these activities.

The committee will assess whether the breadth and quality of existing research, on oysters and other introduced species, are sufficient to support risk assessments of three management options: (1) no use of nonnative oysters, (2) open-water aquaculture of triploid oysters, and (3) introduction of reproductive diploid oysters. Where current knowledge is inadequate, the committee will recommend additional research priorities.

BOX 2.2
Study Sponsors

- U.S. Environmental Protection Agency
- National Oceanic and Atmospheric Administration
- U.S. Fish and Wildlife Service
- Maryland Department of Natural Resources
- Virginia Department of Environmental Quality
- Virginia Sea Grant
- Maryland Sea Grant
- Connecticut Sea Grant
- National Fish and Wildlife Foundation
- Scientific and Technical Advisory Committee

Standish Allen (VIMS) provided an overview of what is known about *C. ariakensis*, both in research conducted at VIMS and in the oyster's native habitat in China. The committee then organized a workshop to be held at its second meeting to bring together scientists, managers, policymakers, watermen, and environmental groups to discuss the many different biological, environmental, regulatory, social, and economic issues that are essential components of informed decision making on the proposed introduction into the bay of a nonnative species. The committee met twice more to prepare the findings that are reported here.

Report Organization

This report is organized as a review of the information available to support a detailed analysis of the risks and benefits of introducing an exotic oyster into the Chesapeake Bay and considering the potential expansion of this oyster into other waters along the eastern seaboard. The remainder of Chapter 2 describes major features of the Chesapeake Bay ecosystem, highlighting the sources of change that may contribute to the current depleted status of the native oyster population. Chapter 3 provides a review of invasive species science and presents case studies of introductions of nonnative oysters around the world. An overview of the biology of both *C. virginica* and *C. ariakensis* is given in Chapter 4, with special reference to the similarities and differences relevant to the survival and propagation of these species in the bay. Chapter 5 presents an analysis of the social and economic value of oysters in the Chesapeake Bay and describes differences between the oyster industries in Virginia and Maryland. Chapters 6 and 7 review the status of native oyster restoration efforts and the aquaculture industry respectively. The regulatory authority governing the decision to introduce a nonnative species into marine waters has not been well defined in the United States. An examination of the existing regulatory framework for introduction of nonnative species at the state, federal, and international levels is presented in Chapter 8. In Chapter 9 the committee presents an assessment of the various risks that may be associated with the three management options defined in the Statement of Task and includes background on risk assessment methodology, underscoring the great uncertainty in all areas. Finally, Chapter 10 presents the committee's conclusions and suggestions for research that would enable decision making for future use of nonnative oysters in aquaculture or to establish another oyster species for harvest in the Chesapeake Bay and neighboring waters.

CHESAPEAKE BAY LIMNOLOGY AND OCEANOGRAPHY

Physical Description

The Chesapeake Bay is a long, narrow, relatively shallow estuary (see Figure 2.2 and Table 2.1) created more than 10,000 years ago when the Susquehanna River valley was drowned by the sea-level rise following the Wisconsin ice age (Boicourt et al., 1999). The bay extends from Cape Charles, Virginia, at its mouth northward for more than 300 km to Havre de Grace, Maryland, at its head. There are three main tributaries of the bay: the Susquehanna River, the Potomac River, and the James River. The Susquehanna River provides approximately half of the freshwater input and with the other two primary tributaries accounts for 70 to 80% of the bay's freshwater input (Boicourt et al., 1999; Seagle et al., 1999).

The Chesapeake Bay can be characterized as a partially mixed estuary with three main factors influencing its circulation: freshwater inflow, the geometry of the basin, and tidal strength (Pritchard, 1967; Boicourt et al., 1999). Synoptic-scale winds (i.e., winds generated by large moving pressure systems) can also be important. Fresh water moves seaward over saltwater that is moving toward the head of the estuary. Mixing is directed to both the upper and lower layers, so water properties are altered in both layers during travel along the axis of the bay. In addition, transport within each layer increases in the seaward direction (Boicourt et al., 1999). Like many rivers and estuaries, the Chesapeake Bay has a characteristic freshwater outflow at its mouth. This transport has been estimated at about five times the entering river flow (Kemp et al., 1999; Malej et al., 1999), which in an average year is typically greatest in the spring and least in the late summer and early fall (Malej et al., 1999). Fresh water in the bay has a mean residence time of 7 months (Malej et al., 1999). The bay's outflow plume is distinct, generally making a wide anticyclonic turn and flowing southward as a coastal current (Rennie et al., 1999; Marmorino and Trump, 2000). However, wind speed and the direction and volume of

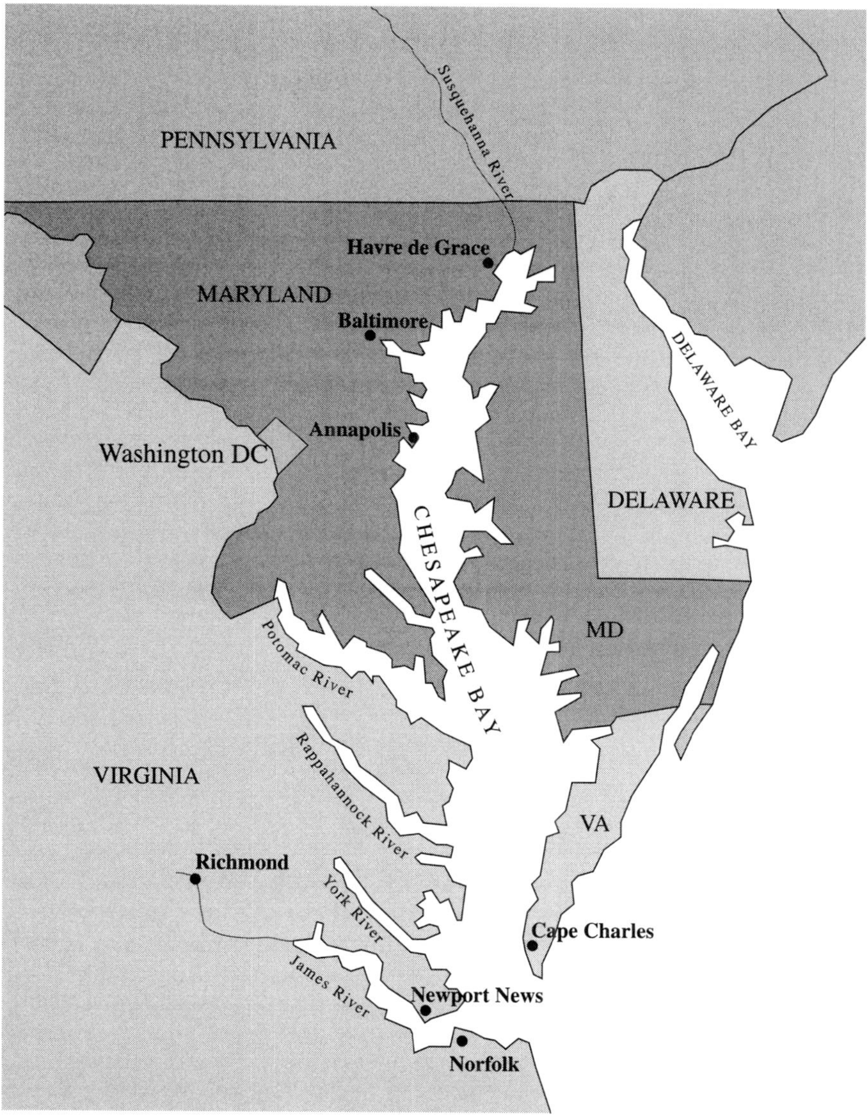

FIGURE 2.2 Map of the Chesapeake Bay.

bay discharge can affect both the direction and speed of this coastal current. The Chesapeake Bay plume propagates more rapidly to the south during moderate-to-weak southeastward winds (downwelling favorable conditions). Under these conditions the current becomes narrower (less

TABLE 2.1 Profile of the Chesapeake Bay

Chesapeake Bay Main Basin		
Length:	195 mi	
Shoreline:	4600 mi	
Volume:	18 trillion gallons	
Area:	Maryland	1726 sq mi
	Virginia	1511 sq mi
	Total	3237 sq mi
Width:	"Cape Charles, VA"	30 mi
	"Annapolis, MD"	4 mi
Depth:	average	25 ft
	greatest (SE of Annapolis)	174 ft
Tidal Range:	Annapolis	1 ft
	head	2 ft
	mouth	3 ft
Surface Salinity:	mouth	30 ppt
	midway to head	15 ppt
	above fall line	00 ppt
	difference between surface and bottom	2-3 ppt

SOURCE: Modified from Maryland State Archives, http://www.mdarchives.state.md.us/msa/mdmanual/01glance/html/ches.html.

than 10 km) and deeper (greater than 8 m). Southward propagation of the plume decreases for moderate-to-weak northwest winds (upwelling favorable conditions). Under these conditions the current is shallower (less than 8 m) and wider, often extending more than 10 km offshore and sometimes detaching from the coast (Rennie et al., 1999; Hallock and Marmorino, 2002).

Within the bay there are continuous gradients of physical, chemical, and biological properties along the axis of the mainstem (Kemp et al., 1999). Most notable is the north-south salinity gradient that is driven by the freshwater input of the Susquehanna River (Seagle et al., 1999; Harding et al., 1999). The salinity ranges from fresh water at the head of the bay to 85% seawater in the lower reaches (Kemp et al., 1999; Harding et al., 1999). Based on factors such as bottom topography, nutrient input, and water clarity, the bay is often divided into upper, middle, and lower regions, each with distinct characteristics (Kemp et al., 1999). The upper bay (above 39°N latitude) is a shallow area (less than 3.5 m) with low relief bottom topography and active sediment deposition. It is turbid and well mixed with high nutrient concentrations (Kemp et al., 1997, 1999). Migrating fish and crabs tend to congregate in this turbidity zone in the spring and fall (Boynton and Kemp, 2000; Simenstad et al., 2000). The midbay is separated from the lower bay just below 38°N latitude and is characterized by relatively clear water and seasonally high nutrient concentrations. The topography of this area includes a narrow central channel and broad flanking shoals. The water in the central channel is well stratified. The lower bay generally has the clearest waters and somewhat lower concentrations of nutrients than the other bay areas (Kemp et al., 1999). The geometry of the Chesapeake Bay is conducive to oxygen depletion, particularly in the deeper areas of the mainstem where stratification suppresses vertical exchange (Boicourt et al., 1999; Kemp et al., 1999). This is significant because the channel, which represents less than 40% of the bay's surface area, contains 75% of its volume (Smodlaka et al., 1999).

Changes in Human Population and Land Use

The Chesapeake Bay drainage basin encompasses an area of 64,000 square miles (166,512 km^2) and extends over parts of six states (Maryland, Delaware, Pennsylvania, New York, Virginia, West Virginia) and the District of Columbia (Smodlaka et al., 1999). The current population in this watershed is estimated at 16 million people (Seagle et al., 1999; Smodlaka et al., 1999), or an average of 98 people/km^2. Archeological evidence suggests that humans first inhabited this area about 10,000 to 12,000 years ago. Significant human impacts on the bay's ecosystem appear to have begun well before the European settlers arrived (Stevenson et al., 1999), although change accelerated during and after colonial times (Cooper and Brush, 1991, 1993). Prior to European settlement, 95% of the Chesapeake Bay watershed was covered by forest (Seagle et al., 1999). By the middle of the 19th century, European settlers had cleared over 55 to 60% of the forested land area (Seagle et al., 1999; Malej et al., 1999). This deforestation resulted in significant increases in nutrient and sediment input into the waters of the drainage basin (Cooper and Brush, 1991, 1993) and caused additional changes to the trophic structure of the bay such that benthic plant productivity decreased and phytoplankton productivity increased (Malej et al., 1999). Today, more than 60% of the watershed area has been returned to forest cover (Seagle et al., 1999). The rest is comprised of urban and suburban development, pasture, and cropland (Seagle et al., 1999). Despite the gradual recovery of forested lands in the bay's watershed, continued human population growth, greater fertilizer use, atmospheric deposition from fossil fuel combustion (especially coal), and increased animal husbandry in the watershed area have added to sediment inputs and increased riverine nutrient inputs 10-fold since the mid-1940s (Krupnick et al., 1998; Russell et al., 1998; Malej et al., 1999). Cooper and Brush (1993), Brush and Hilgartner (2000), and Jackson et al. (2001) provide excellent historical overviews of how the synergistic effects of human activities have led to degraded water quality, loss of oyster populations, and decreased submerged aquatic vegetation in the Chesapeake Bay.

Nutrients and Sedimentation

Compared with other coastal ecosystems worldwide, the Chesapeake Bay estuary receives moderately high inputs of nitrogen and phosphorus (Nixon, 1995; Kemp et al., 1999) and thus is considered to be overenriched in these elements (Malone et al., 1986; Harding et al., 1999). A number of models have indicated that nutrient enrichment is having a major effect on the bay's ecosystem (e.g., Darnell and Soniat, 1981). These effects include increases in the severity, duration, and frequency of hypoxic and anoxic events in the deeper areas of the bay and some of the tributaries (Officer et al., 1984; Seliger et al., 1985) and declining populations of submerged aquatic vegetation (Orth and Moore, 1983; Hurley, 1991), anadromous and freshwater fish (Flemer et al., 1983), and benthic organisms (Kemp and Boynton, 1992). Excess nutrient inputs also enhance phytoplankton blooms and subsequent responses of heterotrophic bacterioplankton (Smith et al., 1992). There is evidence that the trophic structure of the bay has shifted from one that is dominated by autotrophic production (primarily photosynthesis) to one dominated by bacterial production (Jonas and Tuttle, 1990). Most of the nutrients enter the estuary as diffuse nonpoint source inputs associated with river flow, although point source discharges and atmospheric deposition can sometimes be significant (Kemp et al., 1999). The major macronutrients generally decrease in concentration from north to south. The highest nutrient concentrations are near the bay's confluence with the Susquehanna River, and the lowest are at the mouth of the bay (Harding et al., 1999).

The edge-of-stream sediment load for the entire Chesapeake Bay drainage is estimated to be 2.72×10^3 tons/year, with a delivered load of 2.22×10^6 tons. The sediment load is related to river flow and to landcover. Seventy-three percent of the total sediment load is derived from farmland using conventional tillage practices (Seagle et al., 1999). Data from sediment cores show that sedimentation rates in the Chesapeake Bay have increased five- to seven-fold since European settlement, averaging 0.13 to 0.27 cm/year after settlement compared with rates of 0.03 to 0.14 cm/year prior to settlement (Cooper and Brush, 1993).

In the 1970s, the U.S. Environmental Protection Agency (EPA) initiated a decade-long study that was focused on reducing sediment and nutrient inputs into the bay (Horton and Eichbaum, 1991). The first *Chesapeake Bay Agreement* (1983) formally recognized the need for cooperative efforts to reduce pollution (including nutrient enrichment) and enhance bay productivity (Appendix D). A second agreement in 1987 expanded upon the first and committed to a 40% reduction in nitrogen and phosphorus entering the mainstem of the bay by 2000. Although the goal has not been met, nutrient levels have declined despite increased population growth and development in the watershed. The model of the Chesapeake Bay watershed used by the Chesapeake Bay Program (CBP) to estimate nutrient input suggests that nitrogen loads have decreased by 51 million pounds/year between 1985 and 2000 (from 300 million pounds/year to 249 million pounds/year) and that phosphorus loads have declined by 8 million lbs/year (from 25 million pounds/year to 17 million pounds/year) over the same time (Chesapeake Bay Program, 2000). U.S. Geological Survey measurements at nine river input monitoring stations provide more detailed and site-specific data. Nutrient levels computed using actual water quality data collected between 1985 and 1998 at these stations showed no statistically significant trends in either total nitrogen or total phosphorus loads at six of the nine stations. Some of the difference is attributable to high streamflows, which can lead to higher nutrient loads even if nutrient concentrations have maintained status quo or declined. Adjustment of the data to account for this variation resulted in downward trends for nitrogen concentrations (at six of the nine sites) and phosphorus concentrations (at seven of the nine sites; Belval and Sprague, 1999; Sprague et al., 2000). A further analysis of nutrient data from 31 sites in the nontidal portions of the Chesapeake Bay basin from 1985 to 1999 showed that flow-adjusted concentrations of total nitrogen and total phosphorus trended downward at 23 of 31 sites. This suggests that management actions are working in reducing nutrient concentrations (Langland et al., 2001; U.S. EPA, 2002). The *Chesapeake 2000 Agreement* reaffirms the commitment of its signatories to achieve and maintain the 40% nutrient reduction goal agreed to in 1987 and further strives to correct all nutrient- and sediment-related problems in the bay and its tidal tributaries by 2010 (Chesapeake Bay Program, 2000).

3

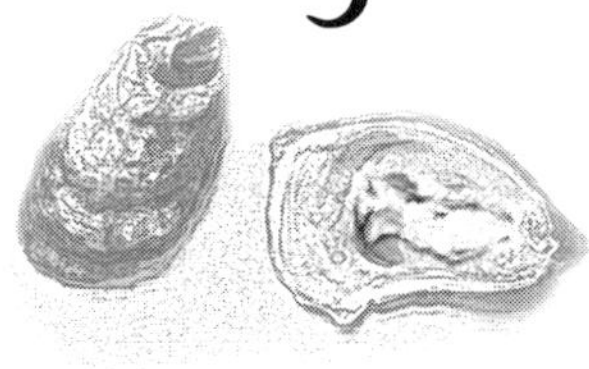

Background on Introduced Species

BRIEF OVERVIEW

Introduced species are organisms that have been intentionally or accidentally transported by human activities into regions where they were not previously found (i.e., nonindigenous; Carlton, 2001). While humans have been transporting marine species for hundreds of years, it has been exceedingly difficult to predict whether a species will become "invasive" (spread from the site of introduction and become abundant; Bergelson, 1994; Crawley, 1989; Kolar and Lodge; 2001; Lohrer et al., 2000; see Appendix J). Most nonnative species do not survive where they are introduced, and few become pests after they are transported to new environments. It has been estimated that about equal numbers of deliberate and inadvertent introductions to the United States of terrestrial vertebrates, fish, and molluscs have turned out to be harmful (U.S. Congress, 1993). Others have suggested that approximately 10% of nonnative species will become "invasive" (Williamson and Fitter, 1996). Low invasiveness is generally explained by a lack of favorable environmental conditions or by the lack of an "open" niche for the nonnative in a new habitat or region (Lohrer et al., 2000). Attempts have been made to predict life history traits of species that will be successful invaders, such as the number and size of propagules (progeny at early life history stages), age at first reproduction, and organism growth rate. Some studies have found correlations between life history traits and invasion success (e.g., O'Connor, 1986; Rejmanek and Richardson, 1996; Williamson and Fitter, 1996) and others have suggested that attributes of the invaded community (i.e., disturbance levels, species diversity) affect its ability to resist nonindigenous species invasion (e.g., Hobbs, 1989; Case, 1990; Tilman, 1997; Stachowicz et al., 1999, 2002a). Often the most reliable predictor is the invasiveness of a species in situations with similar environmental characteristics (Forcella et al., 1986; Gordon and Thomas, 1997; Reichard and Hamilton, 1997).

Recent work has shown that relatively simple approaches may be used to address why some introduced species rapidly spread and flourish while others do not. Kolar and Lodge (2002) studied a number of nonindigenous fish species introduced into the Laurentian Great Lakes and found that different life history traits were important during different phases of the invasion process. Species that grew relatively quickly and had a wide range of temperature and salinity tolerances tended to be most successful during the establishment phase of the introduction. In contrast, species that rapidly spread only after becoming established typically had slower growth rates and a more limited range of temperature tolerance. Those species that eventually were categorized as "invasive" generally had wider salinity tolerances and were capable of surviving lower water temperatures than noninvasive species. These results demonstrate the importance of assessing stage-specific processes and patterns of successful and nonsuccessful invasions. Although the outcomes are habitat specific, the overall approach for predicting invasion success could be applicable to marine invaders. For example, a comparison of the survival patterns of different species as they are transported along a coastline provides useful information on the environmental tolerances of those species—information that is essential in studying the early phases of the invasion process.

Ecosystems that have reduced biodiversity or are under stress from environmental degradation and climate change appear to be more vulnerable to invasions. For example, Stachowicz et al. (2002b) demonstrated that enhanced species diversity directly increases the resistance of subtidal fouling assemblages to invasion and that surveys in a number of coastal habitats in southern New England also revealed an inverse correlation between resident species richness and the number of nonnative species in those habitats. Climate change has its greatest impact where invasive species occur at the southern and northern boundaries of ecosystems (see Root et al., 2003, for a recent meta-analysis). Temperate coastal regions appear to be particularly prone to invasion by nonnative species due to the effect of climate change on invasion by nonnative species. Stachowicz et al. (2002b) have correlated a doubling in the abundance of invasive ascidians in eastern Long Island Sound with a significant increase in seawater temperatures over the past two decades in that region. Sauriau (1991) correlated small increases in seawater temperatures with the spread of a gastropod, *Cyclope neritea*, initially introduced with shipments of the

Pacific oyster, *Crassostrea gigas*, along the Atlantic coasts of Portugal, Spain, and France. Several studies have reported the northward spread of the oyster parasite *Perkinsus marinus* (Dermo disease agent) along the U.S. East Coast in relation to the trend of increasing winter seawater temperatures during the past several decades (Ford, 1996; Cook et al., 1998).

Every major marine, freshwater, and terrestrial ecosystem in the United States contains introduced species, and in some ecosystems it is difficult to find a native species (e.g., Cohen and Carlton, 1998). Nonnative species typically make up a large fraction of the biota of marine and estuarine habitats, largely because of the use of these ecosystems for commerce, fishing, and recreation. Introduced species in coastal habitats arrived via several unintentional human-mediated mechanisms ranging from transport on hulls or in the ballast water of ships to intentional introductions for aquaculture, fisheries, and releases from pet and live seafood trades (e.g., Carlton, 2001).

The human-mediated movement of marine species within and between ocean basins as food items or for aquaculture purposes has occurred for millennia (Carlton, 1999a). For example, there is evidence that Vikings transported the soft-shell clam (*Mya arenaria*) to Europe in the 13th to 14th century (Carlton, 1999b), and the Japanese oyster (*C. angulata*) was brought to southern Europe in the 1500s by Portuguese explorers (Edwards, 1976). The transport of marine species for aquaculture purposes became far more pronounced during the 19th and 20th centuries and today aquaculture has become one of the fastest-growing components of the world's food economy (Naylor et al., 2001). For example, the farming of marine molluscs (e.g., oysters, clams, mussels) is an important industry. In 2000, U.S. molluscan aquaculture was worth more than $75 million (National Marine Fisheries Service, 2002).

The intentional introductions of commercial marine species into coastal areas typically are to augment or replace depleted natural stocks or to diversify the number of species used in aquaculture operations. While the ecological impacts of molluscan aquaculture are relatively small in comparison to other types of seafood farming (Naylor et al., 2000), historically the activity has resulted in the introduction of many nonnative species throughout the world. Oysters have been intentionally transported more often than probably any other marine species (Sindermann, 1990). A wide array of plants and animal species has "hitchhiked" with introduced shellfish. Cohen and Carlton (1995) estimate that about 20% of the introduced species found in San Francisco Bay were the result of shipments of the Eastern and Japanese oysters for aquaculture purposes. The pace of shellfish introductions has greatly diminished with the switch to hatchery-propagated seed on the U.S. West Coast, which permits the application of International Council for the Exploration of the Sea (ICES) protocols that greatly reduce the risk of cointroductions. Other researchers have estimated that the use of introduced species for fisheries activities is the second leading vector (the first is shipping) for the transport of nonnative species throughout the world (Naylor et al., 2001).

In some cases the introduced species have become significant competitors with native oysters. For example, there is evidence that the Japanese oyster, *C. gigas,* has outcompeted native oyster species in New Zealand (Dinamani, 1991a, b), Australia (Ayres, 1991) and France (Goulletquer and Héral, 1991). Some of the species transported with oyster shipments have become important predators and competitors of both cultured and wild molluscan stocks as well as other native species. For example, in the early to mid-1900s, transfers of *C. virginica* from the U.S. East Coast to the U.S. West Coast resulted in the introduction of oyster competitors (e.g., the slipper shell *Crepidula fornicata*) and predators (e.g., the oyster drill *Urosalpinx cinerea*) in those waters. Large-scale transfers of *C. gigas* to the U.S. west coast, British Columbia, Canada, and Western Europe in the early- to mid-1900s resulted in the establishment of a number of oyster predators (Barber, 1997). In addition to affecting economically important species, the invaders can pose significant threats to biodiversity, directly or indirectly alter local community composition, influence the performance of ecosystems, and cause significant economic impacts (e.g., Grosholz et al., 2000; Stachowicz et al., 1999, 2002a). A National Research Council (1995) report ranked invasive species and overexploitation of fisheries stocks as the most important threats to marine biodiversity. Considerable information cataloging and assessing the impacts of introduced species in coastal waters and reviews can be found in Carlton (1985, 1987, 1989), Grosholz (2002), and Ruiz et al. (1997, 1999, 2000).

Notwithstanding the positive impact that introductions of bivalves have had on economic development of local fisheries and aquaculture, there are no documented cases where an intentional introduction of an oyster species has resulted in an overall positive ecological impact on a U.S. coastal ecosystem. However, studies conducted in France have demonstrated that under intensive aquaculture conditions very high densities of the introduced oyster *C. gigas* have resulted in transient ecosystem impacts in several coastal embayments and lagoons by temporarily cropping primary production and increasing nutrient recycling rates (e.g., Bacher et al., 1997a, 1997b; Gangnery et al., 2001). In contrast, *unintentional* introductions of some species of suspension-feeding bivalves into U.S. waters have had pronounced positive and negative ecological impacts on the aquatic ecosystem. For example, millions of dollars are spent annually by Great Lakes cites and industries to unclog water intake pipes overgrown with the nonnative zebra mussel (*Dreissena polymorpha*), and this invader has displaced some native species in certain areas. However, rapid population expansion of zebra mussels also has increased Lake Erie water quality as much as six-fold and has reduced some types of phytoplankton. Phelps (1994) reports that the invasion and subsequent population explosion of the nonnative Asian clam, *Corbicula fluminea*, in the late 1970s in the Potomac River estuary was likely responsible for enhanced water clarity and increased abundance and diversity of submerged aquatic vegetation and aquatic birds in the mid-1980s. In California, where *Corbicula* has become the most widespread and abundant freshwater bivalve, the species is regarded as an economic pest of industrial and municipal water delivery systems (Cohen and Carlton, 1995). In portions of San Francisco Bay, the filtration activities of several different species of nonnative bivalves have resulted in dramatically reduced phytoplankton abundance (Alpine and Cloern, 1992; Cloern, 1996) and may have altered the pelagic trophic structure in parts of the bay (Kimmerer et al., 1994). However, high abundance of the nonnative species has also greatly altered the diversity and biodiversity of the benthic communities (Nichols et al., 1990), and the presence of one species, *Potamocorbula amurensis*, has increased the bioavailablity and trophic transfer of the toxic metal selenium in certain parts of the bay ecosystem (Linville et al., 2002).

The remainder of this chapter reviews case studies of shellfish introductions in France, Australia, New Zealand, and the North American Pacific and Atlantic coasts. In many of these cases the introduction of a nonnative shellfish species was intentional, in response to the severe depletion of native stocks and fisheries. The variety of consequences observed in these case studies, even those employing the same species in different localities or environments, demonstrates the uncertainty of predicting ecological outcomes of shellfish introductions.

CASE STUDIES OF SHELLFISH INTRODUCTIONS

France

Commercial oyster landings by French oyster farmers currently rank fourth in the world, behind China, Japan, and Korea with 150,000 metric tons harvested per year. The landings are valued at $280 million and represent 25% of the total gross value of French seafood production. Landings reached these levels in the early 1990s and have remained stable. Further expansion is limited by the availability of leasing grounds. Two species are cultivated in France: the endemic native flat oyster, *Ostrea edulis*, and the introduced Pacific cupped oyster, *C. gigas* (Goulletquer and Héral, 1997). Culture of the flat oyster has been drastically affected by diseases, with production collapsing to less than 1% of past harvests

(1,500 metric tons a year). The overall French production is stabilized and supplies almost all the yearly consumption. Therefore, exports are rather limited (less than 3%), as are imports from European countries. Presently, the oyster industry is valued at 500 million euros, with more than 4,000 companies and 20,000 direct employees and occupying more than 50,000 acres of state leasing grounds and 10,000 acres of private oyster ponds. Because oyster culture is concentrated in bays and estuaries, usually in rural areas, it has played a critical role in shaping the landscape and stabilizing populations in otherwise economically depressed areas. The oyster industry plays a critical role in coastal zone management and employs environmentally sound practices in wetlands areas. Ninety percent of the production is based on the natural, reliable spatfall that occurs yearly in the southwest part of France. Abnormally low recruitment was observed in only four of the past 30 years. Spat is collected by carefully timed deployment of various types of spat collectors when larval density peaks during the spawning season (July-August). Although their market share is increasing, at present the five private hatcheries only produce about 10% of the total seed (triploids and selected strains) required at the national level to sustain production.

Over the past century the oyster industry experienced major upheavals with *C. gigas* being only the most recent of three species cultured by farmers (see Figure 3-1). Based on past problems, reliance on a single species creates a structural weakness in the industry. This has prompted research to anticipate future problems and assess options for diversifying the culture stock.

Historical Background

Demand for oysters increased significantly during the 19th century, leading to overfishing. Natural flat oyster beds distributed along the French coastline were physically destroyed and overfished, prompting state managers to establish and enforce new regulations based on quotas and seasonal openings. A commercial trade was initiated with Portugal to market the oyster *C. angulata* to meet increasing demand. Beginning in 1860, spat were imported to compensate for the lack of flat oyster spat collected on overfished beds. The first natural spatfall of *C. angulata* occurred in the Gironde estuary after a boatload of Portuguese oysters was jettisoned overboard during a storm. Because some of the oysters were still alive and environmental conditions were highly favorable, this species became established in the area. Later, natural reproduction facilitated invasion of this nonnative oyster from the southwest of the country north to the Loire estuary. At that time a large debate among farmers led to a temporary halt in transfers of oysters into northern Brittany waters. Even-

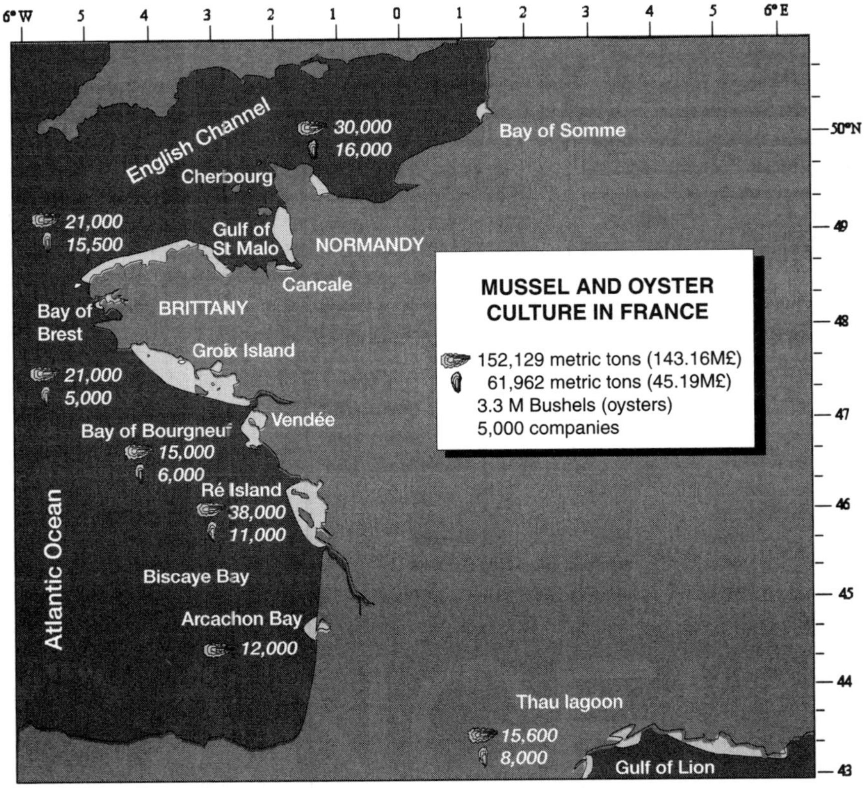

FIGURE 3.1 Mussel and oyster culture in France.
SOURCE: Modified from Goulletquer and Héral (1997).

tually, the Portuguese oysters were cultured outside their natural reproductive range in northern beds (e.g., Cancale) by transplanting seed. After 1920, production increased significantly when spat collectors began to be used systematically. *C. angulata* and *O. edulis* production reached 85,000 and 28,000 metric tons, respectively, in 1960, despite a major disease outbreak affecting the flat oyster in 1920 and low *C. angulata* recruitment in the 1930s. *C. angulata* replaced the native species in the main French rearing areas except Brittany, where oyster farmers were most successful in obtaining spat settlement from *C. angulata* at the same time they were having difficulty obtaining sets of the flat oyster. Although the disease agent was never identified, the massive die-off of flat oysters affected the main European populations simultaneously (Orton, 1924) and therefore has been considered the first oyster epizootic event in Europe.

In the late 1970s, two major disease-causing organisms, *Marteilia refringens* and *Bonamia ostreae*, spread and drastically reduced production of *O. edulis* in almost all rearing areas in France, as well as in Europe. Despite new management practices and an intensive repletion program, *O. edulis* production remains low 30 years later, with no anticipation of a rebound in the foreseeable future. The present approach to rebuilding the population is introduction of a disease-tolerant strain from an ongoing mass selection research program. The strain has been selected for more than 20 years and shows significantly higher survival and growth rates compared to natural populations. However, several questions remain to be addressed, including the potential genetic consequences of using a limited brood stock to restore wild populations. It should be noted that the parasite *B. ostreae* is an exotic species accidentally introduced to Brittany with a small batch of flat oysters originating from California (Grizel, 1997; Cigarria and Elston, 1997).

Concomitant with the increase in *C. angulata* production through the 1960s, abnormal mortality and reduced growth rates were both reported in the main rearing areas (Marennes-Oléron Bay and Arcachon Bay). This resulted from overstocking such that the carrying capacity of the ecosystem was exceeded, a consequence of poor management. From 1966 to 1969 the first outbreak of "gill disease" affected already-weakened oyster populations, leading to a decline in *C. angulata* production. Drastic mortalities from a second viral disease occurred between 1970 and 1973 and caused the total and rapid disappearance of this species from the French coast. Thirty years later, attempts to reintroduce *C. angulata* are still unsuccessful due to high mortality rates. The *C. angulata* collapse led to a major economic crisis, affecting 5,000 oyster farmers and generating a $90 million yearly loss. Therefore, an urgent solution was sought to prevent unemployment and sustain the ailing shellfish industry. A massive introduction of *C. gigas* originating from Japan and British Columbia was organized to provide spat to the industry and brood stock to establish new reefs and sanctuaries along the French Atlantic coastline.

Introduction of the Japanese Oyster *C. gigas*

The Pacific oyster, *C. gigas,* is native to Japan, far-eastern Russia, Korea, Taiwan, and China and has been introduced to all continents except Antarctica (Mann, 1979). Spat of *C. gigas* were first imported from Japan in 1966 by an oysterman looking for growth rate improvements at a time when *C. angulata* production was declining significantly because of overstocking. The nonofficial aspects of these imports and the concomitant increased mortality of *C. angulata* prompted the state agencies to ban further imports and to assume control of the issue. Scientists from Institut français de recherche pour l'exploitation de la mer (IFREMER) were in charge of developing the expertise to (1) evaluate the quality of oyster beds in Japan with regard to zoosanitary and sanitary status, fouling organisms, and oyster quality and (2) study how an introduction could be done to revitalize the industry. Experts concluded that *C. gigas* appeared healthy, and no relationship was established between the pathology observed in France, the mortality of *C. angulata,* and the status of oysters in Japan. Moreover, the environmental conditions sustaining *C. gigas* populations in Japan were considered similar enough to those on the French oyster beds to expect natural colonization after the introduction. Eventually, the introduction of *C. gigas* was approved because the whole industry was collapsing. The introduction was performed through large imports and massive release into the environment in two stages aimed at first building sanctuaries to establish brood stock and then supplying the oystermen with spat.

Sanctuaries—"RESUR" (meaning "resurrection")

After evaluating the sanitary quality of *C. gigas* beds in British Columbia, adult oysters were imported to constitute brood stocks. Histological controls were used for each transfer to ensure oyster zoosanitary status. This operation lasted from 1971 to 1973. The last additional batch was introduced in 1975 in the Marennes Oléron Bay. The sites were free from dying *C. angulata* populations, and environmental conditions were highly favorable (large unused carrying capacity, available habitat) resulting in successful establishment of populations in three areas, all located south of La Rochelle. A large natural spatfall following the introduction resulted from high larval survival rates under favorable temperature and salinity conditions. Presently, the sanctuaries are not maintained because the rearing and wild populations are sufficient to sustain spat supply. In contrast, the carrying capacity is now limiting, prompting the state managers to reduce those sanctuaries.

Spat Supply

From 1971 to 1977, spat was imported from Japan to sustain aquaculture. Each imported batch of spat was inspected and certified for origin, health status, and presence of predators. For disease, histological analyses were performed and an index of meat quality was established. Each batch was immersed in fresh water to reduce the risk of importing species such as the flatworm *Pseudostylochus* (Table 3.1).

The brood stock and spat that generated the present population may be more precisely estimated to be 456 metric tons of brood stock and 3,394 million spat. It should be emphasized that the latter represents 7 years of French hatchery production (500 million spat/year). Therefore, if a simi-

TABLE 3.1 Quantities of Spat and Brood Stock Imported During the Massive Introduction Effort in France from 1971 to 1976

C. gigas stage	1971	1972	1973	1974	1975	1976	TOTAL
Adult (metric tons)	117.5	236.5	173	0	35	0	562
Spat (estimated in millions)	1,226	2,413	365	965	34	5	5,008

SOURCE: Modified from Maurin and LeDantec (1979).

lar introduction were to be done again, hatchery production would technically be able to sustain such an operation under full compliance with the ICES protocol.

Reasons for the Success of the Introduction

The extensive introduction of *C. gigas* was so successful that production exceeded the record for *C. angulata* in less than 10 years. There are several reasons for the rapid increase in production. *C. gigas* has the biological advantage of resistance to the disease-causing organisms affecting *C. angulata* (viruses), *O. edulis* (parasites *B. ostreae, M. refringens*), and clams (*Perkinsus* spp.). Environmental conditions favored *C. gigas* expansion to such an extent that the natural spat supply quickly became sufficient for the entire industry. The ecological cost of this introduction has been limited because few areas were colonized outside the traditional rearing grounds. As early as 1975, spat originating from the sanctuaries and cultured populations were sufficient to sustain the industry. Yearly production is now estimated to be around 15 trillion spat in both Marennes-Oléron Bay and Arcachon Bay. No competition occurs between the Portuguese oyster, *C. angulata,* and *C. gigas* or with other species. Although hybridization has been observed experimentally, it was not a problem because of the very low brood stock of *C. angulata* and the differences in spawning pattern. Where oyster habitat was fully available and carrying capacity was high, optimal *C. gigas* growth was obtained (>70 g in 18 months). No inbreeding problem has been observed despite the relatively limited number of individuals contributing to the population. At least four invertebrate and three algal species were accidentally introduced during the massive operations even with stringent management. The introduced species have limited biomass and distribution and have negligible impact compared to other invasive species such as the limpet *C. fornicata* or species brought in ballast waters (Goulletquer et al., 2002). Similarly, *Haplosporidium sp.* was likely introduced with *C. gigas* as the healthy carrier, but it has remained at low prevalence and does not harm the current shellfish populations. It is unlikely that *Marteilia spp.* were introduced with *C. gigas* as once suggested, based on new genetic information on *M. maurini* and *M. refringens* as well as the worldwide status of the genus *Marteilia* (Berthe et al., 2000; Berthe, 2002). After 30 years of *C. gigas* production, the main management question is how to optimize stocking biomass with available carrying capacity and avoid overstocking effects. In specific areas, overstocking led to decreased growth and increased mortality rates. Where natural spatfall occurs (southwest coastline), management practices have been developed to limit expansion of natural beds that can compete for food with farmed populations. Sanctuaries are no longer in use because the farmed population is sufficient to sustain yearly recruitment.

Human Component

Economic pressure to carry out the species introduction was very high at that time and would be similarly high if the situation were to happen again. A critical step for the success of the introduction was that the government assumed control rather than allow unregulated importation of oysters by farmers. The massive introduction required a defined schedule and preliminary scientific investigations to maximize benefits and limit associated risks. Because *C. angulata* and *C. gigas* share many similarities, the existing aquaculture practices were directly applicable to *C. gigas* without major changes. Scientific advice to optimize recruitment (e.g., timing for spat collector deployment) contributed to the success of the introduction. The new oyster benefited from an overall acceptance by consumers, as in previous disease outbreaks when a new species was substituted. Oyster consumption has increased with production and a slight decrease in price.

Shellfish Introduction into French and European Waters: The Case of Manila Clam

During the 1970s, the shellfish industry was looking to diversify and carried out experimental trials using the Manila clam, *Ruditapes philippinarum*. Rather than a massive old-fashioned introduction, the first clams (150 individuals from Seattle) were obtained from hatchery production to reduce the risk of introducing disease and exotic species. The Manila clam showed larger growth rates than the native species *R. decussatus*, a high value, and a high acceptance by consumers. Because no natural spat were available, hatchery spat and scientific rearing techniques were used to increase production under a variety of environmental conditions. In 1985, clam culture practices using diploids were fully operational and production reached hundreds of metric tons in several European countries. In 1986, abnormal shell calcification affected several intertidal clam-rearing areas, causing abnormal mortality rates (Goulletquer et al., 1989). The Brown ring disease was later attributed to a vibrio infection (*Vibrio tapetis;* Paillard et al., 1989). The origin of Brown ring disease is unknown, but large international transfers of clams may have facilitated spread of the disease. Although no cases of the disease were observed in ponds, most of the intertidal clam culture areas were impacted, drastically reducing the landings. It should be noted that the endemic species *R. decussatus* has not been affected by *V. tapetis*. However, the main factor affecting clam culture was the species' invasive pattern. *R. philippinarum* extensively colonized natural beds in several areas in Europe, leading to an extensive public fishery (65,000 metric tons in 2000) that overtook aquaculture production. Since complete geographic overlap does not occur between the native and exotic clams, both populations are exploited. However, hybridization may occur in specific geographic areas.

Summary

- French oyster production (*O. edulis* and *C. angulata*) suffered from overfishing and five disease events during the 20th century that led to the decision to introduce two exotic species (*C. angulata* and *C. gigas*).
- Scientific and management efforts to reverse the declines of indigenous species were uniformly unsuccessful.
- The introduction of exotics facilitated the survival and expansion of the industry, generating employment and economic benefits.
- Ecological impacts from those introductions have been limited, although the ICES protocols were not followed (ICES, 1994).
- The French shellfish research effort is now focused on disease and its prevention, shellfish ecosystem management to optimize stocking biomass, and genetic studies to diversify and develop selected (and disease-resistant) strains to avoid future population declines (Figure 3.2).

Australia

In Australia shellfish have been commercially harvested since the early 1800s (Biosecurity Australia, 2002). Present production is mainly from aquaculture. Recent research on disease-resistant selected strains and new varieties has promoted expansion of hatchery products. The

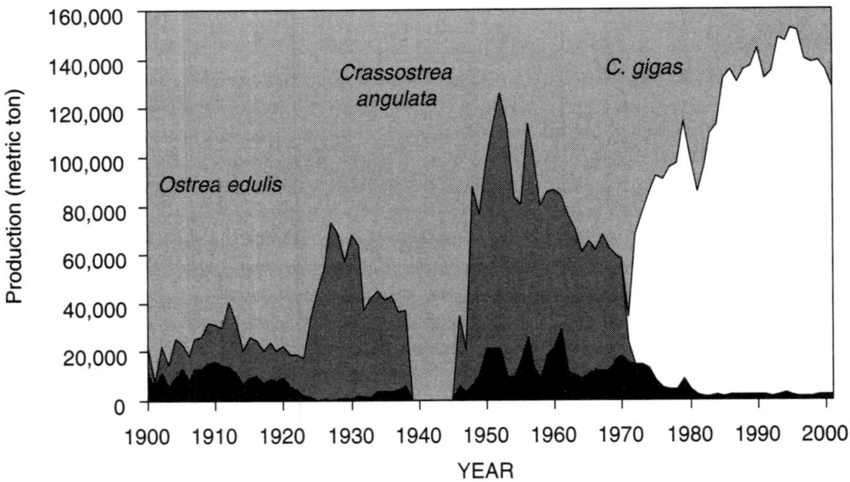

FIGURE 3.2 French oyster landings
SOURCE: Modified from Goulletquer and Héral (1997).

edible oyster species currently grown in Australia include the Sydney Rock oyster, *Saccostrea glomerata* (*S. commercialis*); the Pacific cupped oyster, *C. gigas;* and the flat oyster, *O. angasi*. The Sydney rock oyster is a native species cultured in estuarine areas and rivers, as well as embayments along the coast of New South Wales and Southern Queensland. Since 1988, landings of *S. glomerata* have declined from 7,919 metric tons to 5,000 metric tons in 1997. The seasonal occurrence of the parasite QX (*M. sydneyi*) has reduced the tidal areas where the Sydney rock oyster can be grown. Most of the production relies on natural spatfall and is located in New South Wales with 4,300 hectares of leasing grounds distributed among 41 estuaries. More recently, QX disease has shown a dramatic increase. Tests carried out in 2002 demonstrated that QX has doubled its range, affecting more than 10 major estuaries in New South Wales. Over $1 million has been spent in the past 10 years to come up with disease-resistant strains and fast-growing oysters.

The Pacific cupped oyster (*C. gigas*) is exotic to Australia. The first recorded importation in 1940 was seized and destroyed on the recommendation of the New South Wales state fisheries (Ayres, 1991). However, consignments were imported in 1947, 1948, and 1952 and planted in South and Western Australia and Tasmania. Only oysters planted in Northern Tasmania became established and were spread by larval transport. In 1955, stocks were moved to Victoria and South Australia for aquaculture. Those stocks survived and spread progressively northward to New South Wales and became established. In Tasmania the success of hatchery production led to a decline of the original local industry based on the flat oyster, *O. angasi*. The Tasmanian industry now relies entirely on hatchery production of *C. gigas* and represents 70% of total Australian production.

In New South Wales, Pacific oysters were first found in 1967, and to a greater extent in 1973. The Fisheries Department acted to eradicate the species. However, in 1985 significant numbers of *C. gigas* were found in Port Stephens, the main *S. glomerata* oyster-rearing area, likely resulting from an intentional introduction effort. Since then *C. gigas* has spread rapidly, and all efforts to eradicate the species have been unsuccessful. Although considered a marine pest in this traditional rearing area, the *C. gigas* landings have increased, from 1,516 metric tons in 1988 to 6,000 metric tons in 1997. Joint cultivation of *S. glomerata* with *C. gigas* is unworkable because *C. gigas* overgrows the native rock oyster.

Australian oyster production has increased (10,574 metric tons during the 1998-1999 season), although significant decreases in Sydney rock oyster landings have occurred, mainly due to diseases losses and reduced spat settlement due to *C. gigas* spat overset.

New Zealand

Exploitation of rock oyster *S. glomerata* beds under government control began in 1877 in the northern part of the North Island based on a licensing system (Dinamani, 1991b). Although several attempts were made to develop oyster culture at the turn of the 20th century, the results were disappointing. Between 1908 and 1960, government efforts were focused on maximizing spat settlement surface for recruitment and on managing natural oyster beds, including removing predators, to sustain oyster development. In 1964, the Rock Oyster Farming Act was passed for the purpose of establishing farms on leased areas. It was followed in 1971, by the Marine Farming Act to facilitate the culture of any marine species under a leasing and licensing system. Natural spatfall was irregular, and spat collector deployment suffered from space competition with other species. Although rock oyster production reached more than 500 metric tons in the mid-1970s, it began to be overwhelmed by the Pacific oyster, *C. gigas*, which was first found in 1971 (Dinamini, 1991a). Irregular spatfall during the 1970s prompted farmers to consider hatchery production. When *C. gigas* was introduced to New Zealand, there were some concerns about its impacts on the native species. *C. gigas* colonized much larger areas than the endemic species, especially in northern New Zealand where it outgrew *S. glomerata* on oyster leasing grounds (Smith et al., 1986). The *C. gigas* introduction displaced the active fishery based on the native species. By 1978, rock oyster production was eliminated and the total present oyster landings reached 3,500 metric tons in 2001 (Hay, 1999). No disease has been associated directly with this introduction, but it is having an impact on the coastal ecology; the native species has become so rare as a result of the spread of *C. gigas* that harvest of the native oyster is prohibited in some places. There are reports that the rock oyster can be found higher in the exposed intertidal zone than *C. gigas* (Hay, 1999), providing some refuge for the native species.

Summary

Positive economic benefits have been realized from introduction of the Pacific oyster into several areas of Australia and New Zealand, but there have been negative impacts on the native oyster *S. glomerata*. The fishery for the native oyster has declined or been closed in parts of New South Wales and New Zealand where *C. gigas* is considered a marine pest.

Western North America

The oyster industry and resource management agencies in the western United States and Canada have introduced a variety of nonnative shellfish species over the past 134 years. A review of these introductions provides insight for evaluating both the negative and positive consequences associated with the proposed introduction of *C. ariakensis* to the Chesapeake Bay. In addition, it provides an example of a shellfish industry largely dependent on hatchery-produced seed that is relevant to the development of aquaculture in the Chesapeake Bay.

Historical background

Native Oyster Beds Depleted by Fishery

The oyster industry on the West Coast of North America started with harvest of natural populations of the native oyster, *Ostreola conchaphila*. Native Americans harvested them where large concentrations were found. European settlers gathered native oysters for food and sale (Steele, 1957). Commercial exploitation began in earnest to supply hungry (and wealthy) gold miners during the California Gold Rush following the discovery of gold at Sutter's Mill in 1848. Beds of native oysters in San Francisco Bay were depleted within 3-years of the Sutter's Mill strike (Gordon et al., 2001; Shaw 1997; Barrett, 1963).

The range of *O. conchaphila* extends from Baja California to Sitka, Alaska, but the most abundant resources were discovered in Shoal-Water Bay (today referred to as Willapa Bay) in the Washington territory. A local settler in Shoal-Water Bay, Captain Charles J. W. Russel, reportedly traveled to San Francisco with several sacks of oysters in 1851. This marked the beginning of 20-plus years of a lucrative fishery, where oyster baskets costing $1 at the source could command as much as $30 in San Francisco's wholesale markets. Oysters were shipped by schooner, with one captain making 74 trips between 1875 and 1880 (Gordon et al., 2001). As many as 200,000 bushels per year were shipped from Shoal-Water Bay to the San Francisco markets (Scholz et al., 1984). The Shoal-Water Bay resource was depleted by the end of the 1870s, with little effort directed at conservation or enhancement.

Washington State Oyster and Tideland Laws

Washington became a state in 1890, and some of the first laws passed by the legislature were related to oysters. On March 26, 1890, a law was passed that provided for the appraisal and disposal of state tide and shorelands. It provided for planting of natural oyster beds, declaring them "natural oyster beds reserved." The beds were not to be sold or leased and were intended to provide seed for oyster farmers and stock for wild fishermen. Outside these beds, individuals could purchase up to 80 acres if developed for planting oysters. In 1903 another law was passed to "care for, protect, reseed, and replant the oyster reserves."

In 1895 two other oyster-related laws were passed by the Washington state legislature that allowed the state to dominate U.S. West Coast shellfish production. The laws, known as the Bush Act and the Callow Act, provided for the sale of tidelands into private ownership, specifically for the purpose of culturing oysters. The legislature recognized that an oyster cultivation industry could be encouraged to replenish the depleted wild stocks of native oysters through provisions for private ownership of the tidelands and shellfish grown there. The land returned to the state if the owner used the property for purposes other than oyster cultivation. In 1919 the Clam Act was passed to allow cultivation of clams and other edible shellfish on privately owned tidelands. The practice of selling state tidelands for shellfish culture stopped in the 1970s; however, land already purchased remained in private ownership. Because Washington's shellfish growers held clear title to the tidelands, they were able to obtain working capital through bank loans that were used to acquire more land that was then leveled and diked. With dikes, seawater remained over the oysters at low tide and served as a buffer against lethal temperature extremes.

After passage of the Bush and Callow Acts in 1895, the Washington oyster industry began to aggressively culture the native oyster. Efforts were particularly successful in southern Puget Sound, where more than 1,000 acres of tidelands were leveled and diked for native oyster culture (Lindsay and Simons, 1997).

Another consequence associated with private ownership of the tidelands is that Washington state shellfish growers have become aggressive defenders of water quality in the watersheds in which they farm. Sulfite waste liquor from a South Sound pulp mill, though never conclusively determined to be the cause, likely hastened the demise of the native oyster culture industry and stimulated the transition to the hardier Pacific oyster. Oyster growers brought suit against the pulp mills in the 1930s and 1940s, and litigation over water quality issues continues today with regard to nonpoint pollution sources.

Shellfish growers in other West Coast states rely on being able to lease tidelands or bedlands from their respective states. Washington state growers also lease tidelands from the state; however, the bulk of the production is from privately owned lands. In considering restoration of the oyster resources in the Chesapeake Bay, changing laws to allow private ownership may not be practical. However, long-term leasing for aquaculture could have a similar stimulatory effect on restoration efforts as the Bush and Callow Acts had in Washington state.

California *C. virginica*

Northern California oystermen made the first nonnative oyster introduction to the West Coast. Jealous that their Washington territory counterparts had taken over the San Francisco oyster markets and aided by the completion of the Union-Central Pacific transcontinental railroad, Californians brought the Eastern oyster, *C. virginica,* to San Francisco Bay. The October 22, 1869, issue of the San Francisco tabloid *Alta California* announced the arrival of the first shipment.

Early records of *C. virginica* imports to San Francisco Bay are sketchy and difficult to interpret since they are recorded under different measures, including pounds, gallons, bushels, barrels, shucked meats, and shells. An 1875 issue of the *Manufacturer and Builder* observed: "Every year some 500 carloads of small oysters are transported across the continent, to be brought up in the Pacific." The article further noted, "There appears to be a limit to the growth of any kind of Eastern oyster in Pacific waters; after a certain period, a year and half at the utmost, for some reason as yet not well understood—perhaps the meat becoming too large for the shell planted—the oyster dies."

Between 1887 and 1900, 33,480 bushels of Eastern oysters were shipped annually, constituting 80% of the total oyster production from San Francisco Bay during these years (Gordon et al., 2001). The availability of more popular Eastern oysters in San Francisco markets severely depressed the value of local oysters shipped from Washington. Prior to completion of the transcontinental railway, San Francisco buyers were paying $16 per sack of 1,000 native oysters. By the mid-1870s that price was down to $4 per sack and by the 1880s it was down to $2.50 per sack or about half the price being paid for Eastern oysters (Gordon et al., 2001).

From 1872 until the early 1900s the Eastern oyster dominated West Coast oyster production, peaking in 1899 with an estimated 1,134 metric tons of shucked oyster meat (Conte et al., 1996). Urbanization of the San Francisco Bay area degraded water quality, and by 1908 Eastern oyster production had fallen 50%. The industry stopped importing Eastern oyster seed in 1921, and the last of the San Francisco oysters were commercially harvested in 1939 (Conte et al., 1996).

Washington *C. virginica*

Washington oystermen who wanted to participate in the Eastern oyster business had to wait for a northern spur of the transcontinental railroad. After an unsuccessful planting of 80 barrels in 1895, the state fish commissioner's staff planted an assortment of *C. virginica* from Connecticut, the Chesapeake Bay, New York's East River, Massachusetts, and Maryland's Princess Bay with a $7,500 appropriation from the Washington state legislature in 1899. With proper handling and planting in appropriate locations in Shoal-Water Bay, the oysters thrived.

Within a decade Washington's *C. virginica* industry employed several hundred people and was valued at around $1 million. Production climbed from 4,000 gallons in 1902 to 20,000 gallons in 1908. As a 1906 issue of the *Willapa Harbor Pilot* said: "Many will remember and smile at the great interest, not to say excitement, manifested over the shipment of 80 barrels of eastern oysters to this bay in 1895, while now the importation of from 50 to 100 carloads annually excites little comment" (Gordon et al., 2001).

Transplantation of Eastern oysters to the West Coast in the late 1800s and early 1900s marked a significant transition in the industry from an extractive to a culture-based fishery. Washington's production of *C. virginica* continued to grow until 1919 when the stock collapsed due to an unknown cause. Some speculate that a seed shortage in 1913 coupled with a large harvest in 1918 led to the collapse (Gordon et al., 2001). Today there are no natural established populations of *C. virginica* in Washington state although the industry continues to cultivate them in limited numbers with hatchery-reared seed.

British Columbia *C. virginica*

In 1903, *C. virginica* was introduced to British Columbia in Boundary Bay, Esquimalt, and Ladysmith harbors and Hammond and Nanoose bays. The only site with good yield was Boundary Bay, where *C. virginica* continued to be introduced until around 1940. Because of poor results obtained with transplanted seed, 3- to 4-year old oysters were transplanted instead and kept for about a year in Boundary Bay prior to marketing. A mortality of roughly 25% occurred even with the larger oysters (Quayle, 1988). Successful breeding of *C. virginica* occurred only in the estuaries of the Nicomekl and Serpentine rivers that flow into Boundary Bay.

Washington *C. gigas*

The first documented introductions of the Pacific oyster, *C. gigas*, to Washington took place in the early years of the 19th century. The first large commercial planting was in 1919 by the Pearl Oyster Company in Samish Bay. Four hundred cases of mature adult *C. gigas* (approximately 800 bushels) from the Miyagi Prefecture were shipped in early April 1919 (Lindsay et al., 1997). The oysters were covered with rice matting and given frequent showers of seawater until they arrived 18 days later. Many were dead on arrival, but the young oyster spat attached to the adults survived.

The discovery that spat were hardier than adults shifted the importation strategy to shipments of spat. The wooden cases for shipping seed could contain 12,000 to 30,000 spat averaging 6 to 18 mm in size depending on the size of shell cultch. Pacific oyster seed imports began slowly, reached a peak of nearly 72,000 cases in 1935 and then declined. Oyster seed imports ceased during World War II but began again in 1947 and continued yearly with the exception of 1978 and 1979 (Chew, 1984).

During the summer of 1936, Pacific oysters bred successfully in Dabob Bay on the Hood Canal, southern Puget Sound, and Willapa Bay. The resulting adult oysters provided a large brood stock as well as an important source of market oysters to carry through World War II when Japanese seed shipments were halted. This 1936 spawning by many accounts was huge, particularly in Willapa Bay. Al Qualman in his personal history, *Blood on the Half Shell* (1990), wrote: "There was no place to run, no place where the water would not be filtered and refiltered a thousand times over by the fantastic numbers of hungry little mouths." The overabundance of oysters that resulted from this massive spawning forced the industry to expand its markets. Canning proved to be the answer because it extended shelf life and allowed oysters to be shipped by rail or airplane. The War Department recognized the nutritional value of oysters and in October 1943 approached the industry about supplying the troops. In response, Willapa Bay production numbers nearly doubled from 1943 to 1946 and reached a historical peak of 43,000 metric tons.

After the war the industry was almost driven from the canned oyster market by competition from Korea and Japan. At the time the wholesale price for a carton (48 cans, 10 ounces each) of Willapa Bay oysters was about $14, while cartons of Korean oysters were selling on the open market for $5.65, including shipping costs and import duty (Gordon et al., 2001). With this price advantage, foreign imports might have wiped out the West Coast industry's fresh-shucked and live markets were it not for provisions in the Lacey Act that banned the importation of any insects or other wildlife that might threaten crop production. Because Korean oysters were diagnosed as disease carriers, the importation of fresh-shucked or live oysters into the United States was banned. This ban protected the domestic industry by reducing the level of foreign competition in the fresh-shucked and live markets, thereby ensuring a market niche for the domestic industry.

Subsequent warm years in 1942, 1946, and 1958 resulted in excellent spat sets again in Willapa Bay, Dabob Bay, and southern Puget Sound. In 1942 the Washington Department of Fisheries initiated efforts to study spawning to better predict spatfalls for the industry. Spatfall predictions by the department continued into the 1990s, with many growers putting down cultch every year to supplement Japanese seed and later hatchery seed. Today, natural sets in these bays continue to provide a substantial percentage of Washington state's oyster production.

Production of Pacific oysters on the West Coast has been measured traditionally in terms of gallons. Between 1937 and 1993 in Washington production ranged from 458,000 to 1,553,000 gallons. This would translate to roughly 12,700 to 43,000 metric tons of live oysters, assuming a gallon of shucked oysters weighs 8.75 pounds and live oysters are seven times heavier than shucked meats. Washington state oyster production was estimated by the Pacific Coast Shellfish Growers Association in 2000 at 35,000 metric tons live weight. Most of the production consisted of *C. gigas*. There are no statistics documenting what percentage of this amount is attributed to natural production in the Hood Canal and Willapa Bay versus hatchery-reared seed.

British Columbia *C. gigas*

C. gigas was first introduced to British Columbia around 1912 or 1913, when a few were planted in Ladysmith Harbor and Fanny Bay. By 1925 there was evidence of successful breeding in the harbor. The first significant importation occurred in 1926 with approximately 2,000 adult oysters from Samish Bay, Washington, and 20 cases of seed from Japan. Small numbers of naturally set oysters were found as a result of spawnings in Ladysmith Harbor in 1926, 1930, and 1931. The first major spawning of Pacific oysters in British Columbia occurred in 1932 in Ladysmith Harbor, where natural spawning continues periodically. Subsequently, plantings of imported Pacific oyster seed took place throughout the Strait of Georgia, and in 1942 massive spawning in these areas resulted in a "catch" of oyster seed on many beaches particularly in the northern Strait of Georgia and Gulf Islands. As in Washington, this large set in 1942 carried the industry through World War II when Japanese seed shipments halted.

In 1958, widespread successful spawning of the Pacific oyster throughout the Strait of Georgia basically supplanted the need for importing Japanese seed. Pendrell Sound was established as a seed catching area (no leases allowed), and importation of Japanese seed ended around 1961. As in Washington state and the rest of the North American West Coast, the British Columbia industry today has transitioned to a significant reliance on hatchery-produced Pacific oyster seed. According to the British Columbia Shellfish Growers Association, 1999 production of Pacific oysters totaled 5,800 metric tons.

California *C. gigas*

When the Eastern oyster industry collapsed in the 1920s and 1930s, the industry and state explored the option of using *C. gigas*. In 1929 the California Department of Fish and Game (CDFG) and the industry made experimental plantings of *C. gigas* in Tomales Bay and Elkhorn Slough. In the 1930s additional experiments were carried out in Drakes Estero, Bodega Lagoon, and Morro, Newport, and San Francisco bays. Humboldt Bay was not included in the areas planted with *C. gigas* because CDFG was working there to reestablish the native oyster, *O. conchaphila* (Conte et al., 1996).

Many of these *C. gigas* plantings were successful, and seed shipments continued through the 1930s, halted during World War II (1940-1946), and then expanded again in 1947. Seed was inspected prior to shipment from Japan by CDFG personnel for organisms considered harmful to California waters. The CDFG lifted restrictions on *C. gigas* seed in Humboldt Bay in 1953 (Conte et al., 1996), and for the next 30 years seed shipments of *C. gigas* from Japan revived the California industry. Today approximately 90% of California's oyster production comes from Humboldt Bay and Drakes Estero. The Pacific Coast Shellfish Growers Association estimates California's 2000 *C. gigas* production at 4,500 metric tons.

Washington *C. sikamea*

After World War II, Washington's supplies of Pacific oysters were depleted. The seed oyster industry in Miyagi and Hiroshima were set back by damages sustained in the war. In 1946, the first seed in 6 years was shipped, containing 30 cases of juvenile *C. sikamea* oysters from the Kumamoto Prefecture. The Kumamoto oysters are less likely to spawn and recruit in the northwestern United States because they originate from a more southern and warmer prefecture than *C. gigas*. In 1960 and 1961, 1,004 and 1,200 cases, respectively, were imported (Steele, 1964). Careless brood stock management allowed hybridization with *C. gigas*, such that pure stocks of the Kumamoto oyster were hard to find in the early 1990s, when molecular markers for distinguishing the two species were developed (Banks et al., 1993; Hedgecock et al., 1993). Genetic testing of brood stock and more careful management has allowed the industry to rebuild productive Kumamoto oyster stocks. Kumamoto oysters are petite (50 cm) and have a deep cup that has been extremely popular in the live oyster market. Production continues to increase, and the industry is currently considering expanding the gene pool with oysters from Kumamoto Prefecture in Japan.

West Coast Transition to Hatchery-Based Production

During the 1980s there was a significant shift in the entire U.S. West Coast shellfish industry away from Japanese seed to seed produced in West Coast hatcheries and nurseries. Today, virtually all of the oysters grown in California, Oregon, and Alaska are reared from larvae produced in Washington and Oregon hatcheries. Washington and British Columbia have areas where natural spatfall occurs. In these areas (Willapa Bay and Hood Canal in Washington, Pendrell Sound and the Strait of Georgia in British Columbia) the industry captures seed and harvests these naturalized oysters. In areas with natural recruitment, the resource supports a substantial recreation harvest.

Currently, many West Coast companies are expanding production of single "cultchless" oysters for live or processed "half shell" markets and reducing "cultched" oyster production for the shucked market. This transition has been market driven. Taylor Shellfish Company, one of the largest producers of oysters on the West Coast over the past 8 years, has reversed production from 80% shucked, 20% live to 20% shucked, 80% live.

The transition to hatchery-produced seed has facilitated the use of genetically enhanced oyster stocks. Since the early 1990s, oyster yields have increased due to the success of crossbreeding and selective breeding programs. Growers have transitioned to triploid oyster production for higher growth rates, improved yields, and firm, glycogen-filled oysters during the summer when reproductive oysters are depleted from spawning. Of the 37.5 billion eyed larvae currently produced in West Coast hatcheries, about 12 billion are triploid (Nell, 2002).

Also, hatchery technology has allowed the West Coast shellfish industry to expand and diversify through culture of more species and provision of reliable seed resources to ensure harvests in available growing areas. Hatcheries provide a stable source of seed at predictable prices, enabling sound business planning, bank financing, and market development. Finally, hatchery production of seed has substantially reduced the risk of unintentional species introductions that were common in the early years when seed on cultch was transported within the United States and internationally. Species diversification helps buffer temporary price drops and losses from species-specific diseases.

Unintentional Species Introductions Attributed to the Intentional Introduction of *C. gigas* and *C. virginica* to the West Coast of North America

Lack of adequate regulatory protections associated with the movement of live *C. gigas* from Japan and *C. virginica* from the East Coast resulted in the introduction of 27 species of Asian and Atlantic molluscs to the Pacific Coast of North America (Carlton, 1992b). In 2000 a survey for nonnative species in Elliot Bay (Seattle), Totten and Eld inlets (southern Puget Sound), and Willapa Bay suggests that oyster seed shipments from Japan and the U.S. East Coast may be responsible for 35 of 40 exotic species collected (Cohen et al., 2001).

The consequences of an unintentional species introduction became apparent to West Coast oystermen with the introduction of *Ceratostoma inornatum*, the Japanese oyster drill, in cases of oyster seed from Japan. The drill is particularly damaging to newly seeded Pacific oyster crops. The Japanese drill also attacks exposed Manila clams, and predation by this drill has hampered recent efforts to restore beds of native oysters. An adult *C. inornatum* requires about 2 weeks to drill through a 5-cm-length Pacific oyster and only a day or so for seed oysters up to 2.5 cm in diameter (Quayle, 1988). A major inspection program was established in 1947 with state biologists from Washington and California checking oyster seed shipments in Japan for pest species prior to seed shipment. Additional inspections were made on arrival from Japan. Regulations have subsequently been adopted in West Coast states to prevent accidental introductions of nonnative species or transfer of disease. Washington initiated regulations in 1945 to prohibit transfer of Japanese drills among oyster plantings in the state. These regulations, together with the transition to hatchery-produced seed, now prevent the unintentional introduction of nonnative species due to the activities of the shellfish industry.

A couple of unintentional species introductions have been economically valuable. Most notable is the Manila clam, *Venerupes philippinarum*. The Manila clam was introduced into the Puget Sound during the 1930s and 1940s with shipments of Pacific oyster seed from Japan (Chew, 1989;

Quayle, 1938, 1988). Since its introduction it has become established throughout the Puget Sound region. It reproduces successfully in a number of areas in Puget Sound, Willapa Bay, and British Columbia. Production in 2000 in Washington state alone was estimated by the Pacific Coast Shellfish Growers Association to be in the neighborhood of 7 million pounds valued at $14 million. Second only to the native geoduck clam, the Manila clam is one of Washington's most valuable commercially harvested shellfish.

The Eastern soft shell clam, *Mya arenaria*, is another unintentional species introduction of economic value. The soft shell clam was apparently brought to the West Coast in shipments of the Eastern oyster in the early 1900s. *M. arenaria* became established in a number of West Coast bays at higher tidal elevations. The species has done particularly well in Skagit Bay and Port Gardner in northern Puget Sound, where in the 1980s the natural population was sustaining annual harvests in the 135 to 180 metric tons range. These were all being shipped to East Coast markets. Recently, the resource has been largely inaccessible to commercial harvest because pollution has forced the Washington State Department of Health to downgrade the areas.

Nonbeneficial or harmful species introductions have been more common. Several examples of detrimental unintentional species introductions are described below.

Urosalpinx cinerea, the Atlantic oyster drill introduced to the West Coast with shipments of Eastern oyster seed, was primarily especially a problem when Eastern oysters were raised. It preys on Pacific oysters as well and thus remains a problem for the West Coast shellfish industry. Its distribution extends from British Columbia, Canada, to Newport Bay in Southern California. It is a predatory snail that feeds on barnacles and mussels as well as oysters. Both the Eastern and Japanese oyster drills may prove problematic to current efforts to restore populations of the native West Coast oyster, *Ostreola conchaphila*.

Pseudostylochus ostreophagus is a flatworm that is thought to have been introduced with Japanese oyster seed shipments. It is predatory on juvenile mussels and oyster seed.

Mytilicola orientalis is a parasitic copepod originating in Japan. It occurs in the lower intestine of oysters and mussels. It appears to cause no damage other than reduced condition of the meat tissue. *M. orientalis* was probably brought to Europe in shipments of *C. gigas*. Its occurrence in Pacific oysters was first reported by His (1977). However, another copepod (*M. intestinalis*) was already present, and the impact *of M. orientalis* was only evident at high infestation rates. *M. orientalis* is considered a parasite because it sometimes induces metaplasia in oysters (Deslous, 1981). The zoosanitary monitoring network in France and Europe does not report infestations of *M. orientalis* and the overall impacts of this introduced species are not considered significant (Leppäkoski et al., 2002).

Crepidula fornicata was introduced with the importation of Eastern oyster seed. It has become well established in Totten and Eld inlets in Southern Puget Sound and has also been documented in Willapa Bay (Cohen et al., 2001). This species competes for food with oysters, clams, and other filter feeders that consume phytoplankton. *Crepidula* has not caused problems where phytoplankton resources are abundant such as in southern Puget Sound. However, in the right circumstances, this species could become invasive. In France, populations have become invasive and the oyster industry must control them due to the competition for limited phytoplankton resources.

Sargassum muticum, a Japanese seaweed commonly referred to as wireweed, was introduced with Japanese oyster seed. It first gained a solid foothold in British Columbia waters between 1941 and 1945 (Quayle, 1988). It is thought to have been introduced to the Puget Sound in the 1930s with Pacific oyster seed. A highly invasive species, today it has established itself from southern Southeast Alaska to Baja California, Mexico (O'Clair et al., 2000). It requires rocky bottom in the lower intertidal and has been known to attach to Pacific oysters with its holdfast. The mature alga has thick fronds reaching up to 2 m in length. During storm events, rough seas may dislodge the seaweed and attached oyster causing mortality if they wash ashore. The Monterey Bay Aquarium Research Institute examined the epibiont community associated with *S. muticum* in northern Puget Sound (San Juan Islands) from May to September 1997 (Osborn, 1999). This examination found that epibiont diversity and abundance increased in areas invaded by *S. muticum* because of the increased habitat, productivity, and complexity that *S. muticum* provides. Based on the positive effects on the epifauna, their recommendation was to not attempt to eliminate the species but to further study potential negative effects on water movement, light penetration, sediment accumulation, and anoxia at night.

Zostera japonica, Japanese eelgrass, is thought to have been introduced with shipments of Pacific oyster seed from Japan. First reported in Willapa Bay, Washington, in 1957, it is colonizing midintertidal tideflats (+0.6 to +1.2 m above mean lower low water), which previously lacked macrophyte cover (Bigley and Harrison, 1986). Its range currently extends from southern British Columbia to Coos Bay, Oregon (Posey, 1988). Posey researched the effects of *Z. japonica* invading Coos Bay and found a general positive effect on local species diversity and abundance. In Boundary Bay, British Columbia, where *Z. japonica* is well established, it was found to be a much more suitable food for dabbling ducks than the native *Z. marina* (Posey, 1988). In concluding his paper assessing the *Z. japonica* introduction, Posey offers advice relevant to the *C. ariakensis* question. He suggests "consideration should be given to the possibility that an established exotic may not necessarily be detrimental to native organisms."

In Washington state, *Z. japonica* is present throughout Puget Sound, Hood Canal, and Willapa Bay. Anecdotally, shellfish growers claim it is currently on the increase. It causes problems when it heavily infests Manila clam beds primarily by reducing drainage and obstructing clam harvest. It is interesting to note that the sand/mud substrate in parts of Willapa Bay was not stable enough to allow Manila clams to recruit and survive until *Z. japonica* invaded the area (W. J. Taylor, Taylor Shellfish Company, Shelton, Washington, personal communication, 2003).

Spartina alterniflora, or Smooth Cord Grass, is thought to have arrived on the West Coast when it was used as packing in freight cars carrying loads of Eastern oysters to Willapa Bay. It established itself in the bay over 100 years ago but only became invasive in recent years. Today it is estimated to be filling in the higher intertidal mudflat regions of Willapa Bay at an average rate of 1 acre per day. As Spartina invades shallow mudflats, it traps sediments that build up and allow it to expand farther out away from shore. The Spartina is taking over critical habitat for shellfish, salt marsh plants, migratory birds, juvenile fish, and other wildlife. Massive efforts are being undertaken to eradicate it at great expense. Spartina has also infested parts of Puget Sound but was more likely deliberately introduced by duck hunters as a duck blind or by cattle operations as an alternative food that would grow well in the salty soil recovered by building dikes in the bays.

Although many nonnative species were probably introduced to the West Coast with foreign oyster seed, only one significant disease is thought to have resulted from such shipments. *Mikrocytos mackini*, a parasite that lives in the glycogen storage cells of the Pacific oyster and causes Denman Island Disease (Elston, 1990), is suspected to have been introduced, but its foreign origin has not been confirmed. First reported in 1960 in Henry Bay on Denman Island, British Columbia, it has since been noted at other locales in the Strait of Georgia. In 2001 it was discovered in oysters in Sequim Bay in Washington state, triggering emergency regulations restricting live oyster movements from potentially infected areas. Mortality rates up to 53% in a single season have been reported, but severity fluctuates from year to year (Elston, 1990). The disease mainly affects older oysters that occur at lower tidal elevations.

Summary

- Driven by depleted natural stocks of the native oyster, *O. conchaphila,* and aided by the transcontinental railroad, massive introductions of *C. virginica* were made initially in California (1869 to 1921) and later in Washington state (1894 to 1919) and British Columbia (1903 to 1940). The Eastern oyster naturalized (became self-populating) for a time in Willapa Bay, in southwest Washington, and Boundary Bay, British Columbia. *C. virginica* continues to be farmed with hatchery-produced seed on a small scale on the West Coast.

- *C. gigas* was introduced first to Washington state (1904), then British Columbia (1912 to 1913), and later California (1929). Massive amounts of seed attached primarily to shells imported annually between 1919 and 1979 from Japan, with a short break during World War II. Successful breeding was noted first in Ladysmith Harbor, British Columbia, in 1932 and Dabob Bay, Southern Puget Sound, and Willapa Bay, Washington, in 1936. In notably warm years successful breeding continues to occur in certain areas of Washington and British Columbia; however, the Pacific oyster fails to breed over most of the West Coast of North America, limiting its ability to invade new areas.
- There were few or no protections to prevent unintentional species or disease introductions with oyster seed imports from the East Coast or Japan until 1945. As a consequence, numerous nonnative species were introduced and became established. Many of these introduced species cause ecological and economic harm. At least two introduced clam species have resulted in economic benefit through the establishment of lucrative aquaculture and fishing industries.
- State regulations restricting interstate and international seed importation coupled with the transition to hatchery-produced seed have dramatically reduced, though not eliminated, the risk of unintentional species or disease introduction.
- Hatcheries and nurseries capable of supporting the majority of the North American West Coast oyster industry (estimated 48,000 metric tons annually) evolved over a period of 20 to 30 years. Roughly a third of this production is triploid.
- The transition to hatchery production on the West Coast has facilitated diversification of the industry and supported breeding programs for superior aquaculture stocks.
- Laws allowing the private ownership of tideland in Washington state have served as a critical foundation for a stable and expanding industry.

Eastern and Gulf Coasts of North America

The Pacific oyster, *C. gigas*, and the European or flat oyster, *O. edulis*, have been introduced to the East Coast of North America multiple times (see Table 3.2). Before 1950 some of these introductions were documented in the scientific literature by the scientists who deliberately introduced the new species. This included scientists associated with state fish and wildlife departments. These introductions were small scale and made with the purpose of testing whether the species would be suitable for commercial production. Survival, growth, and reproduction were monitored. Since 1950, two experimental introductions of *C. gigas* in Maine (Dean, 1979) and Massachusetts (Hickey, 1979) have been reported officially. Others reported have been anecdotal accounts of *C. gigas* introductions (Andrews, 1980; Mann et al., 1991; Carlton, 1992b).

Importation of limited numbers of *O. edulis* into New England in the 1940s and 1950s led to the establishment of small, naturally reproducing populations in Maine and Rhode Island, where the species is now also grown in aquaculture. In contrast, introductions of *O. edulis* into the Canadian Maritime Provinces apparently did not naturalize (Medcof, 1961).

In contrast to the establishment of *O. edulis*, no naturalized populations of *C. gigas* have been reported to exist on the East or Gulf Coasts of North America, despite the known and suspected introductions (the latter being undoubtedly more numerous than the literature indicates), including one or two large-scale plantings in Chesapeake Bay (Table 3.2). With a single exception, all of the reported introductions of *C. gigas* have been to the mid-Atlantic, New England, and Canadian Maritime Provinces. A single published account exists for the southeastern United States and Gulf of Mexico and describes the testing of *C. gigas* in Louisiana and Alabama. In this environment the oysters were emaciated and heavily infested with a polychaete worm (*Polydora* sp.) that caused mud-filled blisters on the inner shell; they eventually died (Kavanagh, 1940).

It is probable that water temperatures are too low for successful reproduction of *C. gigas* in the northeastern North America. This is not true of the mid-Atlantic, however, where experimentally held *C. gigas* underwent gametogenesis and spawned in laboratory tanks that received ambient water (Barber and Mann, 1994). It is possible that the numbers of oysters introduced into the mid-Atlantic were simply too low to establish a wild population. It should be noted that the establishment of *C. gigas* in France occurred after an estimated 562 metric tons of adults and several billion spat were imported and planted between 1971 and 1975 (Grizel and Héral, 1991). Further, naturally occurring populations of *C. gigas* on the West Coast of North America developed only after numerous introductions of seed from Japan (Carlton, 1992b), although reproduction is definitely limited by low temperatures in this region.

TABLE 3.2 Reported Introductions of Nonnative Oysters to the East and Gulf Coasts of North America Prior to the Recent Testing of *C. ariakensis*

Date	Species	Location	Comments	Reference
Early 1930s	*C. gigas*	New Jersey	One bushel, placed in Barnegat Bay, grew rapidly during first 2 weeks, then stopped and gradually died out without further growth. The poor performance may have been due to the relatively low salinity (12-20 ppt), with low dissolved oxygen and stagnation.	Nelson (1946)
1930s	*C. gigas*	Louisiana and Alabama	At least two introductions of several "cases," presumably from the U.S. West Coast, were made into Mobile Bay, Alabama, and later transferred to Louisiana. All died. A smaller lot imported later and placed in a flow-through system for closer study rapidly lost meat condition and acquired heavy mud blisters (*Polydora* sp.).	Kavanagh (1940)
1949	*C. gigas*	Massachusetts	Six bushels of spat from Washington state, grown in Barnstable Harbor on Cape Cod over 1.5 years, grew to 5 inches with little evidence of mortality or spawning in the field.	Turner (1949)
1949	*C. gigas*	Maine	6 bushels of spat from Washington State suffered 40% mortality from starfish but otherwise no losses.	Turner (1949; 1950)
1949	*C. gigas*	Maine	A "few bushels" were introduced; gonadal maturation occurred, but no spat were found.	Dow and Wallace (1971) cited in Dean (1979)
1949 and 1955	*O. edulis*	Connecticut and Maine	In 1949 9,000 oysters in three year-classes were brought from Holland: 3,000 to Booth Bay Harbor, Maine; 6000 to Milford, Connecticut.	

			In 1955 another batch of 4,000 from Holland were imported into Maine. Survival was good in both Connecticut and Maine. Successful reproduction and persistence of population was observed in Maine, where small populations remain. Some individuals were later found in Rhode Island, including at borders of Connecticut and Massachusetts.	Loosanoff (1955), Welch (1963), Zabaleta and Barber (1996), R. Rheault, (Moonstone Oysters, Pt. Judith, RI pers com 2002)
1957, 1958, 1959	*O. edulis*	Prince Edward Island and New Brunswick, Canada	About 5,000 individual wild-set oysters were imported each year from England. The shipments were examined macroscopically for parasites and hitchhikers. The oysters did well during the summer but suffered heavy losses during severe winters.	(Medcof, 1961)
Pre-1950	*C. gigas*	Long Island Sound	No information was available other than "numerous attempts at introduction."	Davis (1950) and personal communication in Dean (1979)
Early 1960s	*C. gigas*	Delaware	An unknown quantity of oysters was held in trays in Rehoboth Bay for several years without serious mortality or apparent reproduction.	Andrews (1980)
1971-1973	*C. gigas*	Maine	Oysters were imported for limited controlled experiments on the growth of juveniles in Maine waters. Fast growth and good gonadal development but no spawning was reported.	Dean (1979)
1976-1977	*C. gigas*	Massachusetts	Some 100,000 seed (6-8 mm) from West Coast hatchery were placed in an enclosed pond off Buzzards Bay. About 80% were lost shortly thereafter during a hurricane and presumed buried in mud as they were never found. The oysters grew well, and spawning occurred "more or less continuously" during the summer of 1977. Larvae were found in the water, but no spat were ever recovered even though cultch was placed in the pond.	Hickey (1979)

continued

TABLE 3.2 (continued)

Date	Species	Location	Comments	Reference
1970s?	*C. gigas*	Maryland	"Recently, *C. gigas* from the west coast of North America was planted in Maryland waters (of Chesapeake Bay] by a seafood dealer which resulted in a specific law in that state prohibiting the species. The oysters were recovered as completely as possible by scuba diving."	(Andrews, 1980)
~1988-1990	*C. gigas*	Maryland or Virginia	Thousands of bushels were reported to have been planted in Chesapeake Bay.	(Carlton, 1992a)
1990s	*C. gigas*	Lower Chesapeake Bay	Testing of sterile triploids at three salinities for growth, survival, and resistance to Dermo and MSX diseases was conducted by the Virginia Institute of Marine Science. They were deemed unsuitable for Chesapeake Bay because of poor performance in low salinity and the acquisition of heavy mud (*Polydora* sp.) blisters.	(Calvo et al., 1999)

CHAPTER SUMMARY

- Estuarine habitats frequently contain large percentages of nonnative species because these areas are heavily used for shipping, fishing, and recreation. Introductions of nonnatives occur mainly through human activities. Nonnative shellfish have been transported from region to region for hundreds of years, and oysters have been intentionally transported more than any other marine species.
- The incentive for introduction of commercial marine species has been to replace depleted natural stocks or to diversify the types of species used in aquaculture. Although molluscan aquaculture typically has relatively small environmental impacts relative to culture of fish and shrimp, diseases and "hitchhikers" have been transported along with the shellfish. A number of these unintentional introductions have caused severe impacts on shellfish production and ecosystem structure and function.
- While there are no documented cases that an intentional introduction of a marine bivalve species in U.S. waters has caused pronounced environmental change, there are a number of well-documented cases of how unintentional introductions have resulted in both positive and negative impacts.
- It is exceedingly difficult to predict whether a marine species has the potential to become an "invasive" or a "nuisance" species. Species that exhibit fast growth and high reproductive rates and are tolerant of a wide range of environmental conditions generally tend to be viewed as candidate "invasive" species.
- Ecosystems with reduced biodiversity or that are environmentally degraded may be more vulnerable to invasion. In addition, the southern and northern boundaries of faunal provinces may be more susceptible to invasion by nonnative species as a consequence of climate change.
- A review of a number of case studies of intentional shellfish introduction indicates that there are both benefits and risks dependent on the region. In France, introduction of the Pacific oyster, *C. gigas,* led to revitalization of the oyster industry, but in New Zealand and Australia, the Pacific oyster outcompeted the commercially viable native oyster in some areas.
- The oyster industry on the U.S. West Coast relies almost exclusively on nonnative species. The success of this industry is largely based on culture of several types of molluscs and regulations that allow private ownership of tidelands.

4

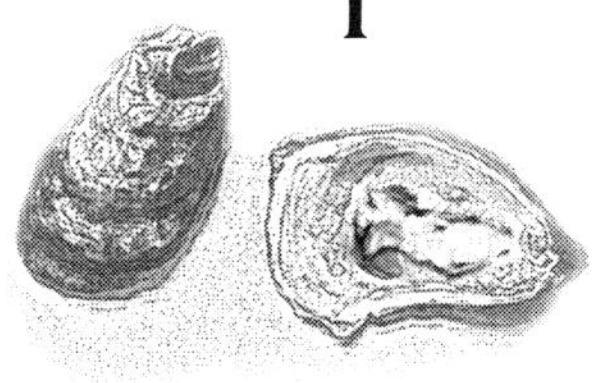

Oyster Biology

GENERAL BIOLOGY OF OYSTERS

Oysters are members of the family Ostreacea, class Bivalvia, in the phylum Mollusca. Under the current systematic schema, most commercially important species are classified in three major genera: *Ostrea, Saccostrea,* and *Crassostrea* and a number of minor genera (Carriker and Gaffney, 1996). Adults are intertidal and subtidal bottom dwellers found worldwide. Most oyster species form the basis of local fisheries or aquaculture operations.

Oysters differ from other bivalves in having a highly irregular shell form. The shape of the shell is typically dictated by environmental constraints, and they are capable of growing over or around adjacent objects, including other oysters. Oysters are plankton feeders; they use their gills to filter microalgae and probably bacteria. During feeding, they relax their single adductor muscle, allowing the two valves of the shell to open slightly. In an action called "pumping," specialized cilia on the gill draw water into the shell cavity (Newell and Langdon, 1996). Other gill cilia trap particles and funnel them toward the palps—large liplike structures, also covered with cilia that surround the mouth and on which particles are sorted. Some particles, such as microalgae, are sent into the mouth; others, such as sediment, are usually rejected and deposited as "pseudofeces" just outside the shell. Filtration rates are a function of several environmental factors, including temperature, salinity, and suspended particulate concentration. Rates increase with size, although per unit weight, small oysters filter more water than do large individuals

(Shumway, 1996). Powell et al. (1992) reviewed the literature on filtration rates for numerous marine bivalves and found that the relationship between filtration rate and size was similar for all species examined, including several oysters.

Oysters do not regulate their body temperature or the salinity of their body fluids; thus, their metabolic activity is closely tied to the temperature of their surroundings, and the salt content of their blood is the same as that of the ambient water (Shumway, 1996). The ability of oysters to tolerate different environments is species specific. For instance, the European oyster, *Ostrea edulis*, grows in relatively cool, clear, water of high salinity (Yonge and Thompson, 1976). *Crassostrea* species, in contrast, are more typically inhabitants of estuaries in which they tolerate wide fluctuations in temperature, salinity, and turbidity.

The oyster's energetic investment in reproduction is prodigious, with individual females capable of producing many millions of eggs. Oysters typically become reproductively mature as males and may become female in subsequent seasons. Reproductive activity is seasonal and in temperate regions is generally dictated by temperature. Spawning occurs predominantly during the warm season, although other factors, such as phytoplankton blooms, may also play a role. Members of the genus *Crassostrea* shed their gametes directly into the water where fertilization occurs, and larval life is spent entirely in the water column. In contrast, fertilization and partial larval development in *Ostrea* take place in the interior of the oyster's shell. Females release eggs within the shell cavity, and fertilization occurs when sperm shed by nearby male oysters get drawn into the female cavity. The larvae develop partially among the female's gill filaments, which turn dark and become gritty as the larvae produce shells and become pigmented. The female's unpleasant appearance and texture at this time are the principal reason that eating *Ostrea* species is avoided in the summer (months without "R," or May through August) when reproduction occurs. The larvae of *Ostrea* species are eventually expelled from the female's shell cavity and complete their development in the water column. Oysters that are brooders produce smaller numbers of offspring than nonbrooders.

The waterborne larval stage of oysters allows them to disperse from the immediate site of the parental stock, enhances genetic mixing, and allows the colonization of new locations. The larvae are both dispersed and concentrated by water currents and wind. At the end of the larval life, usually 2 to 3 weeks, the oysters "set." Unlike clams, which can settle into mud and can shift around as adults, oyster larvae cement themselves to a clean, hard substrate and lose their mobility (Yonge and Thompson, 1976). The substrate may be another oyster, a piece of shell, a pebble, a tree root, or any other solid, clean surface. The concentrating effect of wind and currents, and the fact that larvae prefer to settle where there are other oysters, results in large assemblages on suitable substrates. The mangrove oyster, *C. rhizophorae,* for instance, congregates on the roots of mangrove trees in shallow water. Species that inhabit deeper water tend to form aggregates known as "reefs" or "beds." The Eastern oyster, *C. virginica,* is particularly well known for the large, three-dimensional reefs that it builds as successive generations of oysters settle on each other.

DISEASES OF OYSTERS

Because the terminology is often confused or confusing, this discussion of oyster diseases begins with definitions of key terms. Disease may be caused by an infectious agent or by other factors such as poor diet, exposure to a harmful substance, or a genetic defect. Infection and disease are not synonymous. *Infection* refers to the establishment of a foreign organism (infectious agent or parasite) in the tissues of another organism, called the *host. Disease* indicates damage to a body part, organ, or system such that the affected organism no longer functions normally. Infection does not necessarily lead to disease. Many infectious agents cause localized tissue damage but relatively little overall harm to their hosts. Infectious agents capable of causing disease are termed *pathogens*. Some pathogens are so virulent that they cause disease and mortality in susceptible hosts regardless of the physiological state of the host. Examples include *Haplosporidium nelsoni* and *Perkinsus marinus* (the disease agents), which cause MSX and Dermo diseases, respectively, in the Eastern oyster, *C. virginica* (the host). Other pathogens are described as *opportunistic*. Host organisms that are otherwise "healthy" can prevent infection by, or control proliferation of, opportunistic pathogens through structural (e.g., shell or epithelial barriers) or biological (physiological activity or the internal defense system) mechanisms. Opportunistic pathogens, however, may proliferate and cause disease if the host is compromised in some manner so that it can no longer effectively defend itself or if the number of opportunistic pathogens in the environment is large enough to overwhelm host defenses. Examples are the various bacterial and fungal species that infect and cause mortalities of cultured molluscan larvae and juveniles (Elston, 1984). Similarly, the herpes viruses associated with mortalities of larval and juvenile stages of a number of molluscan species in commercial hatcheries and nurseries are thought to be promoted by culture conditions, especially high temperature and high density (Farley et al., 1972; LeDeuff et al., 1996; Arzul et al., 2001).

For disease to occur, a potential pathogen must find a susceptible host in a favorable environment. A parasite may infect one species without causing apparent harm but can cause catastrophic disease outbreaks when it infects another species. The "other species" in such a case can be a resident host infected by an introduced pathogen or an introduced host infected by a resident pathogen. Pathogens may also be present in an environment that inhibits their proliferation. Under these conditions they remain undetectable, either by causing no observable effect (such as death of the host) or because they are too few to be found by standard diagnostic assays. Climate warming, for instance, is hypothesized to have favored outbreaks of Dermo disease from existing undetected foci of *P. marinus*-infected oysters in the northeastern United States and thus resulted in the apparent range extension of *P. marinus* (Ford, 1996; Cook et al., 1998).

While all commercial molluscan species examined so far are infected by some parasites, oysters have more reported lethal diseases than any other commercial species (Bower et al., 1994; Ford, 2001; see Table 4.1). As a matter of fact, the molluscan diseases listed as "of concern" by the Office International des Epizooties, an international veterinary body concerned with animal health, are primarily those affecting oysters of various species and are all caused by water-borne protozoan parasites that invade through the gut or external epithelium and proliferate inside the soft tissues, killing the oyster when the parasite burden becomes high. Transmission of some parasites, such as *P. marinus* and *Bonamia ostreae* (cause of the disease bonamiasis in *Ostrea edulis*), is directly from oyster to oyster. The mode of transmission, and indeed the complete life cycle of others, such as *H. nelsoni* and *Marteilia refringens* (cause of the disease marteiliosis in *O. edulis*), remains unknown, although a recent study provides evidence of the involvement of a copepod in the life cycle of *M. refringens* (Audemard et al., 2002).

Oyster "mass" mortalities have been recorded at least since the early 1900s. Those not attributable to predation, siltation, or freshwater influxes were simply ascribed to unknown causes (Orton, 1924; Roughley, 1926; Sindermann and Rosenfield, 1968), although one such case was later ascribed to a pathogen (Farley et al., 1988). Another early disease outbreak, which killed large numbers of *C. virginica* in Prince Edward Island, Canada, in 1913 to 1915, has been attributed to an infectious agent (Needler and Logie, 1947). The disease agent is still present but has yet to be identified.

Not until the discovery of *P. marinus* and *H. nelsoni* in the late 1940s and late 1950s, respectively, were specific infectious agents clearly identified as the cause of any bivalve mortality. Shortly thereafter, pathogens were associated with catastrophic mortalities of two oyster species (*C. angulata* and *O. edulis*) in France. A virus identified in the gills of *C. angulata* was thought to be the cause of at least some of the mortalities that wiped out commercial production of this species in France in the 1970s (Comps, 1988). The loss of *C. angulata* prompted the importation of

TABLE 4.1 Important or Common Parasites and Diseases of Oysters

Disease/Condition	Causative agent	Host(s)	Region affected	Comments
Herpes virus disease outbreaks	Herpes virus	Numerous bivalve species including oysters	Worldwide	Typically found associated with mortalities of larvae and juveniles in commercial culture; has been found in adults and in wild larvae and juveniles, but without observed mortality.
Juvenile Oyster Disease (JOD)	Probably bacterial	*Crassostrea virginica* juveniles grown in culture	Northeastern United States	Causative agent unknown, but transmissible; probably has bacterial cause, but may also involve other factors. Caused mortalities from New York to Maine during the 1990s. Problem subsided in most regions in late 1990s.
Summer Mortality	*Vibrio splendidus* bacterium (and various other factors)	*C. gigas*	France	Associated with mortality of juveniles, but adults also suffer. Probably has various causes.
Nocardiosis	*Nodardia crassostreae*	*C. gigas*	West Coast of Unitred States	Associated with summer mortalities.
Maladie du pied	*Ostracoblabe implexa* (fungus)	*O. edulis, C. gigas, Saccostrea cucullata*	Europe, Canada, India	Fungus grows in shells causing "wart-like" protuberances on inner shell. May weaken oyster and diminish marketability.

European oyster haplosporidiosis	*H. armoricanum* (protozoan)	*O. edulis* and *O. angasi*	Northern Europe	Very low prevalence and no significant impact on population.
Marteiliosis (Aber Disease)	*M. refringens* (protozoan)	*O. edulis* *O. angasi, Tiostrea chilensis* (= *T. lutaria*)	Western Europe	In Europe, causes epizootic mortalities in *O. edulis*; other species have proved susceptible when challenged experimentally, but are not known to be affected in their native ranged.
QX Disease	*M. sydneyi* (protozoan)	*Saccostrea glomerata (=S. commercialis)*	Australia	Causes epizootic mortalities.
Dermo Disease	*P. narinus* (protozoan)	*C. virginica, C. gigas, C. ariakensis*	East and Gulf Coast of United States	Causes epizootic mortalities in *C. virginica. C. gigas* and *C. ariakensis* become infected, but do not develop lethal infections.
Bonamiosis	*B. ostrea* (protozoan)	*O. edulis* (European oyster) and other species *Ostrea*: *O. angasi, O. denselamellosa, O. puelchana, Ostreola conchaphila* (= *O. lurida*), and *Tiostrea chilensis* (= *T. lutaria*) *C. ariakensis*(?)	Western Europe; Maine, United States; Northwestern United States	In Europe, causes epizootic mortalities in *O. edulis*; other species have proved susceptible when challenged experimentally, but are not known to be affected in their native ranges. Not known to caused mortalities in *O. edulis* in the United States.

continued

TABLE 4.1 (continued)

Disease/Condition	Causative agent	Host(s)	Region affected	Comments
Australian Winter Disease	*Mikrocytos roughleyi* (protozoan)	*S. glomerata*	Australia	Mortalities caused by *M. roughleyi* apparently first reported in 1926.
Denman Island Disease	*Mikrocytos mackini* (protozoan)	*C. gigas*	British Colombia, Canada	Can be controlled through appropriate aquaculture practices.
MSX Disease	*Haplosporidium nelsoni* (protozoan)	*C. virginica, C. gigas*	East Coast of the United States (*C. virginica*) Pacific Asia and United States (*C. gigas*)	Causes epizootic mortalities in *C. virginica. C. gigas* becomes infected but no mortalities reported. Also found in *C. gigas* in California.
SSO Disease	*H. costale* (protozoan)	*C. virginica*	East Coast of the United States	Restricted to higher salinity locations compared to *H. nelsoni*,
Malpeque Disease	Unknown	*C. virginica*	Atlantic Canada	First outbreak in 1915-16; oysters in affected areas appear to have developed resistance.
Hemic neoplasia (uncontrolled proliferation of blood cells)	Etiology unknown, but reported to have genetic or environmental links	Many species of oysters and other marine bivalves	Widespread	Contagious; may be associated with mortality.

Summer mortality	Various	*C. gigas*	West Coast of United States Japan	Probably a multifactorial cause.
Infection by Rickettsiales- or Chlamydiales-like organisms	Intracellular bacteria-like	All species of oysters examined and most other marine bivalves also	Global	Has been associated with mortality of marine bivalves, but not necessarily as the causative agent—more probably opportunist. Found more frequently in dense associations of bivalves (e.g., nurseries or culture parks).
Other parasites	Trematodespestodes, nematodes, ciliates, and flagellates	All species of oysters examined and most other marine bivalves also	Global	Not deleterious.

SOURCE: Susan Ford, Haskin Shellfish Research Laboratory, Rutger University, Port Norris, New Jersey.

C. gigas as a replacement species (Grizel and Héral, 1991). The *C. angulata* mortalities were followed by disease outbreaks in the European oyster, *O. edulis*, caused by two newly discovered protozoan pathogens, *M. refringens* and *B. ostreae*. The resulting mortalities caused a precipitous decline in *O. edulis* production in France in the 1970s, and accelerated the use of *C. gigas*, which is not susceptible to the diseases caused by these two pathogens or the viral "gill disease."

Infections of marine molluscs by other agents, including parasitic worms, protozoans, Rickettsiales- and Chlamydiales-like organisms (RLOs and CLOs), bacteria, and viruses are not uncommon (Bower et al., 1994), and "new" cases continue to be described as more and more host species are grown in culture and examined by an ever-increasing number of scientists. Some are found when mortalities, often of cultured molluscs, are investigated. Culture conditions, in which the molluscs are grown at high density and often using poor animal husbandry practices, favor the proliferation and transmission of opportunistic pathogens, which can then cause or exacerbate disease and mortality in the cultured organisms (Meyers, 1979; Elston, 1984; Bower, 1987; Bricelj et al., 1992; Lacoste et al., 2001). Various bacterial species and the herpes virus are examples of pathogens most commonly associated with disease outbreaks in hatcheries and nurseries (Hine et al., 1998; Renault et al., 2000, 2001; Arzul et al., 2001). Others are encountered during routine surveys or health examinations required for the shipment of molluscs across governmental boundaries. Most occur at low prevalence and intensity and appear to cause no harm to the host. For instance, certain microorganisms, such as the intracellular bacterialike RLOs and CLOs, have been found in all bivalves examined so far, typically without evidence of being harmful. They have often been associated with mortality (Gulka and Chang, 1984; Le Gall et al., 1988; Norton et al., 1993; Villalba et al., 1999; Moore et al., 2000), although it is probable that they are opportunistic pathogens rather than the original cause of death.

Although not exhaustive, Table 4.1 lists the most important or common diseases and parasites reported for oysters, along with the host species and regions where they are found.

CRASSOSTREA VIRGINICA

Life History

C. virginica, the Eastern oyster, inhabits estuarine waters from the Canadian maritime provinces to the Gulf of Mexico, with reports of the species as far south as Brazil and Argentina (Carriker and Gaffney, 1996). Adults are intertidal and subtidal dwellers, typically found in assemblages called reefs, bars, or beds that range in size from a few acres to hundreds. The general morphological, physiological, and life history characteristics of oysters described earlier apply equally to *C. virginica*. This section provides more detailed characteristics of *C. virginica*.

Reproduction of *C. virginica* is seasonal and largely influenced by temperature. Gametogenesis begins in the spring and spawning occurs from late May to late September in the mid-Atlantic, with the season contracted or extended to the north and south, respectively (Shumway, 1996; Thompson et al., 1996). *C. virginica* are either male or female (the reported incidence of hermaphroditism is <0.5%) but may change sex over the winter when they are reproductively inactive. Small oysters (10 to 20 mm) sometimes develop gametes, almost always sperm. Under favorable growth conditions in the mid-Atlantic, this may occur during the late summer after setting, although it is uncertain whether such individuals actually spawn or produce embryos because they do not ripen until after the normal spawning period. In the southeastern United States and the Gulf of Mexico, sexual maturity is typically reached about 3 months after setting, and the prolonged reproductive period in this region increases the probability that these juveniles do participate in the overall reproductive effort of the population.

Males are more sensitive to spawning stimuli, such as temperature and food, than females and tend to spawn first. The presence of sperm in the water stimulates females to release eggs, which are then fertilized externally. Gametes deteriorate within a few hours of spawning and can be rapidly diluted by water currents; thus, the proximity of oysters to one another increases the chances of synchronous spawning and successful fertilization. In the first 24 hours, oyster larvae develop a large ciliated structure, the velum, which acts as both a swimming and food-gathering organ. Initially, the shell is secreted as a single event at about 24 to 48 hours; thereafter, growth occurs through accretion to both thicken and extend the shell. The larval stage lasts for about 2 to 3 weeks, depending on food availability and temperature. Larvae appear to migrate vertically, particularly at later stages, tending to concentrate near the bottom during the outgoing tide and rising in the water column during the incoming tide, thus increasing their chance of being retained in the estuary (Kennedy, 1996; Shumway, 1996). Larval mortality rates are estimated to be close to 99%.

As is the case with all oyster species, *C. virginica* larvae must eventually find a clean, solid surface on which to cement themselves. Oyster shells meet those criteria if they are not covered with silt or heavily fouled by other epifaunal organisms (although the larvae settle on any type of hard substrate, such as pilings, rocks, and ship bottoms). The suitability of oyster shell for setting, the concentrating mechanisms of wind and water currents, and the gregarious nature of setting all lead to the formation and persistence of oyster reefs. Without continuous setting and growth of juveniles on the reef, it will usually become covered with silt.

Certain sites within estuaries are known to reliably obtain good "sets" of young oysters. These are locations where clean shell is often spread by management agencies or industry members to "catch" the set, after which it is typically moved to grow-out areas. Despite the knowledge of where the larvae consistently settle, the parental stock for these sets is rarely, if ever, known. The potential obviously exists for oysters to be carried long distances during their larval life, both within and between estuaries; the current state of knowledge is insufficient to predict where larvae originating from oysters in a particular area will be transported or to estimate the likelihood that larvae from one estuary will be carried, along the coast, to another estuary. Oyster larvae are common in summer in water samples collected in East Coast estuaries (Kennedy, 1996); however, investigators sampling nearshore waters off New Jersey for surf clam larvae report seeing only one or two *Crassostrea* sp.-like larvae during several years of sampling in the 1970s and 1990s (M. Tarnowski, Maryland Department of Natural Resources, Annapolis, personal communication, 2003; J. Grassle, Rutgers University, Port Norris, personal communication, 2003). Further, there is little evidence for a broodstock size/recruitment relationship for oysters, and very large sets can occur even when the stock of oysters is very low, as occurred in Delaware Bay several years after the MSX disease epizootic and heavy oyster drill predation had severely reduced the Chesapeake Bay population (Fegley et al., 1994). For the Gulf of Mexico, Livingston et al. (1999) reported rapid repopulation of Apalachicola Bay after the oyster population was decimated by two hurricanes in 1985. A widespread heavy set occurred in Chesapeake Bay in 1997 even though the oyster population was severely depleted. Unfortunately, most of the oysters suffered disease-caused mortalities before they reached marketable size.

From a few hundred microns in size at the time of setting, *C. virginica* grow to sizes exceeding 150 mm. They are typically marketed in the United States when they reach about 75 mm (about 3 inches). Growth rates vary with temperature, food, turbidity, and salinity. In the mid-Atlantic, market-sized *C. virginica* are at least 2 years old but more typically are harvested when 3 to 4 years old. To the south, *C. virginica* may grow to marketable size in 12 to 15 months. Growth rates of oysters held in floating aquaculture trays are typically much greater than those of oysters on the bottom. The average life span is about 6 to 8 years; the maximum is probably about 25 years.

Oysters provide food for numerous predatory species, including flatworms, crabs, oyster drills, starfish, and certain finfish. Mortality, mostly due to predation, is high on newly set oysters (spat), typically exceeding 40% in the first week (Haskin and Tweed, 1976; Newell et al., 2000) and sometimes reaching 100% (Roegner and Mann, 1995). Mortality rates diminish as the spat grow but remain high during the first several months after setting. Newell et al. (2000) found that over a period of 14 years less than 0.5% of potential set (measured on clean experimental surfaces during the summer) was found on the bottom by the beginning of winter in a subestuary of central Chesapeake Bay. Higher survival was found in upper Delaware Bay over a 30-year period during which the percentage surviving until winter ranged from <1 to 25 but was mostly <10 (Haskin and Ford, 1986). Predation, especially by oyster drills and crabs, continues on older oysters (MacKenzie, 1970). Ford and Haskin (1982) estimated that predators annually killed 12 to 18% of oysters more than a year old on lower Delaware Bay leased grounds—about half as much as died from other causes, mostly MSX disease.

In addition to providing habitat for numerous other species, a principal ecological function provided by oysters is considered to be their ability to remove, or filter, particles from surrounding water (Newell, 1988). During feeding and respiration, oysters "pump" large quantities of water. A large *C. virginica* (10 to 12 cm) has been found to transport up to 460 liters per day over its gills (Galtsoff, 1964), although this is probably an extreme value. Powell et al. (1992) reviewed the literature on filtration rates for numerous marine bivalves and found that the relationship between filtration and size was similar for all species. The rates that when used in a numerical growth model most closely fit field observations for *C. virginica* were about 0.5 to 0.6 liters per hour per gram dry weight, or about 75 liters per day, for a 10-cm oyster. Filtration rates are a function of several environmental factors, including temperature, salinity, and suspended particulate concentration (Shumway, 1996). Rates increase with temperature, decrease with salinity below about 8 ppt, and also decrease with particle load (Powell et al., 1992). The highest rates occur during the summer. In the mid-Atlantic and the northeastern United States, *C. virginica* are physiologically nearly inactive during the winter; in the Gulf of Mexico they are active much of the time.

Temperature and salinity, and their interaction, are undoubtedly the two most important environmental factors governing survival, growth, and reproduction of *C. virginica*. Many investigators have attempted to define the temperature and salinity tolerance limits and optimum ranges for *C. virginica*, with considerable divergence in results (Shumway, 1996). Differences in methodology (laboratory versus field observations), acclimation conditions (Davis, 1958; Davis and Calabrese, 1964), and geographically associated genetic traits (Barber et al., 1991; Dittman et al., 1998) all contribute to observed variations in optimum temperature-salinity ranges, making it difficult and risky to define limits that apply to all populations. In addition, food and turbidity can confound the interpretation of field observations, especially in the case of salinity, as food availability is often limiting at low-salinity sites.

Despite the variability, it is clear that *C. virginica*, having evolved in an estuarine environment and having an extremely wide geographic distribution, can tolerate a broad range of both temperatures and salinities (Shumway, 1996). For instance, adults can survive freezing during winters in the Northeast and summer temperatures averaging 36°C in Florida and the Gulf of Mexico, with at least short-term temperatures close to 50°C at intertidal sites. They grow in locations with salinities averaging 5 to 40 ppt, although major oyster aggregations are typically found at 10 to 30 ppt. They can survive for weeks in winter in nearly fresh water. Spat appear to have the same tolerance limits and optimal ranges as adult oysters, but those of larvae are more restricted and, in the case of salinity, depend on the conditions in which their parents developed gametes. In the laboratory, normal larval development has been observed at temperatures of about 15° to 30°C and salinities of 15 to 33 ppt. Over most of their range in the United States, *C. virginica* spawn when summer temperatures reach between 18° and 25°C; subsequent larval development occurs at somewhat higher temperatures. Larval growth rates increase rapidly with increasing temperature; the fastest rates occur near 30°C.

Although the species has a wide geographic distribution, clear genetic differences have been identified among geographically separated populations of *C. virginica*, based on both molecular and physiological evidence. *C. virginica* display geographically and genetically distinct growth, reproduction, and disease susceptibility traits. Enzyme polymorphism (Buroker, 1983) and DNA analyses show a genetic discontinuity between Gulf of Mexico and Atlantic populations (Reeb and Avise, 1990; Cunningham and Collins, 1994). Earlier studies indicated that oysters from higher latitudes spawned at lower temperatures than those from more southern locations (Stauber, 1950; Loosanoff and Nomejko, 1951), a difference later demonstrated to have a genetic basis (Barber et al., 1991). Dittman et al. (1998) subsequently observed that Long Island Sound oysters grew faster than Delaware Bay oysters. Disease resistance also has a regional genetic component. The greater resistance of southern oysters to the pathogen *P. marinus* was first noted by Andrews and Hewatt (1957) and demonstrated to be heritable by Bushek and Allen (1996). The existence of geographically distinct genetic traits underscores the potential problems of moving individuals, even within the same species, between regions and environments in which they have evolved.

Ecological Value of *C. virginica* in Chesapeake Bay

Suspension-feeding bivalves are common inhabitants of coastal and estuarine habitats throughout the world. Since these species are capable of capturing large amounts of particulate material from the water column and transporting it to the benthic system, they are implicated as major biological agents in affecting nutrient cycling and benthic-pelagic coupling in shallow-water marine ecosystems (Wildish and Kristmanson, 1997; Dame, 1996). Suspension feeders also provide other types of ecosystem functions. For instance, their filtering activity may enhance water clarity, increasing light penetration and trapping contaminants entering coastal waters (Cloern, 1982; Newell, 1988). Materials ingested by the bivalves are expelled as feces and pseudofeces, which other groups of organisms living in or on the seafloor may bury or remobilize. Sedimentation of materials by filter feeders can change organic matter decomposition rates and anaerobic or aerobic decomposition processes.

Reef-building bivalve species (e.g., oysters, mussels) also function as "ecosystem engineers" (*sensu* Jones et al., 1994) through the creation and maintenance of unique habitats that are used by other species as predator refuges and feeding and nesting sites (e.g., Lenihan and Peterson, 1998; Coen and Luckenbach, 2000). As a consequence, the reefs can enhance ecosystem productivity and biodiversity. The structures also stabilize sediments, reduce coastal erosion, and alter the hydrography of shallow-water marsh creeks and embayments (Kirtley and Tanner, 1968; Dame, 1976; Meyer et al., 1997). Some suspension feeding bivalves can attain densities of 4,000 to 10,000 individuals per square meter and, when particularly abundant, have the potential to directly influence resource availability to pelagic species in the ecosystem by causing state changes in abiotic and biotic materials (e.g., Carpenter and Kitchell, 1988; Kimmerer et al., 1994; Dame, 1996).

Over the past 7,000 years the dominant epifaunal suspension-feeding bivalve inhabiting the Chesapeake Bay ecosystem has been the Eastern oyster, *C. virginica* (Hargis, 1999). During this period, the oysters created large reefs throughout most of the tributary estuaries in the bay with the highest abundances often associated in areas of reduced predation, either in low-salinity (<15 ppt) waters or shallow tidal creeks. But as the *C. virginica* population declined over 100-fold over the past 150 years, the amount of oyster reef habitat has been reduced by more than 50% (Rothschild et al., 1994). Loss of oysters is generally attributed to overfishing, habitat degradation, disease pressures, and the interaction among these factors, while destruction of reef habitat is the direct result of fishing practices used to harvest the oysters.

Habitat and Resource Provision

Oyster reefs provide essential habitat for the maintenance of oyster populations through provision of substrate for larval settlement, refuge from predators and near-bottom hypoxia, and vertical relief above the seafloor, which acts to reduce sediment deposition. Reef formation occurs through a chemically mediated cue (Tamburri et al., 1992; Zimmer-Faust and Tamburri, 1994) in which the planktonic oyster larvae gregariously settle on hard substrates. While there is some controversy regarding the source of the settlement cue (i.e., exudates from juvenile and adult oysters: Keck et al., 1971; Veitch and Hidu, 1971) or microbial biofilms (Fitt et al., 1989, 1990; Weiner et al., 1989), the consequence of gregarious larval settlement is that oyster reefs develop as multiple generations of oysters settling one upon another (Coen and Luckenbach, 2000). Reefs can vary in size and shape (hundreds to thousands of square meters) and in subtidal regions can be up to 4 m in height (Ingersoll, 1881; Winslow, 1882; DeAlteris, 1988).

Reefs with higher profiles above the seafloor appear to promote enhanced oyster productivity. Low-profile reefs, frequently a result of harvesting practices (e.g., Marshall, 1954), enhance sediment deposition on the reef surface (DeAlteris, 1988; Seliger and Boggs, 1988). Increased sedimentation reduces the nutritional value of ingested materials, leading to reduced growth and reproduction and increased physiological stress from clogging of the oyster's filtering mechanism (Mackenzie, 1983). High siltation levels on reefs also impair habitat quantity and quality for settling larvae and attached juveniles (Bahr, 1976). Experimental studies in North Carolina have demonstrated the importance of the interaction of reef morphology (i.e., height above the seafloor) and the abiotic environment (i.e., water flow, dissolved oxygen regime, sedimentation) on oyster productivity (Lenihan and Peterson, 1998). For instance, oyster mortality associated with bottom-water hypoxic/anoxic events was correlated with reef height. In addition to reef morphology, increased amounts of interstitial spaces within artificially constructed reef were found to enhance juvenile oyster survival (Bartol and Mann, 1999), presumably by providing the young life stages with a refuge from predators. While these studies indicate there is a complex interaction between the physical nature of an oyster reef and the physical environment surrounding it, there are few studies examining the direct and indirect impacts of reef destruction over the past century. Experimental studies, similar to those of Lenihan and Peterson (1998) and Coen and Luckenbach (2000), provide a framework for unraveling some of these complexities.

Despite the fact that oyster reefs have been a conspicuous element of the benthic landscape in the Chesapeake Bay for thousands of years, there is surprisingly little information on how reefs provide habitat for other species. The most conspicuous feature is that reefs add habitat complexity relative to the surrounding sediments. Species that enhance structural complexity can be important determinants of population and community dynamics (e.g., coral reefs: Hixon and Beets, 1993; seagrass beds: Heck and Orth, 1980; kelp forests: Tegner and Dayton, 1987; salt marsh plants: Kneib, 1984). However, it is not well understood whether oyster reefs simply act as physical attractants or actually enhance the overall productivity of species that co-occur on the reefs.

A number of community-based surveys of finfish and macroinvertebrates inhabiting oyster reef habitat have been conducted along the south Atlantic and Gulf coasts. For example, Dame (1979) recorded 37 macrofaunal species on an intertidal oyster reef in South Carolina, and Wells (1961) found a total of 303 species inhabiting several reefs in North Carolina. Others (e.g., Arve, 1960; Bahr and Lanier, 1981) have noted that oyster reefs provide habitat for economically important fish (drums, rockfish, speckled seatrout) and invertebrates (shrimp, blue crabs). O'Beirn et al. (1999) recorded a wide diversity of species in studies characterizing changes in the composition of benthic plants and animals following the construction of several artificial reefs in lower Chesapeake Bay. Coen et al. (1999) assessed finfish and decapods crustacean species that are found in association with oyster reefs in the bay. Using criteria developed by Breitburg and colleagues (1995, 1999), species were classified into three groups: those that used oyster reefs as their primary habitat (reef residents), those that were generally found on the reefs (facultative residents), and species that were more far-ranging and tended to forage on or near the reefs (transient species). Summarizing data collected in Maryland and Virginia waters, Coen et al. (1999) found five fish species were considered to be reef residents, four were deemed facultative residents, and 48 were considered reef transient species. All of the decapods (three species) associated with reef habitat were considered transient species.

Breitburg and Miller (1999) noted some of the resident reef fish species in the bay (particularly the naked goby, *Gobiosoma bosc*) were significant zooplankton predators that in turn became prey for larger transient species (e.g., striped bass, *Morone saxatilis*). High densities of striped bass were regularly found swimming within 1 m of artificial reefs constructed by the authors. Breitburg et al. (1999) also recorded large numbers of larval stages of some resident fish that tended to use the artificial reefs as a hydrodynamic refuge by actively congregating on the down-current side of the structures. Laboratory studies indicate that grass shrimp (*Palaemonetes pugio*) use oyster reef habitat as a predator refuge (Posey et al., 1999), while other species have been found to temporarily occupy natural and artificial reefs as refuges from hypoxic/anoxic events (Seliger et al., 1985; Breitburg, 1992; Lenihen and Peterson, 1998). Mann and Harding (1998) and Breitburg and Miller (1999) have characterized the trophic links between oyster reefs and selected fish species in the bay. Clearly, additional work is needed to examine the trophic and population dynamics of reef-dwelling assemblages and the contribution reef habitat plays in supporting the productivity of those assemblages.

Coastal and estuarine ecosystems (e.g., seagrass beds, salt marshes, mangrove forests) are frequently viewed as nursery habitats for a diversity of fish and invertebrate species because of their effects on productivity and biodiversity (Chambers, 1992). While little attention has been paid to the potential value of oyster reefs as nursery habitats (Beck et al., 2001), a number of resident reef fish species have been shown to use specific microhabitats in reefs as nesting sites. Coen and Luckenbach (2000), for example, list 11 fish species common to the bay that potentially could use oyster reefs as sites for reproduction. Of these species, Coen et al. (1999) consider reef habitat essential (i.e., a habitat that is necessary to the fish for breeding, feeding, and/or growth to maturity) for five species in the bay (naked goby, *Gobiosoma bosc*; striped blenny, *Chasmodes bosquianus;* feather blenny, *Hypsoblennius hentz*; skilletfish, *Gobiesox strumosus*; and oyster toadfish, *Opsanus tau*).

Pelagic-Benthic Coupling and Nutrient Cycling

Oysters are active suspension feeders, acquiring food by filtering organic materials from the water column and depositing copious quantities of particulate waste materials on the seafloor. Studies in South Carolina and France (Dame et al., 1989; Zurburg et al., 1994a, 1994b) have identified reefs as intense sites of organic matter decomposition and sources of inorganic nutrients. The oysters are functioning as an important feedback loop by catalyzing the flux of particulate nutrients from the water column to sediments through biodeposits of feces and pseudofeces. Deposition fuels remineralization of organic material through complex bacterial biogeochemical transformations and a reverse flux of organic nutrients from the sediments back into the water column. Added to the positive feedback loop of remineralization of inorganic nutrient pools is the reduced storage of nutrients in phytoplankton biomass as a consequence of bivalve grazing. This process forms another positive feedback that influences inorganic nutrient availability (Prins et al., 1995, 1997) and may also stimulate phytoplankton and macrophyte primary production. Remineralization of the nutrient pools may also influence phytoplankton development as differences in regeneration rates can change nutrient dynamics (Smaal and Prins, 1993). In addition to the positive feedback loops, Newell et al. (2002) demonstrated in laboratory experiments that oyster biodeposits can serve as sites for the removal of nitrate from the ecosystem. Accumulation of phytoplankton-rich oyster biodeposits accumulate on the sea-floor stimulates the conversion of nitrate to nitrogen gas by anaerobic bacterially mediated processes.

By promoting the flux of organic materials from the water column to the sediments, oysters help retain organic matter that may otherwise be lost from the system. Deposited organic matter supplies food to a variety of other species living in and on the reef habitat. In particular, deposit feeders and grazers (i.e., annelid worms, amphipod crustaceans) and microphytobenthos utilize organic matter found in the oyster biodeposits that support a greater diversity and abundance of benthic species than would otherwise occur in the surrounding sedimentary habitat.

Oysters and other suspension feeders can help determine the fate of contaminants through filtration, metabolic modification, and biodeposition. In estuarine waters, trace metals and synthetic organic compounds adhere to particulates and become concentrated by filter feeders such as bivalves. Bivalve tissue often reflects the current (weeks to months) contaminant burden of an ecosystem, whereas sediments reflect the long-term contaminant loading (e.g., Huanxin et al., 2000).

Ecosystem-Level Effects and Trophic Interactions

Studies over the past three decades in a number of estuarine and coastal systems strongly suggest that when benthic suspension feeders are sufficiently abundant, their feeding activities regulate the amount of phytoplankton biomass (see Smaal and Prins, 1993; and Dame, 1996, for reviews). In a theoretical analysis, Officer et al. (1982) concluded that suspension feeder grazing control of phytoplankton is only possible when bivalve biomass is high (~100 g total fresh weight per square meter), water depth is relatively shallow (a few meters), and systemwide bivalve filtration rate is of the same magnitude as the time constant for phytoplankton growth. The strongest evidence for the ability of natural populations of suspension feeders to control phytoplankton populations generally comes from systems in which bivalve biomass has been artificially enhanced or through the introduction and subsequent proliferation of nonnative species. For example, blue mussels and oysters stocked in experimental enclosures or cultured on off-bottom racks or ropes have been shown to reduce chlorophyll levels (e.g., Riemann et al., 1988; Tenore et al., 1982). Also, several species of exotic bivalves introduced into San Francisco Bay and zebra mussels in the Laurentian Great Lakes have been shown to be responsible for controlling phytoplankton stocks (Cloern, 1982, 1996; MacIssac, 1996).

Studies examining the impacts of benthic suspension feeder grazing on phytoplankton are based on in situ experiments and laboratory experiments extrapolated to provide estuarywide estimates of filtering activity. The estimates are used to develop secondary productivity models, which are coupled with primary productivity and hydrodynamic models to develop ecosystem models. The models can be calibrated with field data and used as management tools to assess interactions between environmental parameters and shellfish populations (Bacher et al., 1997a; Dame, 1993; Héral, 1993).

The feeding activities of oysters remove materials in the water column besides phytoplankton that can reduce turbidity and allow more light to reach submerged aquatic plants (e.g., seagrasses, benthic microalgae). Historically, seagrass beds have been an important habitat in the bay, but beds have declined to a fraction of the previous abundances (Orth and Moore, 1983; Brush and Hilgartner, 2000). Seagrass beds are habitats for supporting many species of mobile and sessile benthic species (Orth et al., 1984) and are valuable nursery habitats for economically important shellfish and finfish. Benthic microalgae provide an important food source for benthic herbivorous meiofauna and macrofauna (Miller et al., 1996).

The reduction of oyster populations in the bay over the past 150 years is hypothesized to have removed an important control over phytoplankton blooms and to have led to the damaging consequences of nutrient enrichment and a change in the bay's trophic structure. Newell (1988) estimated that oyster densities prior to 1870 were abundant enough to filter the entire volume of the bay every 3.3 days. As a consequence of more than a century of oyster harvesting, habitat degradation, and disease, the reduced densities of oysters are now estimated to perform the same task in 325 days (based on Newell's calculation). Using a carbon flux model, Ulanowicz and Tuttle (1992) expanded on Newell's work and hypothesized that an increase in oyster abundance in the mesohaline region of the bay should increase benthic diatom production and decrease phytoplankton production and stocks of pelagic microbes, ctenophores, and particulate organic carbon. While the authors cautioned that the magnitude of their predictions were not absolute, they postulated that the combined effects of decreased planktonic primary productivity and increased benthic primary productivity might reduce eutrophication in the bay. Because oysters compete with zooplankton for food, the model also predicted that the numbers of gelatinous predators that feed on zooplankton (e.g., ctenophores, medusa) would dramatically decline with increased oyster abundance.

Most of the information that the bay has changed from a "benthic-dominated" to a "pelagic-dominated" system with the reduction in oyster abundance is based on very limited or anecdotal evidence. For example, Baird and Ulanowicz's (1989) ecosystem model of the mesohaline portion of the bay revealed that zooplankton grazing on summertime primary producers was greatly reduced by the predation activities of gelatinous zooplankton (ctenophores and sea nettles). As a result, some of the ungrazed phytoplankton fuels water column microbial activity, and the remainder sinks to the bottom and is utilized by a variety of deposit-feeding worms, amphipods, and crabs. The authors conclude their work with the question: "Was this always the case?" While records indicate that benthic filter feeders were far more abundant in the past (e.g., Newell, 1988) and presumably exerted a stronger control on water clarity, Baird and Ulanowicz (1989) note the lack of historical data on phytoplankton, zooplankton, and benthic faunal abundance to confirm whether there were significant changes in phytoplankton or other factors that could have resulted in trophic restructuring in the bay. Coupled with the 100-fold decline in oyster abundance and loss of oyster reef habitat have been dramatic increases in anthropogenic nutrient loading and associated oxygen depletion in the bottom waters during the summer (Seagle et al., 1999). Typically, more eutrophic ecosystems have shorter food chains, few levels of biological organization, lower diversity, and higher phytoplankton productivity. Some reports claim there is a lack of functional redundancy in the bay, and unlike other systems no other suspension-feeding animals have effectively replaced the effects of oyster grazing. There is also anecdotal information that when oyster reefs were abundant along the edges of the deep-water channels of the bay, they may have affected surface water mixing processes by enhancing bed roughness (Newell and Ott, 1999). As a result, less saline and higher oxygenated surface waters may have been mixed downward to deeper regions of the bay and may have helped to ameliorate the effects of hypoxic conditions in the bottom waters.

Gerritsen et al. (1994) developed a model of bivalve suspension feeding by including water mixing parameters, bivalve abundance, and filtering capacities and applied it to the Maryland portion of the Chesapeake Bay. Benthic faunal surveys in the bay (conducted in 1986 and 1987) indicated that despite severely reduced abundances of oysters, 97% of benthic organism biomass consisted of suspension-feeding bivalves (*Corbicula fluminea, Macoma* spp., *Mulinia lateralis, Mya arenaria, Rangia cuneata, Tagelus plebeius*). Exceptionally high biomass in the upper shallow reaches and oligohaline regions of the bay were due to the proliferation of nonnative species (*Corbicula fluminea* and *Rangia cuneata*). The authors found that most of the primary production appeared to be consumed by the combined actions of the bivalves, zooplankton, and deposit feeders. Phelps (1994) suggested that the major factor influencing reestablishment of submerged aquatic plants in the Potomac River was increased light penetration due to cropping of phytoplankton by filter-feeding bivalves. For the middle bay and in deeper regions, however, Gerritsen et al. (1994) found that less than half of the annual primary production was consumed. The primary factors limiting consumption of primary production by suspension feeders in these portions of the bay were related to low flow rates. They concluded that consumption could be increased by raising suspension feeders in the water column where flow rates are higher.

Diseases

Disease has been among the most important influences on the population dynamics of *C. virginica* in the United States over the past half century. Two major diseases, MSX and Dermo, both caused by waterborne parasites, have severely reduced the abundance of Eastern oyster populations along the East Coast of the United States (Ford and Tripp, 1996). Although overharvesting had been diminishing the population abundance for decades (Rothschild et al., 1994), diseases have become more widespread and intense during this period (Burreson and Andrews, 1988), and there is no question that disease-caused mortality is largely responsible for the dramatic declines in oyster landings observed since the early 1980s. (Figure 2.1 shows the decline in landings relative to the appearance of the diseases MSX and Dermo).

P. marinus *(Dermo) and* H. nelsoni *(MSX)*

The first oyster disease to be recognized in the United States was Dermo disease, caused by *P. marinus*. Although it was discovered in the late 1940s in the Gulf of Mexico, it had probably been present throughout the southeastern United States and the Gulf of Mexico for many decades at least (Ray, 1996). Between its discovery and 1990, *P. marinus* was prevalent only in waters south of Delaware Bay. Since then epizootic outbreaks associated with a pronounced winter warming trend have been recorded as far north as Maine (Ford, 1996).

The second disease, MSX, is caused by *Haplosporidium nelsoni*, a parasite introduced to the East Coast of the United States from Asia, where it infects the Pacific oyster, *C. gigas* (Burreson et al., 2000). *H. nelsoni* prevalence in *C. gigas* is very low and results in no reported mortalities, but it is lethal to *C. virginica* and began causing epizootic mortalities in Delaware and Chesapeake bays in the late 1950s and early 1960s (Ford and Tripp, 1996). The pathogen is now present along the entire East Coast of the United States, although its major impact has been from Virginia north to Maine. The Long Island Sound oyster industry, which had been spared recurrent disease outbreaks, has recently suffered heavy losses, mostly due to MSX disease (Sunila et al., 1999). The devastating effect of both diseases during the past decade and a half is associated with a period of above-average temperatures and repeated droughts, both of which favor the parasites. Not only is the current depressed level of oyster production in the Chesapeake Bay strongly linked to recent environmental extremes, but native oyster restoration efforts have been conducted under the same conditions, which so obviously favor the parasites over the oyster throughout most of the bay.

Oysters can become infected by *H. nelsoni* or *P. marinus* shortly after they set; however, infection levels in spat are typically very low because the small volume of water pumped by these tiny oysters makes their chances of encountering infective stages extremely small compared to larger oysters. Once infected, however, parasite burdens in spat can become very high. There is no evidence that either parasite can be transmitted directly from parent to offspring via infected gametes or during spawning (Ford et al., 2001). Of the two parasites, *H. nelsoni* generally infects and kills sooner than does *P. marinus*. *C. virginica* can suffer very heavy *H. nelsoni*-caused mortality during their first year of exposure, whereas *P. marinus* typically requires 2 or 3 years to attain full epizootic status. Nevertheless, each parasite is capable of killing 90 to 95% of susceptible *C. virginica* within 2 to 3 years.

The parasites have different environmental limits and different methods of transmission. *P. marinus* is transmitted directly from oyster to oyster, with most new infections occurring in the late summer when previously infected oysters die and release parasites into the water (Ragone-Calvo et al., 2003a). Transmission of *P. marinus* is thus dependent on the density of, and proximity to, infected oysters. This parasite develops the heaviest infections and kills most readily at salinities >10 ppt and temperatures >20°C, but it survives at much lower salinities (3 ppt) and temperatures (<5°C) (Chu and La Peyre, 1993; Chu et al., 1993; Ragone-Calvo and Burreson, 1994).

The life cycle and means of transmission of *H. nelsoni* are unknown. The spore, a life stage common to other members of the same phylogenetic group and one presumed to play a role in transmission, is extremely rare in adult Eastern oysters, although it develops readily in juvenile oysters (Barber et al., 1991; Burreson, 1994). The scarcity of spores suggests that adult Eastern oysters are a dead-end host for *H. nelsoni* and that there may be another host involved in the life cycle. None has been identified, however, and the possibility of a direct life cycle with juvenile oysters being the source of infective stages cannot be discounted. *H. nelsoni* is rare in *C. virginica* living at salinities <10 ppt; in fact, exposure to low salinity can eliminate the parasite from infected individuals (Andrews, 1983; Ford, 1985). This parasite begins proliferating at temperatures of about 10°C and causes most mortalities from early summer into autumn. New infections are acquired during the same period but are not dependent on proximity to infected oysters.

The temperature and salinity constraints on *P. marinus* and *H. nelsoni* have largely influenced the historical and present distribution and prevalence of infections in *C. virginica*. Within Chesapeake Bay, salinity has had the predominant effect. For many years after they were originally discovered in the estuary, both parasites were prevalent mostly in the high-salinity Virginia portion of the bay. Even during the drought of the mid-1960s, only the southern portion of Maryland's Chesapeake Bay experienced significant mortalities. It was not until the 1980s and the beginning of repeated periods of below-average river flows and elevated salinities (Burreson and Ragone-Calvo, 1996) that both parasites began infecting large numbers of Maryland oysters, causing heavy mortalities and much-reduced harvests. At the same time, the parasites spread upriver in the major tributaries of the Virginia portion of the bay, killing oysters on reefs that had previously been protected from the diseases by low salinity.

A trend toward warmer winters has further favored the parasites (Cook et al., 1998). *P. marinus* infection intensity in oysters declines over the winter due to the death of the parasites, but some parasites survive and proliferate again as temperatures rise in the spring. When winters are short and mild, more parasites survive and are available to begin proliferating in the spring. Lethal infections are reached earlier in the summer and result in earlier transmission of parasites from dead oysters, leading to another round of infection development, death, and transmission in the same season. Winter temperatures also affect *H. nelsoni*, although the mechanism is not known because the method of transmission is not known. Cold winters are followed by years of low *H. nelsoni* infection activity (Ford and Haskin, 1982). Mathematical modeling of *H. nelsoni* infection cycles predicts that several successive cold winters would substantially diminish infection prevalence and suggests that climate warming, with concomitant relaxation of the cold winter effect, may be responsible for some recent disease outbreaks in the northeastern United States (Hofmann et al., 2001) and perhaps more recently in Nova Scotia, Canada (Office International des Epizooties, 2002).

Few viral or bacterial infections have been reported in adult *C. virginica*, and no disease outbreaks associated with these organisms have been reported. A disease of suspected bacterial origin, Juvenile Oyster Disease, was a serious problem in *C. virginica* nurseries in New York and New England for a number of years in the 1990s but has since subsided in most locations. Bacterial and fungal disease outbreaks sometimes develop in hatcheries where larvae are reared at high densities but have not been reported in nature, where larval densities are far less. A lethal viral disease of larvae was detected on the U.S. West Coast in a hatchery rearing the Pacific oyster, *C. gigas* (Elston and Wilkinson, 1985), and herpes virus has been associated with mortalities in hatcheries and nurseries rearing a variety of commercial molluscs in numerous places around the world (LeDeuff et al., 1996; Renault et al., 2000, 2001; Arzul et al., 2001). Viruses have not been reported as the cause of problems for *C. virginica;* however, this may be due more to the difficulty of diagnosing viral infections than to their absence in this species or to a lack of viral involvement in the "unexplained mortalities" that occur from time to time in hatcheries and nurseries rearing *C. virginica.*

Interaction of Disease and Habitat

The possible influence of oyster habitat on disease is sometimes mentioned in arguments for reef restoration. A paper by Lenihan et al. (2001) is cited to demonstrate that oysters at the base of reefs become less heavily infected with *P. marinus* than do oysters at the crest of reefs. However, in this study, which took place in a North Carolina river, the oysters were sampled only once, shortly after they had become infected and when infections were still very light. The investigators were unable to continue sampling over time as infections intensified. Another report (Volety et al., 2000) followed oysters at intervals of 2 to 4 weeks over a 2-year period on an artificial reef in a tributary of Chesapeake Bay. *P. marinus* and *H. nelsoni* infection levels were determined in oysters near the top of the reef and those near the bottom. When data for all sampling times were pooled, oysters at the bottom of the reef had statistically fewer and lighter infections of both parasites than those near the top, but most differences in *P. marinus* levels occurred during the first year when infections were relatively few and very light. By the second year, nearly all samples were 100% infected, with the average intensity being moderate to heavy. No *H. nelsoni* was detected until year two when infection intensities were very low at both locations. Nevertheless, they were clearly fewer and less intense in oysters near the reef crest. The two studies suggest that oysters at the base of reefs may experience heavier infection pressure, perhaps because of a greater concentration of infective stages that may accumulate, along with other suspended particles, at the base of reefs (Lenihan et al., 2001). But the Chesapeake Bay study indicated that the differential in *P. marinus* infection levels had largely disappeared by the second year of exposure.

In Florida, Quick and Mackin (1971) found that *P. marinus* prevalence did not change and that intensity actually decreased with increasing depth. The results of the Chesapeake study for *H. nelsoni* are also at variance with a study in Delaware Bay, which found no difference in the acquisition and intensification of that parasite in oysters suspended in the water column compared to those on the bottom (Ford and Haskin, 1988). Nor do oysters in intertidal locations have fewer infections than subtidal oysters (Gibbons and Chu, 1989; Littlewood et al., 1992; Burrell et al., 1984). The variable results of these studies, which may be linked to the particular site and methods involved, simply are not consistent enough to support the notion that position in the water would play a significant, long-term role in determining disease levels, especially during the second or third years of growth required to reach market size on reefs in the Chesapeake Bay.

Disease Resistance

Disease Resistance in Wild Oyster Populations

The potential for oysters to develop resistance to a disease was demonstrated in the early 1900s after an outbreak of "Malpeque Bay disease" in eastern Canada (Needler and Logie, 1947). Years after the epizootic, native *C. virginica* had good survival whereas those transplanted into the area suffered high mortality, indicating that the disease agent, never conclusively identified, was still present and that natural selection had produced a resistant local population. In the 1950s, comparison of South Carolina and Chesapeake Bay oysters exposed to *P. marinus* showed that the former consistently developed fewer and lighter infections than did the latter, leading to speculation that South Carolina oyster stocks may have experienced greater selective pressure than those from the Chesapeake (Andrews and Hewatt, 1957). More recently, a number of studies have involved Gulf of Mexico oysters, which have been under selective pressure from *P. marinus* for many decades, if not longer. Gulf stocks consistently show fewer and lighter *P. marinus* infections than do mid-Atlantic or northeast stocks under the same disease challenge (Bushek and Allen, 1996). Gulf oysters, however, are extremely susceptible to *H. nelsoni*, which has never been reported in the Gulf of Mexico. The Gulf oysters therefore experience overall higher mortalities than mid-Atlantic/northeast stocks, which have undergone varying selective pressure to both *H. nelsoni* and *P. marinus* over the years. In fact, the wild Delaware Bay oyster population appears now to be highly resistant to *H. nelsoni*. The development of a moderate degree of resistance was documented after the initial epizootic in the late 1950s but did not increase further because most of the surviving oysters were on the upper bay beds where they were protected from lethal infections by low salinity (Haskin and Ford, 1979). Infection prevalences remained high (50 to 80%) even though mortalities were reduced. Since the late 1980s, however, *H. nelsoni* infection prevalences in native oysters have been very low (<30%) in Delaware Bay, and the parasite has caused essentially no commercially significant oyster mortalities. At the same time, imported non-

Delaware Bay stocks that have not undergone selective mortality still become heavily infected, and molecular detection methods indicate that the parasite is still present and widespread in the bay. Although the putative high resistance of present-day Delaware Bay stocks has not been systematically and rigorously tested, the evidence available suggests that the current level was achieved only after extensive *H. nelsoni*-caused mortalities occurred on seed beds in the upper bay during two drought years in the mid-1980s. Had the initial epizootic mortalities during the late 1950s been as severe in the upper bay as they were in the mid-1980s, the high degree of resistance apparent now probably would have developed earlier.

A number of papers suggest that some localized oyster stocks in the Chesapeake show selective survival despite disease pressure (Andrews, 1968; Burreson 1991; Ragone-Calvo et al., 2003b), and Farley (1975) reported that some oysters display a histological appearance generally interpreted as evidence of resistance. However, there has been no systematic effort to document resistance in Chesapeake Bay native oysters. The development of resistance through natural selection depends on the degree of selective mortality that a population has experienced, and until recently many Chesapeake Bay populations were protected by low salinity from acquiring and developing lethal infections of either *P. marinus* or *H. nelsoni*. As long as unselected oysters survive and reproduce, it is unlikely that the overall level of resistance in the bay will improve measurably. Disease-caused mortalities in the past 2 years have extended into many subestuaries of the upper bay and may increase resistance in the overall population. Whether mortality events will be extensive enough to significantly lessen future epizootics is unknown, as are the time and the number of such widespread mortality episodes that might be needed to establish enough resistance for recovery of the population. Yet another complication is that the repletion program carried out for several decades by Maryland's Department of Natural Resources has transferred oysters throughout the bay, diluting any acquired genetic resistance in local populations.

Selective Breeding Programs for Disease Resistance

In the early 1960s, after the initial *H. nelsoni* epizootics, selective breeding programs were begun at Rutgers University and the Virginia Institute of Marine Science (VIMS). Both programs used survivors as initial brood stock, exposed the progeny to natural infections, and then produced another generation from the survivors of that generation. A generation in the Rutgers program was 2.5 years, by which time the oysters were market size. By the fifth generation, on average, about 70% of the oysters were still living at market size compared to about 7% in unselected controls (see Figure 4.1). Interestingly, the greatest improvement in survival was from the unselected to the first generation; the rate of improvement slowed noticeably after the third generation. The VIMS strains did not achieve as high a level of resistance because selection pressure was not as great at the VIMS exposure site (Haskin and Andrews, 1988). Also, development of resistance may have been less in the VIMS oysters because they were subject to infection by *P. marinus*. In the late 1980s and early 1990s, some of the Rutgers strains were compared to local stocks at sites from Maryland to Massachusetts. The selected strains survived and grew better than the local stocks when *H. nelsoni* was prevalent and performed similarly when it was not. Current tests continue to show that the selected stocks outperform unselected and local controls (see Table 4.2).

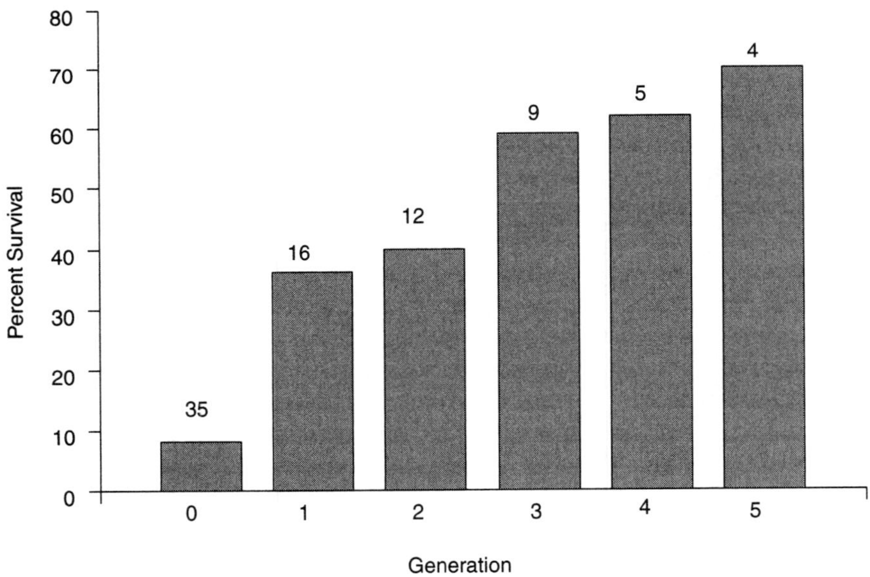

FIGURE 4.1 Percent survival, at age 2.5 years, of *C. virginica* bred for resistance to MSX disease at the Haskin Shellfish Research Laboratory, Rutgers University over five generations of selection, compared to susceptible controls (Generation 0), 1964 to 1988. Numbers above bars represent the number of different groups in each generation tested. One or more susceptible groups were produced each year, along with the selected groups. The continued poor survival (and high infection levels) of the susceptible oysters verified the sustained presence of the disease agent *H. nelsoni*.
SOURCE: Unpublished data from S. Ford, Haskin Shellfish Research Laboratory, Rutgers University, Port Norris, NJ.

TABLE 4.2 Cumulative Mortality of Selected *virginica* Strains and *C. ariakensis* and Local Control Stocks in Various Tests

% Mortality Resistant Strain[1]	Local Control	Site	Test Period	Reference
XB and WHS				
40-45	47	Chesapeake Bay, VA	26 months	SK Allen, VIMS, per com, 2002
40-50	78	Delaware Bay, NJ	26 months	SK Allen, VIMS, per com, 2002
XB and WHS				
50-70	60	Chesapeake Bay, VA	21 months	SK Allen, VIMS, per com, 2002
22	38	Delaware Bay, NJ	21 months	SK Allen, VIMS, per com, 2002
55-58	65	Delaware Bay, NJ	36 months	SK Allen, VIMS, per com, 2002
NE 1 and NE 2	Cape Shore Set			
30-4[2]	80[3]	Delaware Bay, NJ	30 months	[4] X Guo and S Ford, per com, 2002
10-1[5]	85[3]	Cape May Harbor, NJ	30 months	[4] X Guo and S Ford, per com, 2002
DEBY F_1	Mobjack Bay F			
20	82	Great Wicomico River, VA	18 months	Ragone-Calvo et al., 2003b
52	83	York River, VA	18 months	Ragone-Calvo et al., 2003b
36	82	Burton Bay, VA	18 months	Ragone-Calvo et al., 2003b
C. ariakensis	Mobjack Bay F			
14	80	Great Wicomico River, VA	18 months	Calvo et al., 2001
13	100	York River, VA	18 months	Calvo et al., 2001
16	100	Burton Bay, VA	18 months	Calvo et al.,2001

[1]Resistant strain codes: XB = Delaware Bay origin; WHS = Delaware Bay/Long Island Sound hybrid; NE 1 = Long Island Sound Origin, all selected for resistance to MSX and Dermo diseases. NE 2 = hybrid between NE 1 and brood stock from F. M. Flower Company selected for resistance to MSX, Dermo, and juvenile oyster diseases. DEBY = Delaware Bay wild, selected in Chesapeake Bay for resistance to MSX and Dermo diseases

[2]Infection pressure was much heavier at the Delaware Bay site compared to the Cape May Harbor site.

[3]Much of this mortality was due to unknown cause(s), not MSX and Dermo diseases.

[4] X. Guo, Graduate Institute of Environmental and Occupational Health, Medical College, National Cheng Kung University, personal communication, 2003; S. Ford, Haskin Shellfish Research Laboratory, Rutgers University, 2002.

The selective breeding programs were crude in comparison with modern plant and animal breeding practices, and inbreeding problems eventually occurred in some lines (Vrijenhoek et al., 1990; Hu et al., 1993), which were later remedied by an improved breeding strategy (S. K. Allen, Virginia Institute of Marine Science, Gloucester Point, personal communication, 2002). Nevertheless, there has been a clear response to selection in the resulting strains, measured as both reduced infection levels and improved survival (Ford and Haskin, 1987; Ragone-Calvo et al., 2003b).

Unfortunately, the strains were not resistant to *P. marinus*-caused mortality (Burreson, 1991), and mortality was high when this parasite caused an epizootic in Delaware Bay in the early 1990s (S. Ford, Haskin Shellfish Research Laboratory, Rutgers University, 2002). Thus, the selection program evolved to develop Dermo disease resistance in strains already highly resistant to MSX disease. Parallel programs have been conducted at Rutgers and VIMS, and resulting strains have been evaluated at sites from Virginia to Massachusetts.

Oysters "resistant" to MSX and Dermo diseases may still become infected by the causative agents, *H. nelsoni* and *P. marinus*, but infections are slower to develop to the stage when they cause disease and mortality. In fact, a negative diagnosis in selected strains may not indicate a lack of infection, but simply infection below the level of detection by standard assays. Over time, with repeated exposures, infections may intensify to a level that is not only detectable but also heavy enough to cause disease and mortality (Ford and Haskin, 1987). Further, the differential between selected and unselected strains may diminish under conditions of very heavy infection pressure (Haskin and Andrews, 1988). Thus, the differential depends on when the comparison is made and how heavy the disease pressure is. Probably the best measure of performance for an oyster stock is survival at market size because it integrates both survival and growth rates to a commercially relevant time. Under infection pressure from both *H. nelsoni* and *P. marinus* in Delaware and Chesapeake Bays and coastal embayments of Virginia and New Jersey, the best-performing selected *C. virginica* strains suffer 25 to 50% the mortalities of local oysters (Table 4.2). In comparison, mortality of *C. ariakensis* (Chesapeake Bay only) was 15 to 19% that of local stocks. The reduced mortalities were accompanied by lower infection rates of *P. marinus* and *H. nelsoni* (Calvo et al., 2001; Ragone-Calvo et al., 2003b; S. K. Allen, Virginia Institute of Marine Science, Gloucester Point, personal communication, 2002; X. Guo, Graduate Institute of Environmental and Occupational Health, Medical College, National Cheng Kung University, Taiwan, personal communication, 2003; S. Ford, Haskin Shellfish Research Laboratory, Rutgers University, Port Norris, New Jersey, personal communication, 2002).

CRASSOSTREA ARIAKENSIS

Life History

C. ariakensis is reported to be found along the entire Chinese coastline (Tschang and Tse-kong, 1956), southern Japan (Rao, 1987), Taiwan, the Philippines, Thailand, Vietnam, and northern Boreno and Malaysia (Zhou and Allen, 2003). There are also reports of *C. ariakensis* on the northwest coast of India and Pakistan. However, the distributional range outside China and southern Japan has not been genetically confirmed (Allen et al., 2002), and there is considerable taxonomic confusion about the species (Coan et al., 2000; Carriker and Gaffney 1996; Zhou and Allen, 2003). Sometimes *C. ariakensis* has been identified as *Ostrea rivularis*, *C. rivularis*, *C. discoidea* (e.g., Awati and Rai, 1931; Harry, 1985; Rao, 1987) or *C. paulucciae* (Carriker and Gaffney, 1996). As a result, biological information on *C. ariakensis* in its native distribution range is somewhat difficult to unravel because *C. ariakensis* may have been misidentified.

C. ariakensis was inadvertently introduced to Oregon with shipments of *C. gigas* and *C. sikamea* spat from Japan in the 1970s (Breese and Malouf, 1977). Although *C. ariakensis* seed has been repeatedly outplanted on intertidal mudflats or suspended from floating rafts at several sites from Washington to central California (Breese and Malouf, 1977; Langdon and Robinson, 1996), there are no reports of established wild populations existing on the U.S. West Coast (Coan et al., 2000; J. T. Carlton, Williams College-Mystic Seaport Program, personal communication, 2003; R. Malouf, Oregon State University, personal communication, 2003). Apparently, seawater temperatures are not warm enough for the species to reproduce and maintain self-sustaining populations at these sites.

C. ariakensis has limited aquaculture use in Washington and Oregon because of difficulties in obtaining sufficient quantities of seed for large-scale production (Langdon and Robinson, 1996). Field and laboratory experiments in Virginia (e.g., Calvo et al., 2001) and North Carolina have used the U.S. West Coast strain originally imported from Japan. In addition, scientists at VIMS have imported strains collected from the Yellow River estuary (northern China *ariakensis*) and from the Guangxi province near Beihai (southern China *ariakensis*) for use in comparative biological studies (S. K. Allen, Virginia Institute of Marine Science, personal communication, 2003). Laboratory work has also been conducted in France with the strain that was accidentally imported from Japan to the U.S. West Coast. No wild populations currently exist in France.

While *C. ariakensis* has been extensively cultured throughout southern China and Japan for over 300 years (Cai et al., 1979), there is relatively little information on the ecology or biology of natural populations of the species in its native distributional range. In China the common name is *Jinjiang-muli* (meaning "close to river [Jinjiang] oysters [muli]": Zhou and Allen, 2003). Indeed, Cai et al. (1992) reported that the oyster is cultured in estuarine mid- to shallow subtidal regions of Zhanjiang Bay where water temperatures range from 14 to 31.8°C and salinity varies from 7.5 to 30.2‰. Others report that the species is normally found in areas where the salinity varies between 9 and 30‰ (Cahn, 1950) and that it can tolerate salinities less than 10‰ for short periods of time (Amemiya, 1928). Guo et al. (1999) reported that the species can tolerate a wide salinity range, though settlement is most pronounced in low-salinity estuaries and river beds. Extensive Chinese aquaculture of *C. ariakensis* is typically found in intertidal and shallow subtidal estuarine habitats, where concrete stakes are transported to lower-salinity waters for spat collection and then moved to more saline areas for grow-out to marketable sizes (Guo et al., 1999).

Similar to *C. virginica*, spawning activity in *C. ariakensis* is seasonal, and the reproductive cycle is influenced by temperature and regional environmental conditions. For China, Cai et al. (1992) reported that the reproductive season was between April and June and that larval settlement was most pronounced on the shady sides of hard surfaces. Moazzam and Rizvi (1983) reported that larval settlement in saltwater creeks in Pakistan was most pronounced from September through October, though recruitment was also observed during July to August. In a saltwater creek near Karachi, Pakistan, Ahmed et al. (1987) reported that larval settlement occurred from April to October, with highest spatfall recorded in April through July. In the Zhujiang River estuary the reproductive season is June to September and spawning is mainly in June to July; a second spawning may occur if environmental conditions are appropriate (Zhou and Allen, 2003). Asif (1979) reported that gonads generally were first present in individuals 2 to 3 months of age (0.4 to 0.6 cm in shell length) and that protandric hermaphrodites were found.

Langdon and Robinson (1996) report that mature oysters collected from Yaquina Bay, Oregon, became available for hatchery spawning in early to mid-May, and female oysters had some percentage of mature eggs until December. From August to November, females were found with more that 50% mature eggs. There was a higher degree of annual variability in the gametogenic cycle of male oysters, though the highest percentages (>50%) of mature individuals were recorded from June to February. For Dabob Bay, Washington, Perdue and Erickson (1984) reported that gonadal proliferation began in late April to early June and the greatest proliferation occurred in mid-June to October. The authors reported that *C. ariakensis* failed to spawn, however, due to low summer water temperatures. Luckenbach (personal communication, Virginia Institute of Marine Science, Gloucester Point, 2003) has conducted preliminary laboratory studies which suggest that *C. ariakensis* is capable of reproducing throughout the same salinity ranges as *C. virginica* (i.e., 5 TO 35‰). Modeling of larval salinity and temperature tolerances will be needed to predict the dispersal of *C. ariakensis* both within and beyond the Chesapeake Bay as has been done for other species (Lough, 1975; Goulletquer et al., 1994b).

One concern regarding the potential introduction of *C. ariakensis* to areas with native populations of *C. virginica* is that both species may spawn at the same time and place, raising the possibility of interspecific hybridization. Laboratory experiments conducted by Allen et al. (1993) demonstrate that, although fertilization occurs in interspecific crosses, no viable progeny are produced. The larvae do not grow and begin to die a week after fertilization.

Coan et al. (2000) reported that *C. ariakensis* can reach shell heights up to 20 cm, while Carriker and Gaffney (1996) noted that the oyster can be 20 to 24 cm high. Oysters growth relatively quickly in southern Chinese coastal waters and are typically harvested, when raised in extensive aquaculture conditions, within 2 to 3 years (10- to 15-cm shell length) after larval settlement (Guo et al., 1999). In U.S. waters, growth rates appears to vary with environmental conditions. For example, Langdon and Robinson (1996) reported no differences in the growth rate of *C. ariakensis* and *C. gigas* at sites in Puget Sound, Washington, and Coos Bay, Oregon, when planted on intertidal mudflats or when suspended from floating rafts in Yaquina Bay, Oregon. Both species attained shell heights of about 10 cm in roughly 2 years. *C. gigas* did grow faster than *C. ariakensis* when placed in mesh bags supported by intertidal trestle in Tomales Bay, California, (central California *C. ariakensis* grew to only about 6 cm in about 1.5 years, while *C. gigas* attained an average shell height of 10 cm). Calvo et al. (2001) grew triploid *C. ariakensis* in mesh bags in floating trays at six sites in the Virginia portion of Chesapeake Bay; two sites were located in low-salinity (<15‰), two in medium-salinity (15-25‰), and two in high-salinity (>25‰) waters. At deployment, oysters averaged 6.4 cm in shell height (2 years old), and growth was followed for about 9 months. It was found that shell growth was significantly slower (2.6 mm/month) in the low-salinity regime when compared to medium- (4.9 mm/month) and high-salinity waters (6.2 mm/month).

A limited number of studies have examined the predators and competitors of *C. ariakensis* in its natural distribution range. Several species of carnivorous snails and crabs are reported to feed on adult oysters (Zhang and Lou, 1956), while sea urchins and sea stars consume oyster spat (Zhou and Allen, 2003). Barnacles have been noted to compete for food and space with settling oyster larvae (Cai et al., 1992). Calvo et al. (2001) reported that mortalities of oysters suspended in mesh bags above the bottom at several places in the lower Chesapeake Bay varied from 13 to 16% over a 9-month period, though the agent of mortality was not reported. The oysters were contained in mesh bags placed in suspended trays where benthic predators (e.g., crabs, snails) had reduced access to them. Moreover, the oyster leech (flatworm), *Stylochus ellipticus*, considered a significant predator on *C. virginica*, may affect *C. ariakensis* populations (Daniel et al., 1983; MacKenzie, 1997; Sagasti et al., 2001). Because *C. ariakensis* is similar in size and shell characteristics, it is likely that the same suite of benthic predators that prey on *C. virginica* in Chesapeake Bay would also forage on *C. araikensis*. Luckenbach (personal communication, Virginia Institute of Marine Science, Gloucester Point, 2003) thinks that the shell of *C. ariakensis* may not as be strong as *C. virginica*, which may suggest that the effective size refuge from crab predators is larger for *C. ariakensis* than for Eastern oyster.

Investigations of the larval biology of *C. ariakensis* are restricted to laboratory studies, most of which were conducted in highly artificial conditions to assess optimal nutritional and environmental conditions for potential aquaculture purposes. Breese and Malouf (1977) reported that maximum larval growth was obtained at 28°C at both 20‰ and 30‰ salinity. The authors found that larval settlement was highest at 28°C and 20‰, with some settlement occurring at 28°C and 15‰ and 26°C and 20‰. Other temperature-salinity combinations yielded poor larval growth and no settlement. No larvae survived at 32°C. Langdon and Robinson (1996) found that larval settlement at salinities of 15 and 20‰ was greater than 25 and 30‰ and that no settlement occurred at 35‰. A similar pattern was found when examining larval growth rate. The authors found that growth of settled oyster spat over a 2-week period (at 20‰ salinity and on a diet of the diatom *Chaetoceros calcitrans*) was generally fastest at 25° to 30°C, though growth occurred at temperatures ranging from 20° to 35°C. Salinity and temperature conditions were not reported, but Cahn (1950) found that newly settled *C. ariakensis* could be found as early as June in Japan. Preliminary laboratory studies by Luckenbach (personal communication, Virginia Institute of Marine Science, Gloucester Point, 2003) revealed that swimming larvae of *C. ariakensis* tended to concentrate nearer the bottoms of culturing tubes while those of *C. virginica* were more commonly found nearer the surface of the containers. No field observations are available on *C. ariakensis* larval behavior and mortality rates.

Like other oyster species, larvae of *C. ariakensis* must settle on solid surfaces. As mentioned earlier, suitable substrates for successful oyster settlement are those not heavily fouled by sediment or organisms (e.g., Osman et al., 1989). While Ahmed et al. (1987) reported that spatfall was primarily in the lower intertidal zone in Pakistan, a much broader settlement range (low tide to 10-m depth) has been reported in China (Nie, 1991). Larval settlement, 12 to 18 days after spawning (Zhou and Allen, 2003), varies primarily with temperature and other environmental conditions. In southern China, spatfall has been recorded from June to August (Cai and Li, 1990), while in Pakistan larval settlement has been observed from May to December (Moazzam and Rizvi, 1983; Hasan, 1960).

There has been some uncertainty about the reef-building characteristics of *C. ariakensis* in different areas depending on substrate type. Yao (1988) reported finding an oyster reef primarily composed of fossil shells of the species in the tidal zone of Fujian, China. X. Guo (Graduate Institute of Environmental and Occupational Health, Medical College, National Cheng Kung University, personal communication, 2003) noted, "It is common knowledge among oyster (workers) in China that *C. ariakensis* is a reef builder. Areas [with] *C. ariakensis* reefs are north Shandong, Guangzi and Gunangdong [provinces]." There is a brief reference to reef building in the Bohai Sea cited in Wang et al. (1993, referred to by Guo). Lastly, there are some newspaper accounts of using historical reefs found in coastal low-salinity waters of the Bohai Sea in northern China to make concrete and apparently only *C. ariakensis* is found in this region, according to X. Guo. In Japan, however, the oyster apparently is restricted to lower intertidal muddy bottom habitats (Amemiya, 1928; Hirase, 1930). There are several reports that in India and Pakistan the oyster can be found on both hard substrates and muddy creek bottoms (Mahadevan, 1987; Patel and Jetani, 1991; Ahmed et al., 1987).

Oyster filtration rates are typically related to environmental conditions (e.g., temperature, salinity, hypoxia) as well as the quantity and quality of suspended particulate materials (e.g., Higgins, 1980; Shumway et al., 1985; Riisgard, 1988). Zhang et al. (1959) examined the feeding biology of adult *C. ariakensis* relative to a number of environmental parameters, and concluded that feeding rate was highest when temperature and salinity were between 10 to 12°C and 15 to 30‰, respectively. While the oyster is known to be able to tolerate reduced salinity conditions, feeding rate was significantly reduced at less than 5‰ salinity. Zhang et al. (1959) also noted that feeding rate was not influenced by high levels of suspended material. In preliminary laboratory experiments conducted by R. Newell (Horn Point Laboratory, University of Maryland Center for Environmental Science, personal communication, 2002), size-specific filtration rates of *C. ariakensis* appeared similar to those of *C. virginica*. These results are consistent with Powell et al.'s (1992) conclusion that size-specific filtration rates are quite similar for most marine bivalve species.

Disease and Disease Resistance

Observed Diseases and Parasites in Native Range

There is little information on disease outbreaks in *C. ariakensis* over its native range. Most *C. ariakensis* production appears to occur in China, which also encompasses a large portion of this species' range. A recent publication describes recurrent large-scale mortalities (80 to 90% annually) recorded since 1992 in Guangdong province (Wu and Pan, 2000). The mortalities occur from February to May in 2- to 7-year-old oysters. The authors found heavy concentrations of what they considered to be a Rickettsiales-like organism in moribund oysters collected during one mortality episode. Rickettsiales-like organisms are common and ubiquitous in bivalves worldwide and are generally considered benign. In a few instances they have been associated with mass mortalities of other bivalves (Gulka and Chang, 1984; Gardner et al., 1995; Villalba et al., 1999), although it is not certain in any case whether these organisms are the original cause of the deaths or merely contributed to the deaths of otherwise stressed hosts. Further, the bodies identified by Wu and Pan as rickettsia have morphological and staining characteristics atypical of rickettsia and may have been misidentified. Although Wu and Pan do not mention it, toxic algal blooms are thought by others to cause the occasional large-scale mortalities of *C. ariakensis* in China (Zhang and Lou, 1956; Zhang et al., 1995). In a survey of molluscan aquaculture in China, Guo et al. (1999) noted reports of mass mortalities in scallops, abalone, and certain clams but none for oysters of any species.

As part of the International Council for the Exploration of the Seas (ICES) protocol being employed in tests of *C. ariakensis* in Chesapeake Bay, the VIMS has examined tissue sections of 242 individual *C. ariakensis* from three separate stocks originating from the U.S. West Coast and north and south China. Cells characterized as "unusual" were found in three oysters from the northern China site. One of the oysters contained spores and plasmodia consistent in morphology with members of the Haplosporidia, the phylum containing *H. nelsoni*; cells in the other two oysters were not identifiable. Neither the haplosporidian infection nor the other cells were abundant in the affected oysters.

VIMS is currently engaged in further testing of *C. ariakensis*. Samples were obtained from five sites along the Chinese coast in autumn 2002, and another set from the same sites will be collected in spring 2003, with further sampling planned if mortality is observed. Each sample includes 30 small and 30 market-sized cultured oysters and if present, 30 market-sized wild oysters, for a potential total of 900 individuals. The oysters are being examined by tissue slide histopathology for potential pathogens and pathological conditions, and, by polymerase chain reaction (PCR) using DNA primers specific for *Perkinsus* spp. and herpes virus. Using PCR, VIMS researchers have detected herpes virus in adult *C. ariakensis* collected in Japan (Itoki River, Ariaki Bay) but not in two spawns of *C. ariakensis* made in 2002 at the VIMS hatchery (K. Reece, Virginia Institute of Marine Science, personal communication, 2002). It should be noted that herpes virus is already present in the East Coast oyster population, as reported by Farley et al. (1972) who found it in *C. virginica* subjected to elevated temperatures in a power plant discharge pipe in Maine.

French researchers at the Institut français de recherche pour l'exploitation de la mer La Tremblade Laboratory examined tissue sections of 15 female *C. ariakensis* from Hong Kong and found parasites in the egg cells of 7 individuals. The parasite could not be identified with the material available but resembles a Marteilioides-type parasite described as occurring in the egg cells of *C. gigas* in Japan (Comps et al., 1986).

An "Exotic" Disease

As a possible replacement for *C. gigas* in France, in the event the species began to have serious mortalities, French scientists imported *C. ariakensis* (from the Haskin Shellfish Research Laboratory in New Jersey), which they held in a "quarantine" system that treated outgoing, but not incoming, water. A sample was inspected histologically before the importation and found to have no detectable parasites, but after 7 months in the "quarantine" system, the oysters experienced high mortality. Tissues of moribund oysters were examined and found to be heavily parasitized by an organism that resembled, in all morphological and pathological details, *B. ostrea*, which infects *Ostrea edulis*, causing the disease "bonamiasis" in the same region (Cochennec et al., 1998). The authors "strongly suspected" that the *C. ariakensis* became infected through exposure to incoming untreated "waters . . . endemic for bonamiasis." They pointed out that the parasite is the first report of a bonamia parasite in a species of *Crassostrea*; all other findings have been in *Ostrea* spp.

Resistance to MSX and Dermo Diseases

In several tests conducted since 1998, *C. ariakensis* has demonstrated rapid growth, high survival, and low infection levels after exposure to *H. nelsoni* and *P. marinus* in a number of locations in the Virginia portion of Chesapeake Bay and the coastal bays of Virginia's Eastern Shore. The oyster has not been tested in Maryland, but the Virginia sites represented a variety of environments that spanned high-, medium-, and low-salinity regimes. The first trial, conducted by VIMS, extended over 18 months in 1998 and 1999. Mortality of *C. ariakensis* was about 15% compared to 80 to

100% of similar-sized lower Chesapeake Bay wildstock *C. virginica* (Calvo et al., 2001). Infection prevalence of *H. nelsoni* was low during the test, with a maximum in *C. virginica* of only 25%. No *H. nelsoni* was detected in any *C. ariakensis*. In contrast, both species developed high prevalences of *P. marinus*, with peaks of 68 to 84% in *C. ariakensis* and 100% in *C. virginica*. Despite the high *P. marinus* prevalences in *C. ariakensis*, all infections were categorized as "light" (unlikely to be lethal), whereas a substantial proportion of the infections in *C. virginica* became advanced and ultimately lethal. Subsequent trials of *C. ariakensis*, initiated in 2000 and 2001, were conducted by industry and did not include a disease diagnosis component; however, survival and growth of the species were reported to be extremely good. Within a year of the 2000 deployment, the *C. ariakensis* were of market size with little or no mortality, whereas half of the *C. virginica*, against which they were compared, were dead (Thompson, 2001).

One concern about the 1998 to 1999 trial (Calvo et al., 2001) was that the *C. virginica* had been exposed to infection during the year before the trials began, which might have explained their higher infection levels compared to *C. ariakensis*. Infection prevalence and intensity, especially of *P. marinus*, typically increase with exposure in *C. virginica*. Although *C. ariakensis* did become infected with *P. marinus*, there was no consistent trend toward increased prevalence or intensity in this species over the 18-month trial, which included two summer infection periods. To investigate the possibility that longer exposure might increase parasite burdens in *C. ariakensis*, VIMS is now examining residual oysters from the 2000 and 2001 industry trials (E. Burreson, Virginia Institute of Marine Science, personal communication, 2002). Samples collected in June, September, and October 2002, after 2 to 3 years of exposure, showed 20 to 80% prevalence of *P. marinus* in the *C. virginica*, depending on test site, but no more than 20% (all very light infections comprising just a few parasites) in the *C. ariakensis*. No *H. nelsoni* have been detected in any *C. ariakensis* examined. A few instances of hemocytic infiltration into the digestive epithelium and lumena have been found, but no recognizable parasites (E. Burreson, Virginia Institute of Marine Science, personal communication, 2002).

All evidence to date, then, indicates that even though *C. ariakensis* can become infected with *P. marinus* (and perhaps also with *H. nelsoni*, although no infections have yet been detected), infections do not develop to the point where they inhibit growth, fattening, and survival. If *C. ariakensis* actually does become infected with *H. nelsoni*, infections remain so light and localized that they have so far been undetectable by standard diagnostic methods. This result is similar to that demonstrated for *C. gigas* in earlier trials (Calvo et al., 1999). One difference, however, is that *C.*

ariakensis performed well in low salinity, whereas *C. gigas* grew poorly and had high mortalities. Further, *C. ariakensis*, under the conditions of the test, did not develop the heavy mud blisters caused by the polychaete *Polydora websteri* that formed in *C. gigas* tested in Chesapeake Bay and were also reported in earlier trials in Louisiana (Kavanagh, 1940). More recently however, some 2-year-old *C. ariakensis* from the industry trials, which had been deployed at a low-salinity site, were observed to have a heavy infestation of Polydora blisters (S. Ford, Haskin Shellfish Research Laboratory, Rutgers University, personal communication, 2002). Infestation by Polydora is probably a site-dependent phenomenon, as the prevalence and intensity of blisters in the 2 to 3-year-old oysters from four sites being monitored by VIMS have not been noted as being heavier than those of *C. virginica* at the same locations (E. Burreson, Virginia Institute of Marine Science, personal communication, 2002).

SUMMARY

- Oysters, like other bivalve filter feeders, feed on microalgae and are capable of removing large quantities of particulate matter from the water; they have a larval stage that allows them to be dispersed from the immediate site of the parental stock; and they require a clean, solid substrate on which the larvae can settle and cement themselves.
- *C. virginica* spawns during the summer, producing a larval stage that lasts for 2 to 3 weeks, during which time mortality is typically about 99%. Between settlement in mid-summer and winter, mortality of the "spat" due to predators may be nearly as high. Adults can tolerate a wide range of environments: subtidal and intertidal; salinities from <5 to 40 ppt; temperatures from freezing to 36°C, with short-term exposure to 50°C. They display genetically conserved, geographically distinct growth, reproduction, and disease susceptibility traits.
- Two important ecological functions have been attributed to oysters: habitat and resource provision and filtration of suspended particles from the water. The first function results from habitat created by oyster beds or reefs that provides refuge for the next generations of oysters and structure, refuge, and food resources for other benthic organisms and fish. The second ecological function is the removal of organic material, primarily phytoplankton, from the water and its transformation into inorganic nutrients, thus helping to reduce turbidity. Although these functions are often attributed to oysters, it is unclear how much impact oyster reef habitat and oyster filtering capacity have on the Chesapeake Bay ecosys-

tem, especially given other stresses from sedimentation and nutrient overenrichment. Most of the arguments that the loss of oysters from the Chesapeake Bay has changed it from a "benthic"- to a "pelagic"- dominated system are based on very limited or anecdotal evidence.

- Oysters, like other organisms, become infected by some parasites that cause disease (damage severe enough to cause dysfunction of an organism) and by other parasites that cause little overall harm. The parasites responsible for MSX and Dermo diseases are so virulent that they cause disease in otherwise healthy oysters. Other parasites, including many bacteria and at least one virus (herpes), are "opportunists" that proliferate and cause disease in oysters that are stressed by other factors. Opportunistic parasites are most common in hatcheries and nurseries, where larval and juvenile bivalves are grown at high densities and elevated temperatures.
- Over the past two decades, above-average temperatures and repeated droughts have favored the parasites that cause MSX and Dermo. Not only have these conditions caused heavy oyster mortalities on existing reefs, they have impeded restoration efforts by causing mortalities on newly created reefs.
- Resistance to disease has been documented in wild populations of *C. virginica* after heavy selective mortality. Selective breeding has also produced *C. virginica* strains that are highly resistant to MSX disease and moderately resistant to Dermo disease; however, when tested under the same conditions, *C. ariakensis* shows a significantly higher degree of resistance (i.e., lower infection prevalence and intensity). The best *C. virginica* strains have shown 50 to 75% lower mortality than local (unselected) stocks, whereas *C. ariakensis* has shown 80 to 85% lower mortality.
- *C. ariakensis* is reported from Japan and Korea, south along the coasts of China, Malaysia, the Philippines, and westward to India and Pakistan, although genetic confirmation is lacking for most of this region and the species may have been misidentified in some cases. Relatively little biological and ecological data are available on *C. ariakensis* in its native range; however, available information indicates that *C. ariakensis* grows under environmental conditions very similar to those of *C. virginica* and has the same size-specific filtration rates. It is found intertidally and subtidally on reefs as well as on muddy bottom and is cultured in areas where the temperature ranges from 14° to 32°C and salinities from 7 to 30 ppt. The salinity range under which it reproduces is also similar to that for *C. virginica*, although it may require higher temperatures to induce spawning. Stocks imported to the West Coast of the United

States in the 1970s have not naturalized, presumably because the temperature is too low for reproductive success. *C. ariakensis* grows faster than *C. virginica* but has a thinner shell, which may make it more susceptible to predators such as crabs; otherwise, it should be subject to the same predation pressures as *C. virginica*.

- Little information is presently available on parasites or diseases of *C. ariakensis* in its native range. Histological examinations of *C. ariakensis* at various institutes have found a few protozoan parasites in groups known to cause mortality in oysters, including a haplosporidian (MSX-like) and a *Bonamia* parasite, which was associated with mortality of *C. ariakensis* and was probably the same parasite that caused widespread mortalities of the European oyster. Other parasites, including a putative Rickettsiales-like organism and herpes virus, have been found worldwide in numerous marine bivalves.

5

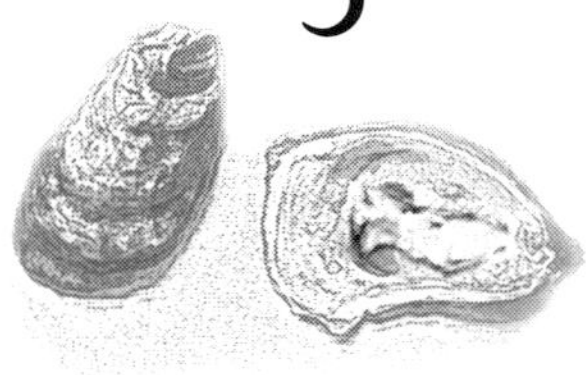

Social and Economic Value of Oysters in the Chesapeake Bay

Crassostrea virginica has occupied a prominent position in the ecology and economy of the Chesapeake Bay. In the early colonial period, oysters were so abundant that reefs posed navigational challenges to vessels. Early fisheries supplied food and fill for road construction. Demand grew and harvesting capacity expanded through the end of the 19th century. Oyster harvests peaked in the 1870s and 1880s with the commencement of dredging on deep channel reefs and then began a precipitous decline, falling by nearly 60% between 1880 and 1930. Harvests held fairly stable from 1930 through the late 1950s in Virginia, before collapsing to the present low levels. Maryland harvests held stable for longer, from 1930 through the 1970s, but have since collapsed. Present harvests from Maryland and Virginia are less than 2% of the 1880 harvest. The decline from 1880 through 1930 is consistent with excess harvesting. The subsequent decline seems closely linked to Dermo and MSX, exacerbated by increased salinity levels throughout the bay associated with a sequence of years with below-average precipitation. In addition, elevated levels of tributyltin oxide, agricultural pesticide residues, and other chemical stressors have been implicated in increased morbidity. Kennedy and Breisch (1983) provide a history of Chesapeake Bay oysters from the mid-1800s through the early 1980s.

Economic analyses of the oyster fishery have included studies of the theoretical merits of leased-bottom versus public-bottom management paradigms (Christy, 1964; Agnello and Donnelly, 1975, 1976, 1984; Santopietro and Shabman, 1992a, 1992b), an ex ante evaluation of the potential consequences of introducing *C. gigas* (Lipton et al., 1992), estimates of the cost efficiency of relaying for depuration or to reduce morbidity (Easley, 1982; Dunning and Adams, 1995; Keithly and Diop, 2001a, 2001b), the strength of leased-bottom claims as a basis for redress against sources of bioaccumulated toxic compounds (Ajuzie and Altobello, 1997), an evaluation of the role that increased seed prices played in the recent fishery decline (Bosch and Shabman, 1989), a model of the return to investment in research into the causes of recent fishery declines (Bosch and Shabman, 1990), and assessments of the current economic status of the oyster fishery (Lipton and Kirkley, 1994; Kirkley, 1997; Lipton, 2002).

Between 1980 and 2001, oyster production and harvest effort declined throughout the bay; however, production did not fall evenly. As shown in Table 5.1, Maryland produced roughly 2.1 million bushels of oysters in 1980, while Virginia produced 1.1 million bushels.[1] By 2001 production from Maryland waters was approximately 348,000 bushels, a decline of approximately 83% from 1980 levels. During the same period, production from Virginia waters declined approximately 98%, to about 23,000 bushels. Virginia's share of the harvest fell from 34% in 1980 to 6% in 2001. Total production, man-days of labor, number of license holders, estimated average income, and number of processing plants are all below their 1980 levels.

As production declined, the per unit value of oysters increased. The nominal value of a bushel of oysters in 2001 was more than three times the 1975 value (see Figure 5.1). After adjusting for the effect of inflation, it becomes apparent that increases in prices paid to harvesters have not been high enough to offset overall production losses; thus, the total value of the oyster harvest in 2001 was significantly smaller than the 1980 value. For 1980 the total dockside value of the Chesapeake oyster harvest was $29.3 million. In 2001 it was $4.3 million (National Marine Fisheries Service [NMFS], 2003), a drop of more than 85% (see Figure 5.2). When expressed in real terms (2001 dollars), the loss in total dockside value was closer to 93%; a drop from $60.1 million in 1980 to $4.3 million in 2001 dollars. Moreover, as oyster abundance has declined, catch per unit of effort has also declined. The adoption of cost-reducing technological changes has been constrained by regulation, suggesting that the cost per unit landed has increased. Harvester net returns have been caught in the double bind of falling revenues and rising costs. Because oyster harvests declined more precipitously in Virginia than in Maryland, real value fell from $40.6 million to $3.8 million in Maryland and from $19.5 million to

[1]There is a wide variation in pounds of oyster meat per bushel across states and regions: in Connecticut a bushel yields 7.7 pounds of oyster meats; New York 7.5 pounds; Maryland 6.48 pounds; Virginia 6.59 pounds; and Louisiana 4.82 pounds (NMFS, 1976).

TABLE 5.1 Chesapeake Bay Oyster Statistics for Maryland and Virginia

	Maryland and Virginia					
SEASON	1980	1985	1990	1995	2000	2001
LANDINGS (BU)						
Maryland	2,111,080	1,142,493	414,445	164,641	380,675	347,968
Virginia	1,074,776	627,052	162,618	55,414	22,623	22,573
Virginia Share	34%	35%	28%	25%	6%	6%
Total for Bay	3,185,856	1,769,545	577,063	220,055	403,298	370,541
MAN-DAYS						
Maryland	154,460	146,994	80,170	28,637	47,595	42,176
Virginia[a]						
LICENSE HOLDERS SELLING OYSTERS						
Maryland	2,246	1,807	1,156	603	1,773	1,293
Virginia	2,622	2,081	1,782	399	295	320
Total for Bay	4,868	3,888	2,938	1,002	2,068	1,613
AVERAGE BUSHELS PER MAN-DAY (BU)						
Maryland	13.7	7.8	5.2	5.7	7.9	8.3
Virginia[b]					3 to 6	
ESTMATED INCOME FROM OYSTERS PER LICENSE HOLDER SELLING ($)						
Maryland	7,900	9,300	8,600	5,400	4,100	5,300
Virginia[c]	3,600	3,500	2,400	2,700	1,600	1,800
OYSTER PROCESSING PLANTS						
Maryland			20			
Virginia			28			
CERTIFIED FACILITIES						
Maryland						
Shucker Packers					29*	
Shellstock Shippers					67*	
Repackers					06*	
Reshippers					16*	
Virginia						
Shucker Packers					62*	
Shellstock Shippers					09*	
Repackers					05*	
Reshippers					08*	

*Data from 2002

[a]Virgina does not track man-days.

[b]The VMRC estimates Virigina harvest per manday at 3 to 6 bushels.

[c]Virginia income estimates may be biased downward because not all license holders may be selling oysters.

SOURCE: Data from J. Wesson, Virginia Marine Resource Commission (VMRC), Newport News, personal communication, 2003; C. Judy, Maryland Department of Natural Resources, Annapolis, personal communication, 2002.

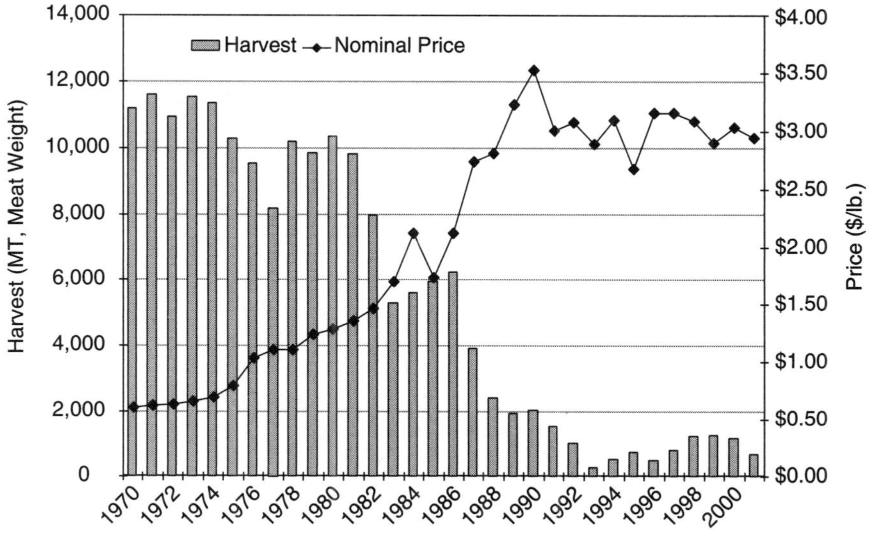

FIGURE 5.1 Chesapeake oysters: harvest and nominal price.
SOURCE: Data from the National Marine Fisheries Service (NMFS, 2003).

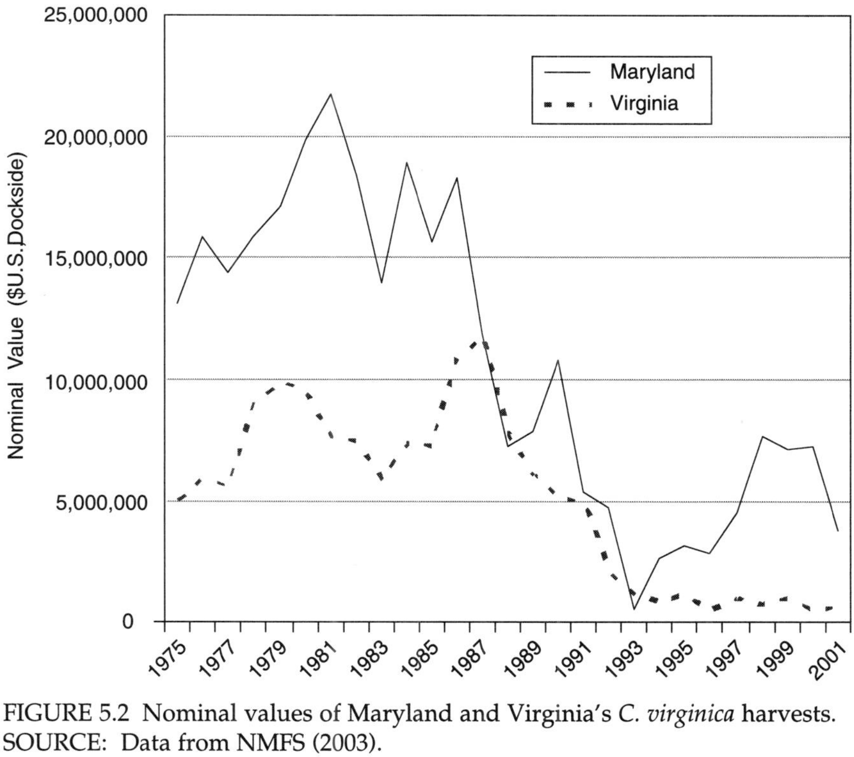

FIGURE 5.2 Nominal values of Maryland and Virginia's *C. virginica* harvests.
SOURCE: Data from NMFS (2003).

$0.57 million in Virginia, declines of 91% and 97%, respectively (see Figure 5.3) (NMFS, 2003). In 1980 the value of Chesapeake Bay oyster harvests was shared, with 64% going to Maryland and 36% to Virginia. By 2001 the split had shifted such that Maryland's share increased to 87% and Virginia's share declined to 13%.

Between 1980 and 2001, effort also declined throughout the bay, but not by as much as production (in relative terms). The number of man-days of effort in Maryland fell 73%, from 154,000 in 1980 to 42,000 in 2001 (Table 5.1). Because the decline in effort has not matched the decline in production, the average number of bushels produced per unit of effort has also fallen. Catch per unit of effort in the Maryland fishery in 2001 was about 39% lower than it was in 1980. Because Virginia does not track the number of man-days of effort for oyster harvests, it is not possible to compare changes in catch per unit effort for Virginia oystermen. Nevertheless, based on reported estimates that Virginia oystermen average 3 to 6 bushels per man-day of effort (J. Wesson, VMRC, Newport News, per-

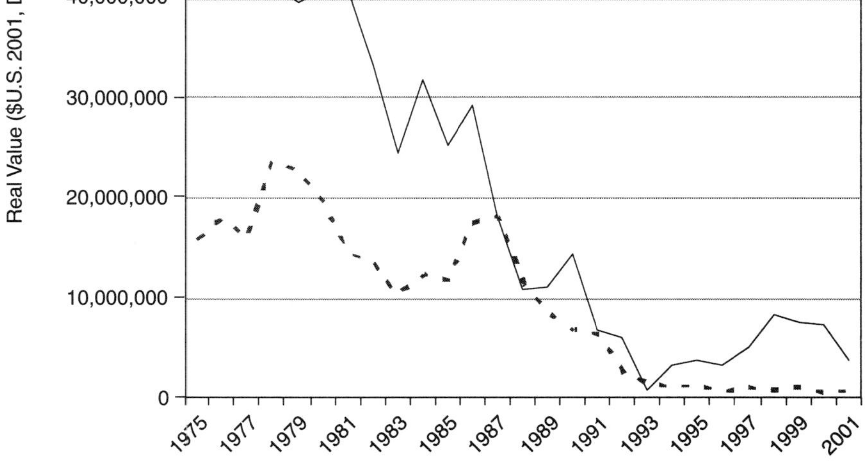

FIGURE 5.3 Real (2001 $U.S.) value of Maryland and Virginia's *C. virginica* harvest. SOURCE: Data from NMFS (2003); U.S. Department of Labor, Bureau of Labor Statistics (2003).

sonal communication, 2003), it is likely that Virginia oystermen catch 25 to 65% less per man-day than their Maryland counterparts.

The number of license holders has dropped 42% in Maryland and 88% in Virginia since 1980. Average revenues per license holder have declined by 50 to 67% from 1980 levels. The average revenue per license holder in Virginia is 25 to 33% of that in Maryland. However, it should be noted that holding a license does not indicate active participation in the fishery. Licenses can be bought to maintain the option of entering the fishery if conditions improve. Since Virginia oyster harvest licenses cost less than $70 in 2001, purchasing a license can be viewed as an inexpensive investment that ensures an option to access the fishery and serves as evidence of interest should the state choose to implement a license limitation program or seek disaster assistance on behalf of oystermen who have been adversely impacted by the decline in oyster populations.

As the Chesapeake oyster harvest has declined, the Gulf of Mexico region has become the dominant source of *C. virginica* harvested in the United States. In 1950 the gulf accounted for less than 20% of the *C. virginica* harvest; the gulf region now provides about 90%. In contrast the

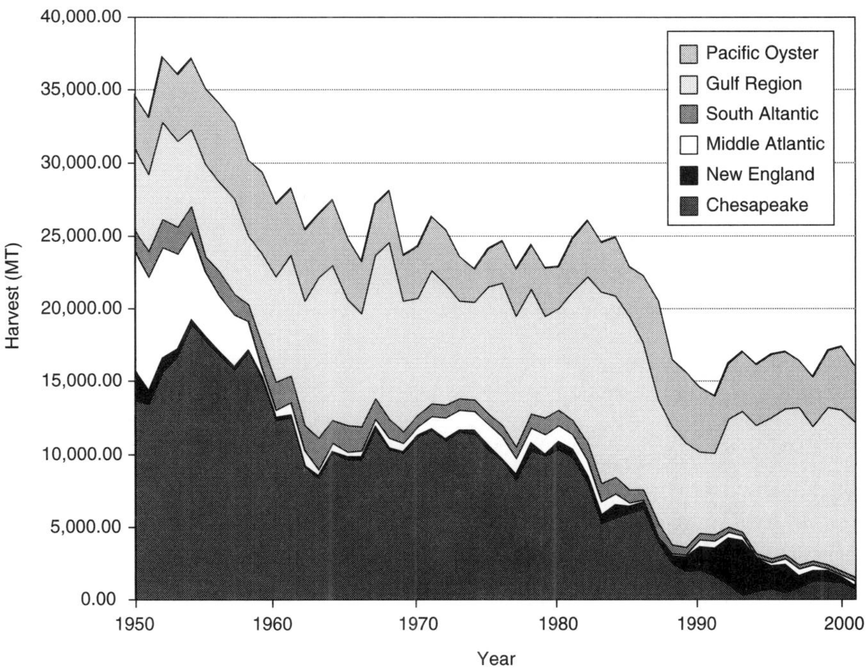

FIGURE 5.4 U.S. *C. virginica* (by region) and Pacific oyster harvest.
SOURCE: Data from NMFS (2003).

Chesapeake Bay region accounted for only about 6% of the 2001 harvest of *C. virginica* (see Figure 5.4). When harvests of oysters on the West Coast (primarily nonnative *C. gigas*) are included, the Chesapeake's share of the U.S. domestic oyster harvest declines to 4%. Production increases outside the Chesapeake region have helped offset losses in the bay. Although total domestic harvest declined consistently throughout the 1980s, it has gradually increased since 1991. In addition, imports of oysters have generally increased since 1994. In 1994, imports of competing oysters (fresh, frozen, and live, primarily from Canada, excluding smoked and canned) were 1,727 million tons. By 2001 oyster imports had increased to 2,437 million tons (NMFS, 2003; see Figure 5.5a). As a result, in 2001 Chesapeake oysters accounted for only 3% of total oyster supply and have ranged between 1 and 5% of total supply for the past decade. As indicated in Figure 5.5b, total oyster supplies have trended upward, and the real price—the price adjusted to eliminate the effect of inflation—of Chesapeake oysters has trended down over the past decade. Total U.S. oyster supply (including fresh, frozen, and live, excluding smoked and canned) increased from 1991 to 2001 by 15%. At the same time, the real price of Chesapeake Bay oysters declined by 24%, despite the fact that the Chesapeake oyster harvest declined by 55% over the period. The Chesapeake

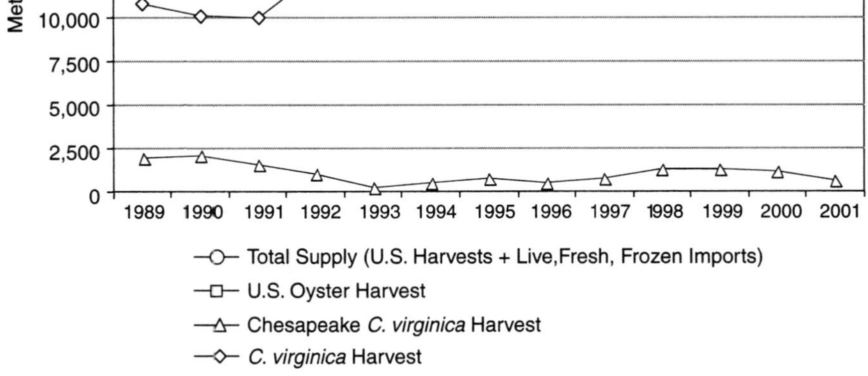

FIGURE 5.5a U.S. Oyster supplies (meat weight).
SOURCE: Data from NMFS (2003).

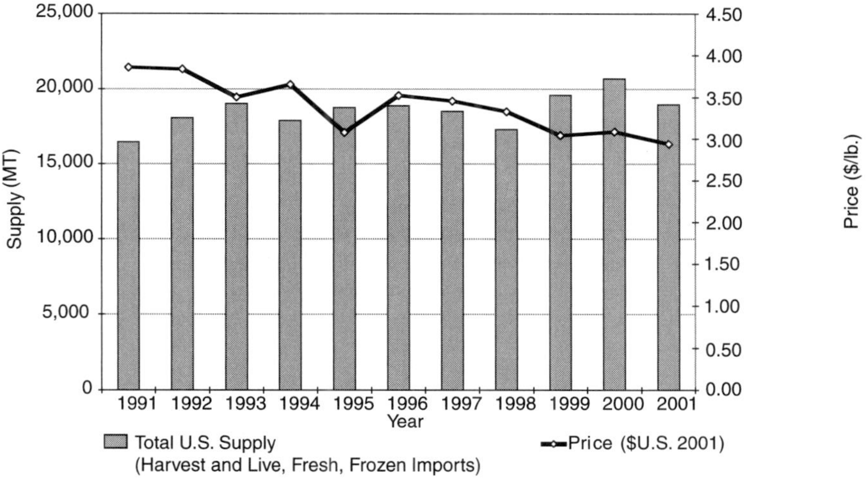

FIGURE 5.5b Chesapeake oyster real prices and U.S. oyster supply.
SOURCE: Data from NMFS (2003).

Bay has gone from being the dominant source of oysters in the 1950s to providing only about 3% of the present-day supply (including imports).

The economic effects of the decline in Chesapeake Bay oyster production extend beyond just the oystermen themselves. Many Virginia and Maryland processors have begun to rely increasingly on oysters from outside the region to meet their needs. In a survey of processors conducted during 1991, it was found that 31% of Maryland respondents and 53% of Virginia respondents handled Gulf of Mexico oysters. In addition, 20% of Virginia respondents and 8% of those in Maryland also handled oysters from the Pacific. This dependence on oysters from outside the region increased through the 1990s. Although Virginia has a smaller harvest, it has a larger processing sector. NMFS data for 1997 indicate that there were 21 processing and shucking plants in Virginia with 389 employees producing 1,670 million tons of shucked oysters (about 25 times Virginia's harvest) valued at $16.3 million. In contrast, Maryland had 11 reporting plants with 249 employees producing 546 million tons of shucked oysters (about 10% greater than Maryland's harvest) valued at $5.4 million (Muth et al., 2000). Between 1974 and 2000, both Maryland and Virginia experienced a 65% drop in the number of oyster-processing plants. In fact, during this period Virginia lost almost as many plants (52) as the total number (58) of plants present in Maryland at the beginning of the period (Lipton, 2002). These losses have affected oystermen, processors, and the people and communities that relied on these processors for employment.

Very little public information is available on the cost efficiency of different aquaculture systems for oyster production. Uncontained systems characteristic of current leased-bottom fisheries require limited capital investment and, consequently, are of lower cost than other production systems. However, uncontained systems are subject to greater losses to siltation and predation. In a comparison of stake and rack production systems, Samonte-Tan and Davis (1998) reported that cost efficiency is dependent on the scale of the production system. While it is often assumed that capital intensity will increase with the scale of production, Samonte-Tan and Davis also found that labor- and capital-intensive rack production was most efficient for small-scale producers, while the comparatively less capital intensive stake production method was most cost efficient for large-scale producers.

MARYLAND AND VIRGINIA STATE SUMMARIES

Despite the decline in the *C. virginica* fishery, it continues to provide economic and sociocultural value to various stakeholder groups in the region. In this section, economic and sociocultural information for Maryland and Virginia is presented in order to establish a comparative baseline from which to evaluate some of the potential benefits and risks of introducing a triploid or diploid *C. ariakensis,* into the Chesapeake Bay.

Most of the following discussion focuses on approximately the past 10 years. The selection of this recent time period provides a contemporary picture of the value of the *C. virginica* oyster fishery. Finally, in discussing the value of the bay oyster fishery based on *C. virginica* it is important to include both economic and sociocultural data at the industry- or fisherywide and household or community levels.

Maryland

During the 60 years before oyster diseases became a dominant factor in the 1980s, harvests in the Maryland portion of the Chesapeake Bay supported an annual commercial fishery of approximately 2 million bushels (Chesapeake Bay Program, 2002). Since 1993 the average annual harvest has been approximately 233,000 bushels, with the lowest recorded harvest occurring in 1994, followed by a rebound in harvests to 423,000 bushels in 1997, followed by another decline in 2002 to 148,000 bushels. Reports based on small-scale surveys for the 2003 season suggest that the harvest will be very low, perhaps even below the all-time low in 1994. The particularly low harvests in recent years have resulted from increased oyster mortality due to the prevalence of MSX and Dermo associated with elevated salinity occasioned by drought conditions in the region. For the 1993 to 2002 period, the mean dockside value of *C. virginica* in Maryland was $4.4 million, with a minimum value of $1.3 million coinciding with the record-setting low harvest in 1994, and the maximum value of $7.8 million coinciding with the largest yearly harvest for this period in 1997. During the 1990s the dockside value of oysters varied between $16 and $27 per bushel (approximately $2.50 to $4.25 per pound meat weight). The mean dockside value per bushel of oysters for the 1993 to 2002 period was $19.30/bushel ($3.00 per pound of meat weight). Although the 2002 to 2003 harvest portends to be the lowest ever, Maryland watermen have received well above average prices of between $25.00 and $45.00 per bushel ($3.85 to $6.90 per pound meat weight), prices that have helped offset their financial losses due to the low harvest amounts. Moreover, the higher prices led some watermen to end crabbing early in the 2002 to 2003 season in order to take advantage of early-season high oyster prices. (The fall 2003 market was depressed, leaving many watermen without markets for their crabs and low bushel prices for those watermen who had markets.)

For the period 1993 to 2002, the average number of man-days of effort in Maryland's oyster fishery was 35,513, with a minimum of 12,907 and a maximum of 72,516 days corresponding to 1994 and 1999, respectively. The average number of bushels per man-day was 6.5, with a range of 4.6 to 8.3 bushels. Applying the average harvest value of $19.30 per bushel, the average daily harvest earnings for Maryland watermen for the 1993 to 2002 seasons was $125.45 less expenses.

Hand tongs are the most common gear type used to harvest oysters in the Maryland portion of the bay. During the 2001 to 2002 season, hand tongers harvested approximately 41% of total landings. Over the past 12 years, hand tongers have harvested between 38 and 75% of the catch. Patent tongers and divers harvest approximately 20% each, with power dredges and skipjacks responsible for the remaining harvest (Chesapeake Bay Program, 2002). Description of these different gear types can be found in Kennedy (2002).

In terms of the number of Maryland license holders selling oysters in 1999, the peak harvest period for the past decade, 2,520 Maryland watermen sold oysters, while in 2002 only 915 watermen reported an oyster harvest (C. Judy, Maryland Department of Natural Resources, Annapolis, personal communication, 2002).

In addition to industrywide data on harvest, dockside value, and man-days, it is also important to understand the economic and sociocultural importance of the oyster fishery at the watermen household and community levels. Currently, the commercial (and recreational) oyster season in Maryland begins on October 1 for shaft and patent tonging and diving and November 1 for power dredging and dredge boats (skipjacks) and ends for all on March 28th. Current daily harvest limits are 15 bushels per license for tongers and divers (maximum of 30 bushels per boat), 12 bushels per day per license for power dredging (24 bushels maximum per boat), and 150 bushels per day for skipjacks, which are limited to just 2 days per week.

For Maryland watermen, oystering is second in economic importance to crabbing. Prior to the 1980s, oystering was the most important fishery on the bay, with watermen hard crabbing (or farming) in the summer mainly to provide income in between oystering seasons. Today, oystering is the "off-season" income producer. Maryland watermen have intensified their crabbing efforts and diversified into the harvesting, shedding, and marketing of soft crabs. Today, the commercial crabbing season runs from April 1 until November 15, and most watermen fish 6 days a week during this period as soon as the crabs begin moving until they bury in the bottom for over-winter hibernation or (for mature females) migrate out of the bay to spawn.

Given the seasonal characteristics of the two fisheries and their reversed economic roles compared to earlier times, the oyster fishery takes pressure off the blue crab fishery, allowing Maryland watermen flexibility to shift to oysters in early fall when prices are the highest (pre-Thanksgiving and Christmas holidays), if the blue crab harvest or market prices drop off. The oyster fishery, even at its currently reduced levels, is an economic safety valve for Maryland watermen and can reduce harvest pressure on the blue crab, which is a fishery management priority (additional blue crab harvest regulations have been issued over the past 3 years to help ensure a 15% spawning stock).

The household or community benefits of the income from oystering in Maryland are more important than a simple consideration of industry-wide figures might suggest. Once blue crabs have buried in bottom or moved to lower reaches of the bay to spawn, oystering provides Maryland watermen with their only source of fishing income. This income and "work on the water" are particularly important from October to January. Prices are also higher during this period, and the weather is generally better in the earlier months of this period. There is minimal work involved in rerigging boats with tonging or dredging gear, and the crews are either familybased or a continuation of a captain-mate arrangement used for crabbing (excluding skipjacks, which have crews of 6 to 8 watermen; it should be noted that only 5 to 10 skipjacks continue to dredge for oysters). Thus, watermen can shift to oystering with minimal effort and expense in either time or money.

For Maryland watermen, harvesting oysters is flexible work. While watermen crab daily (excluding Sundays) and most for the full 8 hours allowed by Maryland law, in part because crab mortality increases the longer the crabs are left in the crab pot, most Maryland watermen oyster, on average, only a few days per week. Based on a combination of factors, such as weather conditions, market prices, immediate household income needs, and competing work and family requirements, watermen decide whether to "go and catch a few bushels." This decision is not to imply that Maryland watermen do not take oystering seriously, and if there are oysters to be caught and marketed, watermen will go daily and brave bad weather. However, there is a flexible and varied side to their harvesting of oysters. Without overstating it, it represents a source of work and income that can be tapped to meet immediate household economic needs. It allows watermen to use their boats and gear and to continue, albeit at a reduced level, their cultural tradition of "working the water." Again, the oyster fishery is available at a time when for most watermen there are no other significant income-earning alternatives, and it is a "set" resource that will be available throughout the season. As one Maryland waterman explained in an interview: "Oystering has none of the pressures that crabbing has, where you have to be one step ahead of the crab's movements. Although it is hard work, it is a more relaxed way of fishing. You know where to go, and the oysters are either there or not. But you can always get a few bushels to put some food on the table." Finally, with the arrival of the new year, as weather worsens and market prices drop, Maryland watermen begin to shift their efforts to preparing crab pots and boats for the upcoming crab season.

The loss of the oyster fishery for Maryland watermen would be a severe blow to their efforts to continue the livelihood and traditions of watermen communities. Being able to dredge or tong for oysters even in small amounts is an important part of the cultural heritage of watermen communities, a heritage that is increasingly being celebrated through baywide efforts to support and value the region's cultural resources. Maryland watermen are increasingly participating in partnerships with scientists, environmentalists, and others to restore or replenish oyster habitat (e.g., Maryland's Oyster Recovery Partnership). Maryland watermen bring to these partnerships detailed knowledge of the location, ecological status, and economic potential of oyster reefs throughout the bay and a strong commitment to restoring the oyster to provide a natural resource on which their families and communities can depend.

Virginia

In Virginia, overharvesting reduced turn-of-the-century landings of 6 million to 7 million bushels in the 1930s to between 3 million and 4 million bushels in the 1960s. Following outbreaks of MSX in the 1950s, Virginia's harvest decreased to less than 20,000 bushels by the mid-1990s and has remained at that level ever since (Chesapeake Bay Program, 2002). Commercial harvest from the Virginia public fishery was 16,000 bushels in the 2001 to 2002 season and is expected to be even lower for the 2002 to 2003 season.

A major difference between Virginia's oyster fishery and Maryland's is that Virginia does much more processing and marketing. About two-thirds of the processed oysters in Virginia come from areas outside the state, including the Gulf of Mexico states, Maryland, and Delaware Bay (J. Wesson, VMRC, Newport News, personal communication, 2003). In the 2001 to 2002 season, Virginia processors shucked and repackaged close to a million bushels of out-of-state oysters. In another example, total 1994 oyster landings from Virginia waters were 300,526 pounds valued at $812,000. However, the estimated total sales of oysters and related products that year reached $81 million (Kirkley, 1997). This figure is another indication that the overall demand for oyster products is strong. It is interesting to note that 1993 to 1994 was the record-low year of oyster harvests from Maryland waters (see above). Virginia also "wet stores" oysters brought in from other states.

Many of the household sociocultural and economic factors related to oystering described above for Maryland watermen also apply to Virginia watermen. Winter oystering on public grounds or private beds provides small amounts of much-needed income during late fall and winter; an earlier shift to fall oystering reduces fishing pressure on the blue crab; most oystering is familybased with minimal effort required to rerig for oystering; and oystering continues to be a fishing occupation central to the heritage and identity of watermen communities. However, in Virginia, watermen do not have a unique, statewide organization similar to the Maryland Watermen's Association. Instead, there are 15 watermen associations, each based in a different geographical region.

Harvesting of oysters in Virginia is done on public beds during winter and private beds during summer. Traditionally, watermen would work public bottoms in winter and then in the summer work for processors planting seed and shell, dredging, and harvesting. According to data from the VMRC oyster harvest methods in Virginia have changed only slightly over the past 25 years. From 1976 to 1991 the top three oyster harvest methods, by harvest weight, were patent tongs (44%), the oyster dredge (35%), and hand harvest through diving (18%; VMRC, 2003b).

For the 1993-2001 period, the mean dockside value of C. *virginica* in Virginia was $0.8 million, with a minimum value of $0.6 million in 1996 and a maximum value of $1.1 million in 1993.

Historically, Virginia's production was primarily from privately leased oyster grounds. From the 1950s until the 1973 to 1974 season, typically 66 to 89% of Virginia harvest came from privately leased bottom.

TABLE 5.2 Distribition of Virginia Oyster Harvest, Private vs. Public Beds

YEAR	PUBLIC LANDINGS (BU)	PRIVATE LANDINGS (BU)	TOTAL	% PRIVATE
57-58	586,304	2,926,750	3,513,054	83%
58-59	703,915	3,347,170	4,051,085	83%
59-60	699,420	2,553,275	3,252,695	78%
60-61	781,783	2,237,736	3,019,519	74%
61-62	227,921	1,815,001	2,042,922	89%
62-63	278,830	1,652,880	1,931,710	86%
63-64	576,857	1,223,549	1,800,406	68%
64-65	615,864	1,605,759	2,221,623	72%
65-66	605,982	1,188,633	1,794,615	66%
66-67	226,855	587,105	813,960	72%
67-68	262,996	790,483	1,053,479	75%
68-69	227,577	621,463	849,040	73%
69-70	192,187	818,943	1,011,130	81%
70-71	281,001	836,014	1,117,015	75%
71-72	260,241	928,404	1,188,645	78%
72-73	157,890	394,121	552,011	71%
73-74	374,522	424,277	798,799	53%
74-75	403,737	491,860	895,597	55%
75-76	397,209	475,159	872,368	54%
76-77	312,539	320,711	633,250	51%
77-78	512,687	394,692	907,379	43%
78-79	590,533	441,082	1,031,615	43%
79-80	608,880	465,896	1,074,776	43%
80-81	704,848	472,465	1,177,313	40%
81-82	464,280	326,809	791,089	41%
82-83	329,492	361,792	691,284	52%
83-84	334,749	247,525	582,274	43%
84-85	308,392	318,660	627,052	51%
85-86	328,338	386,665	715,003	54%
86-87	501,075	279,872	780,947	36%
87-88	325,527	194,654	520,181	37%
88-89	165,061	107,612	272,673	39%
89-90	88,635	73,983	162,618	45%
90-91	59,883	52,109	111,992	47%
91-92	53,288	29,079	82,367	35%
92-93	34,355	30,182	64,537	47%
93-94	7,401	28,134	35,535	79%
94-95	18,583	36,831	55,414	66%
95-96	3,709	13,182	16,891	78%
96-97	8,801	14,548	23,349	62%
97-98	4,416	15,834	20,250	78%
98-99	19,620	28,556	48,176	59%
99-00	4,758	17,865	22,623	79%
00-01	1,808	20,765	22,573	92%

SOURCE: Data from VMRC (2003b).

From the 1973 to 1974 season until the 1993 to 1994 season, private-leased bottom accounted for 35 to 55% of Virginia's harvest. Since then, leased beds have accounted for 62 to 92% of Virginia's meager harvests (VMRC, 2003a; Table 5.2). In contrast, the private harvest in Maryland has been less than 4% for the past 25 years. Virginia's utilization of private leased beds places it in a better position to take advantage of any program that utilizes hatchery-based oysters that are grown to market size on private beds using aquaculture systems. The introduction of triploid *C. ariakensis* will most likely employ aquaculture-oriented grow-out systems and therefore will tend to favor private lease owners. In such an environment, state and federal regulators can work with individual watermen to track how much seed is going into an area, how much product is coming out, and whether and with what frequency reversion is occurring in this semi-controlled environment. It may be much more difficult to track and control triploid oysters in a public harvest area. Private bed watermen may also benefit from a diploid *C. ariakensis* program more than public bed watermen because tracking of diploid oyster growth and reproduction may be easier in a private bed environment. It should be noted that this difference is specific to an aquaculture-based system such as currently proposed by the Virginia Seafood Council. A government-supported program to enhance public grounds with nonnative oysters would not differentially favor leaseholders.

A program that utilizes hatchery-based *C. ariakensis* and moves Chesapeake Bay oyster production toward an aquaculture-based regime could change the proportion of oysters being shucked. Aquaculture-based systems are likely to result in greater investment in equipment and in animal growth and health tracking. Watermen using these methods will want to make sure that they get the best return on their investment, and this return is more likely to be found in the half-shell rather than the shucked market. Evidence from the Pacific Northwest shows that aquaculturists tend to prefer the half-shell market over the shucked market. In this region the majority of oyster production enters the half-shell market.

RECREATIONAL AND AMENITY BENEFITS

The Chesapeake Bay supports a wealth of recreational activities and amenity services that are directly or indirectly related to the abundance and distribution of oysters. The benefits of these activities and services can be characterized as flows of use, option, and nonuse benefits through time. Use benefits include benefits net of costs associated with sportfishing for oysters and other fish and shellfish, boating, swimming, bird watching, waterfowl hunting, tourism, and the value of ecosystem services contributed by the bay. Option benefits reflect individuals' willingness to pay for the opportunity to engage in recreational activities and to benefit from amenity services at some future time. Nonuse benefits accrue to vicarious consumers of bay resources and include the benefit of knowing of the existence of these resources, the value associated with bequesting recreational opportunities and amenity services to future generations, the altruistic value of preserving recreation resources and amenity goods for other users, and the value associated with beliefs that maintaining or enhancing bay resources is intrinsically desirable.

Although linkages between the introduction of nonnative oysters and the magnitude of use and option benefits are poorly understood, it is likely that the linkage is through the effect of introduction on water quality, substrate characteristics, and the composition of benthic vertebrates, invertebrates, vascular plants, and algae. The effect of nonnative introductions on the magnitude of nonuse benefits is probably dependent on similar factors and on individual preferences regarding the benefits or disutility of the introduction of an alien species.

Water quality improvements provide direct and indirect benefits. The direct benefits of water quality improvements include increased use, option, and nonuse benefits associated with improved recreational opportunities and enhanced amenity values (Greenley et al., 1982; Smith and Desvouges, 1986). Indirect benefits of water quality improvement arise when the improved water quality reduces the costs of production or increases the value of products that depend on the water body as an input in production or a repository for byproducts of production. For example, improved water quality could lead to increased consumer demand for seafood or reduce the costs of certifying the safety of harvested fish and shellfish (Kahn and Kemp, 1985; Keithly and Diop, 2001a, 2001b; Jakus et al., 2002).

Bockstael et al. (1987a, 1987b, 1988, 1989) found that use benefits associated with swimming, boating, and sportfishing in the Chesapeake Bay were strongly influenced by perceived water quality. The aggregate benefits of a 20% improvement in water quality (defined as a 20% reduction in nitrogen and phosphorus loading and a 20% increase in sportfishing catches) relative to conditions present in 1980 were estimated to be $188 million for public western shore beach use, $26 million for recreational boating, and $8 million for striped bass sportfishing.[2] Aggregate willingness to pay for water quality improvements was estimated to be $213 million for users and $74 million for nonusers. Kahn and Kemp (1985) linked the extent of submerged aquatic vegetation to the abundance of finfish and shellfish populations exploited by commercial and

[2]Monetary values reported here are derived from estimates reported in the original sources by using the Consumer Price Index to adjust to 2002 price levels.

recreational fishermen. Changes in the extent of submerged aquatic vegetation were, in turn, related to watershed-scale variations in nutrient runoff, suspended sediments, and metropolitan sewage outfall. The magnitude of recreational and amenity benefits are also positively correlated with water quality improvements through the effect of water quality on reduced health concerns regarding the consumption of oysters and finfish (Keithly and Diop, 2001a, 2001b; Jakus et al., 2002). In addition, the extent of submerged aquatic vegetation is linked to the abundance of finfish and shellfish populations exploited by commercial and recreational fishermen. Consequently, the extent of submerged aquatic vegetation is positively correlated with recreational use benefits (Kahn and Kemp, 1985). Moreover, changes in the extent of submerged aquatic vegetation are related to watershed-scale variations in nutrient runoff, suspended sediments, metropolitan sewage outfall, and nitrous oxide emissions by utilities, motor vehicles, and nonutility point sources in the bay airshed (Krupnick et al., 1998). The Chesapeake Bay airshed and watershed are depicted in Figure 5.6.

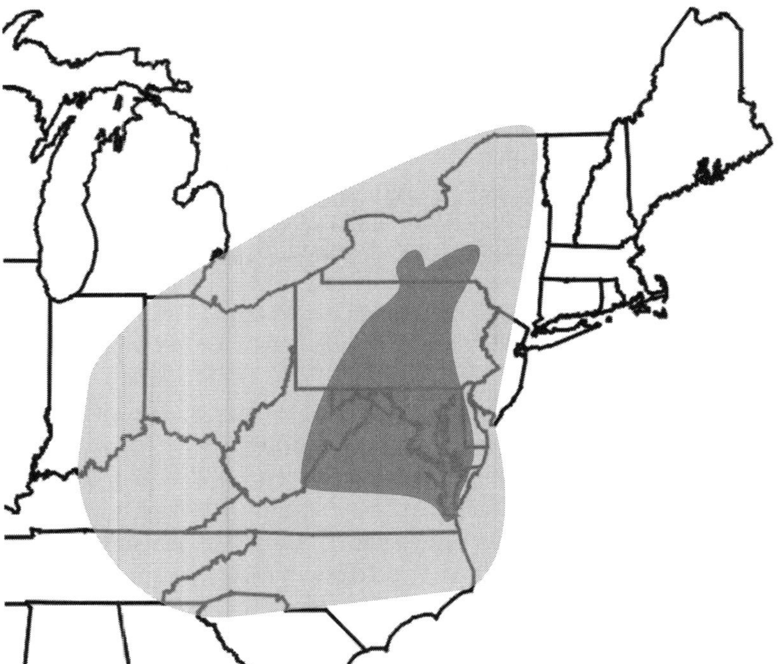

FIGURE 5.6 The Chesapeake Bay regional watershed and airshed. Light shading denotes airshed; dark shading denotes watershed.
SOURCE : Modified from Krupnick et al. (1998).

Changes in water quality, substrate characteristics, and the abundance of waterfowl populations and fish and shellfish stocks affect recreational and amenity benefits in two ways: (1) as water quality improves, the number of participants will increase and (2) as water quality improves, the average net benefit per trip increases. Because the travel costs and the number of alternative recreational areas increase at increased distances from the Chesapeake Bay, it is likely that, although improved water quality would attract increased participation, the incremental increases in participation are likely to be larger for initial water quality improvements and smaller for subsequent ones. Similarly, while improved water quality can be expected to increase the average net benefit of each trip taken, the increase in average net benefit per trip is likely to decline at ever more pristine levels of water quality. The overall effect of water quality improvements is represented graphically in Figure 5.7. The potential gains in use and option benefits depend on current water quality levels, the response of participants to increases in water quality, and the rate at which average net benefits per trip increase in response to improved water quality.

To the extent that introduction of nonnative oysters leads to improved water quality, desirable substrate characteristics, and increased abundance of waterfowl populations and fish and shellfish stocks, the introduction will lead to increased use and option benefits. Because of the diverse character of nonuse benefits, the effect of nonnative introductions

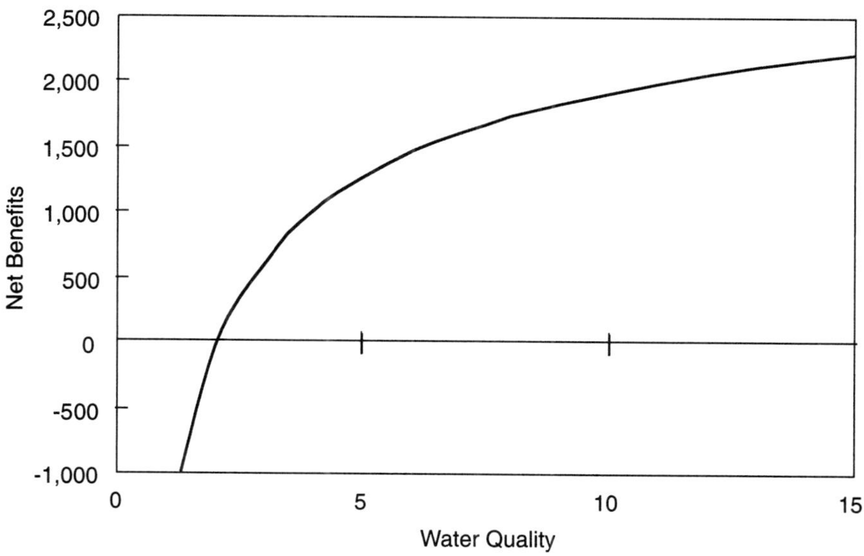

FIGURE 5.7 Relationship of net amenity and recreational benefits of the Chesapeake Bay to relative levels of water quality.

on the magnitude of nonuse benefits is uncertain. Nonuse values could be depressed because of disquietude over the introduction of an alien species or increased in anticipation of the existence of self-sustaining oyster reefs irrespective of species composition.

SUMMARY

- Virginia and Maryland *C. virginica* harvests fell substantially between 1980 and 2001—83% in Maryland and 98% in Virginia.
- In the past decade the Chesapeake Bay region produced 2 to 10% of the U.S. *C. virginica* domestic harvest and only 1 to 5% of the total oyster supply (fresh, frozen, and live, excluding smoked and canned). The bay's share of domestic harvest was roughly 50% in 1980.
- Total U.S. oyster supply (fresh, frozen, and live, excluding smoked and canned) increased by 15% from 1991 to 2001. At the same time, the real price of Chesapeake Bay oysters declined by 24%, despite the fact that the bay's oyster harvest declined by 55% over the period.
- Despite industrywide declines in oysters for both Maryland and Virginia, at the community and household levels oysters continue to provide small amounts of much-needed income during a period when watermen have no other source of fishing income. Moreover, oystering can reduce fall harvest pressure on the blue crab. Continued oystering, even at a reduced level, represents an important part of the cultural heritage of watermen communities in both states.
- Both Virginia and Maryland processors are now dependent on oysters from outside the Chesapeake Bay region. A 1991 survey of processors found that 31% of Maryland respondents and 53% of Virginia respondents handled Gulf of Mexico oysters, while 20% of the Virginia respondents and 8% of those in Maryland handled oysters from the Pacific. Processors' dependence on oysters from outside the region has continued to increase from these levels.
- The structures of oyster production in Maryland and Virginia differ from one another. Over the past decade more than 60% of Virginia's production came from privately leased beds, while less than 4% of Maryland's harvest during the same period came from private beds. This structural difference places Virginia in a better position to take advantage of any solution that might utilize aquaculture methods.
- Given that the Chesapeake Bay region now accounts for less than 3% of the total U.S. oyster supply, doubling or tripling of the harvest in the bay over several years is likely to have relatively minor

price effects. However, if growth in the harvest is extremely rapid as a result of the recovery of *C. virginica* or through the introduction of *C. ariakensis,* significant price declines could result.

- Recreational and amenity benefits associated with the Chesapeake Bay are large relative to the dockside value of recent commercial oyster harvests. Because the magnitude of recreational and amenity benefits is closely linked to water quality, management actions that reduce the level of nutrient inputs or that accelerate the rate of nutrient clearing can be expected to yield increased recreational and amenity benefits. To the extent that increased oyster populations contribute to accelerated nutrient clearing, policies that foster increased native or nonnative oyster populations will yield increased recreational and amenity benefits.

6

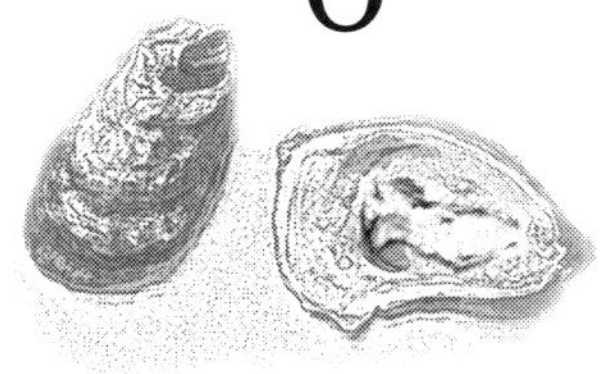

History and Current Status of Restoring Native Oysters Reefs in the Chesapeake Bay

INTRODUCTION

The potential for restoring the Eastern oyster, *Crassostrea virginica,* on oyster reefs to self-sustaining populations appears to be one of the critical issues in restoring the overall integrity and functionality of the Chesapeake Bay ecosystem. Since oyster reefs are an essential component in the estuarine ecology of the bay, restoring reefs to functioning levels is a multifaceted priority for many resource managers. Both Virginia and Maryland have a long history of oyster restoration, and recent restoration strategies are based on information gained over many decades of restoration management (see Box 6.1). Oyster resource management programs have historically been directed toward maintaining a sustainable oyster fishery and producing fishery-dependent revenues. Only recently has there been a shift in management objectives toward rehabilitation of impaired resources and habitat to restore ecological function. Oysters have long been recognized as a keystone species in the bay, and there is growing awareness of the role productive oyster reefs play in providing vital ecological and economic benefits other than fisheries alone. Currently, more emphasis is being placed on the ecological benefits of functioning oyster reefs in estuarine ecosystems, including values related to filtering capacity, structural fishery habitat, species diversity, and trophic dynamics. However, it is also important to understand that restoring productive oyster reef habitat is only one part of a complex problem, and resource managers and researchers must guard against the sentiment that oyster restoration can singularly resolve all of the ecological and environmental problems facing the bay. Successful oyster

BOX 6.1
Chronology of Oyster Restoration Management in the Chesapeake Bay

1914 First experiments with transplanting oyster seed in Maryland
1921 First experimental shell-planting project in Maryland
1922 Maryland initiates shell-planting program
1927 Watermen's Advisory Committee formed in Maryland
1927 Maryland dedicates funding for shell-planting/oyster rehabilitation program
1928 Virginia initiates shell-planting program
1943 Maryland Board of Natural Resources is created
1943 Maryland BNR Oyster Management Plan developed (seed areas and seed planting)
1960 Maryland initiates oyster repletion program
1960 Maryland initiates shell-dredging program (fossil shell)
1961 Department of Tidewater Fisheries is given authority over natural oyster reefs
1963 Potomac River Fisheries Commission is established
1969 Maryland Department of Natural Resources is created by legislation
1988 Virginia convenes Blue Ribbon Oyster Panel
1989 Chesapeake Bay Oyster Management Plan
1990 Oyster Disease Research Program is established (NOAA/Sea Grant)
1991 Maryland establishes surcharge to fund repletion program
1992 Virginia's Blue Ribbon Oyster Panel reports recommendations to VMRC
1993 Virginia uses selected hatchery-reared larvae (disease resistance)
1993 Maryland convenes Oyster Roundtable
1993 Maryland develops the Oyster Roundtable Action Plan
1993 Chesapeake Bay Policy for Introducing Nonindigenous Aquatic Species
1994 Maryland initiates hatchery production of larvae
1994 Chesapeake Bay Oyster Management Plan is revised
1994 Chesapeake Bay Aquatic Reef Habitat Plan is adopted
1994 Maryland Oyster Recovery Partnership (broad partnership)
1999 Virginia Oyster Heritage Program (broad program goals and participants, funding)
2000 Chesapeake 2000 Agreement is developed
2002 Draft Comprehensive Oyster Management Plan

SOURCE: Modified from Tarnowski, 1999.

reef restoration and the recovery of oyster resources does not directly equate with the overall recovery of Chesapeake Bay, but successful oyster reef restoration is a major component of returning the ecosystem to a more productive condition and should be linked with other ongoing efforts to improve conditions in the bay.

Current programs to restore native oyster populations seek to identify successful management strategies and measure performance in terms of functionality. The oyster industry has long held that successful restoration could be measured by increased harvests, a perspective that has influenced fishery management policies for decades. Although increased economic benefit derived from increased landings is a legitimate measure of success, it is not the only criterion for measuring success. Recent research and management practices have also emphasized measuring success by evaluating ecological benefits.

The importance of restoring oyster reefs to functional levels is also a central element in the argument about whether to continue restoration efforts with the native oyster or with a surrogate nonnative oyster. The underlying question is whether the native oyster or a surrogate oyster provides the greatest potential for restoring ecological functionality and stability to the bay. Research and management programs in Virginia and Maryland and neighboring states have provided substantial multidisciplinary information about the potential restoration of *C. virginica*. On the other hand, very little is known about the restoration potential of nonnative oysters, especially *C. ariakensis*. Researchers at the Virginia Institute of Marine Science (VIMS) in cooperation with the Virginia Seafood Council conducted investigations to determine the feasibility and potential of using a nonnative species for open-water aquaculture. The results of these investigations suggest that the Suminoe oyster, *C. ariakensis*, is a promising surrogate for the native oyster.

Virginia's oyster industry, via the Virginia Seafood Council, has expressed the opinion that past efforts to restore *C. virginica* have generally met with failure and that the future of the industry depends on introducing *C. ariakensis* as quickly as possible. A different opinion is held by some of the partners involved in the Chesapeake Bay Program, who recommend continuation of oyster restoration with the native oyster and have expressed concerns that directing funds and efforts away from current restoration programs will be counterproductive. Recent restoration efforts combine coordinated actions by state, federal, and private organizations under the mandates of Chesapeake Bay 2000 to restore and maintain the valuable ecological services provided by native oyster populations while continuing to support local oyster fishing interests.

The following information represents a cursory review of the history of oyster restoration efforts in the Chesapeake Bay to identify the benefits and shortcomings of these programs and to evaluate the potential of future restoration programs.

Need for Restoration

Throughout the last century researchers and resource managers have provided a consensus regarding the depletion of oyster resources in the Chesapeake Bay. Foremost among the causes has been overfishing, a problem first recognized at the turn of the 20th century that has continued to the present (Hargis and Haven, 1999; Rothschild et al., 1994). Concomitantly, overfishing has led to the loss of oyster habitat.

Rothschild et al. (1994) showed that oyster bar acreage in Maryland waters declined by more than 50% from 1907 to 1982, and the quality of existing oyster bars has been diminished to the point where population dynamics, productivity, and yields per habitable acre are substantially reduced. Various methods of harvesting, primarily mechanical harvesting devices, used over the long term have destroyed the structural integrity of oyster reefs and depleted available substrate that is suitable for larval settlement. Rothschild et al. concluded that the effects of fishing manifested through modification of oyster reefs had a much greater influence on the long-term decline of the oyster than degraded water quality and the effects of diseases.

It is important to note that the conclusions of Rothschild et al. focused on habitat destruction prior to the period when extensive oyster mortalities were associated with diseases. Since the 1990s, diseases caused by two protist parasites, *Haplosporidium nelsoni* (MSX) and *Perkinsus marinus* (Dermo), have substantially increased mortality rates among older oysters, contributing to decreased harvests and reproductive potential.

Additionally, environmental disturbances affect oyster reproduction and survival. Mining shell from extant and extinct reefs has substantially reduced reef structure and elevations to levels where recolonization has been unsuccessful. Diminished water quality, resulting from numerous human activities, has adversely affected overall estuarine habitat and environmental health. Eutrophication from excessive nutrient inputs and different types of contaminants and riverine sediment loads have combined to adversely affect growing waters where oysters can survive and reproduce.

It is clear from the history of management in both Virginia and Maryland that poor management decisions, legislation, and failure to react to available scientific information have contributed to resource management problems. Historically, most management efforts were directed at sustaining the oyster industry as opposed to restoring oyster populations over the long term (Haven et al., 1981; Kennedy and Breisch, 1983; Rothschild et al., 1994; Tarnowski, 1999; Hargis and Haven, 1999). Numerous researchers and managers have pointed out that many resource management decisions were based on fishery-driven objectives and that the decision-making process was influenced by social and political interests instead of scientific data (Haven et al., 1981; Kennedy and Breisch, 1983; Rothschild et al., 1994; Tarnowski, 1999; Hargis and Haven, 1999).

Rothschild et al. (1994) suggested that the effects of a diminished oyster population must have changed the ecology of the bay and that the effects should have become evident at the time of maximum stock declines (from 1884 to 1910). More recently, the complex ecological community associated with oyster reefs has gained more attention, and developing functional ecological relationships has become the focus for restoration efforts. Science plays a major role in the decision-making process, as resource managers take a holistic approach to oyster management. Resource managers have now agreed on a more comprehensive approach to oyster resource management and oyster restoration. The holistic approach includes coordinated multifaceted management strategies to restore oyster populations to self-sustaining levels, to provide ecologically valuable reef habitat, to improve ecological services such as water quality, and to provide an economic benefit for resource users.

Rothschild et al. (1994) proposed a four-point strategy to effect recovery of Maryland's oyster resources and to revitalize the oyster fishery involving fishery management, repletion, habitat replacement, and broodstock sanctuaries. Similarly, Hargis and Haven (1999) listed four purposes for restoration other than increasing harvests of seed and market oysters: restoration of broodstock levels, genetic enhancement by allowing for natural selection, restoring the biological and ecological functions of oyster reefs (filtration), and restoring the oyster reef-associated community structure. In combination these elements provide broad benefits from restoration, serve multiple purposes and user groups, and allow for sharing costs among multiple objectives.

Mann (2000) posed several questions regarding oyster restoration in the Chesapeake Bay, including whether revitalization of the oyster fishery should be the prime motivation for restoration of oyster populations. He further asked if restoration of the resource should optimize harvest and economic return or should restoration optimize ecological complexity and stability. Both strategies provide important fishery restoration goals and positive societal benefits but are influenced by biology, economics, perception, and time.

The question remains whether both the economic benefits related to the oyster fishery and the ecological benefits related to productive self-sustaining oyster populations can be generated concomitantly from future resource restoration programs (Chesapeake Research Consortium, 1999). In the near term, restoration efforts intended to support fishery harvests are incompatible with restoration efforts intended to renew ecological functionality. Similarly, restoration efforts focused on ecological objectives are unlikely to ensure economic viability of the fishery. In the long term, restoration of ecological functionality could provide harvestable surplus sufficient to meet fishery needs. Coen and Luckenbach (2000) have proposed that ecologically motivated restoration of oyster reef habitat will be a growing practice and that the challenge is to identify their ecological benefits. In the broadest sense the goals of restoring oyster reef habitat are maintenance of biodiversity, increased finfish and shellfish production, and improved ecosystem function (Coen and Luckenbach, 2000).

HISTORY OF OYSTER RESTORATION IN THE CHESAPEAKE BAY

Restoration of oyster reefs as a practicable resource management strategy has been used in many oyster-producing regions for more than a century. Oyster fishery management in the Chesapeake Bay can be traced back as early as the 1880s when the Baylor Survey was initiated to delineate public oyster grounds (Haven et al., 1981). Oyster resource management in the Chesapeake Bay has been described by Haven et al. (1978, 1981), Hargis and Haven (1988, 1999), and Wesson et al. (1999). Numerous researchers have also reported the precipitous decline in oyster production and provided management guidelines for protecting, conserving, and maintaining oyster stocks and habitat in the bay (Haven et al., 1978, 1981). The complex nature of oyster resource restoration has been described by Kennedy and Breisch (1981), Haven et al. (1981), Bartol and Mann (1997), Southworth and Mann (1998), and Mann (2000). Hargis and Haven (1999) have listed the ecological conditions under which oyster reefs originate and survive and applied them in developing a list of guidelines to plan and conduct reef restoration projects.

Virginia

Virginia's public reef shell-planting program began in 1928 (Hargis and Haven, 1999) when repletion taxes were enacted to set aside monies to fund public oyster repletion programs. Repletion programs (in 1928, 1952, and 1961) were supported by funds generated by state and federal sources but were largely financed by state subsidy and were not self-supporting (Haven et al., 1981). Haven et al. provided an overview of Virginia's oyster repletion program carried out by the Virginia Marine Resources Commission (VMRC) but added that the efforts of the state had not succeeded in reversing the downward trend in oyster production from public grounds or private leases. Harvest reduction was attributed by loss of habitat through many activities (harvesting and shell mining), sedimentation, predation, disease, and poor water quality (Hargis and Haven, 1999).

Haven et al. (1981) included a list of major public management problems facing the repletion program. Poor recruitment of oysters in the James River seed area was identified as a major factor contributing to declines in oyster production. This seed area was one of the principal sources of seed oysters for the private sector growing oysters on leased bottoms prior to 1960. Private growers resorted to importing seed oysters from neighboring states for planting on their leases.

Haven et al. reported that without a reliable source of seed stocks the oyster farming industry would continue to face difficult times and ultimately cease to exist in its present form. In response to declining production, Haven et al. summarized methods to improve growing oysters, and emphasized enhancing natural production. Among the most promising management approaches was depositing reef shell or processed oyster shell on public grounds and private leases to create favorable substrate for larval settlement.

Early replenishment programs in Virginia focused primarily on watermen transplanting seed oysters to enhance harvests (Wesson et al., 1999). Transplanting seed stocks is a complicated practice, with seed being generally moved from high-salinity areas to low-salinity areas. This practice takes advantage of heavier spatfall in high-salinity areas and lower disease prevalence and intensity in lower-salinity areas. In 1994 and 1995 the replenishment program in Virginia received two oyster disease research grants to develop and test protocols that take advantage of higher salinity for setting while reducing the impacts of oyster diseases (Wesson et al., 1999).

Until the mid-1990s almost all shell-planting efforts were directed toward the practice of creating new oyster reefs rather than maintaining existing natural oyster reefs. In 1993 the replenishment program began concentrating restoration efforts on existing reefs that appeared to have favorable contours, but where substrates were depleted of shell and live oysters. The VMRC's Shellfish Replenishment Program initiated a reef-based restoration effort in the Piankatank River in 1993 (Bartol and Mann, 1997), and a contrasting approach was employed in the Great Wicomico River in 1996 (Southworth and Mann, 1998). Initial shell planting on seed-producing grounds of the James River proved successful by doubling the natural spat set on almost all areas that were subjected to shell application (Wesson et al., 1999). A second study of natural reefs indicated that many had been harvested to such an extent that reef elevations were below optimal elevations for recruitment and survival. The practice of creating new oyster bars in areas that did not historically support oyster reefs was shown to be more expensive and less effective than enhancing natural reefs and taking advantage of existing reef elevations in areas where oysters had previously occurred. Reduction in reef elevations was seen as a serious problem, and any large-scale reef restoration requiring substantial reconstruction would be very expensive. Wesson et al. (1999) reported that, when reef elevations are too low, restoration will be unsuccessful unless the entire reef elevation is raised.

The Virginia Blue Ribbon Oyster Panel's plan for managing oysters was adopted in 1992 as a guide to oyster restoration over the next 10 years. The goals of the plan were to achieve no net loss of existing standing stocks of oysters over the next 5 years and to achieve a doubling of existing standing stocks of the native oyster over the next 10 years. In 1994 the Chesapeake Bay Aquatic Reef Plan and the Oyster Fishery Management Plan also specified oyster restoration as a management practice. The Chesapeake Bay Program designated approximately 5,000 acres each in Maryland and Virginia and 1,000 acres in the Potomac River to create new oyster habitat by 2000. Progress toward these goals was made through several projects, including direct application of cultch to improve substrates and facilitate settlement and recruitment, reef enhancement using dredged fossil shell (buried shell), and the construction of elevated reef structures in Virginia's subestuaries of the Chesapeake Bay (Wesson et al., 1999).

Since 1999 the Virginia Oyster Heritage Program has provided the framework for broader participation in reef restoration projects. The program was established as a fund-raising program, but it also provides public relations and educational components. This program funded a 3-year project that included construction, management, and monitoring of restored oyster reefs in sanctuaries and public grounds.

Most recently, the Chesapeake 2000 Agreement (see Appendix E) established oyster restoration goals to increase native oyster populations in the bay by a minimum of 10-fold by 2010 and to develop and implement a strategy to achieve this increase by using sanctuaries sufficient in size and distribution, aquaculture, continued disease research and disease-resistant management strategies, and other approaches to restore native oyster productivity to the bay. The baseline for this goal was the estimated biomass of oysters at the beginning of 1994. An important element in establishing the oyster restoration goal is recognition that the native oyster is a keystone species and that oyster reef communities are essential components in the ecology of the bay. The cost of achieving this goal was estimated at $100 million. It has been calculated that 1,500 acres in the bay need to be restored by 2010 to reach the Chesapeake 2000 Agreement's goal of a 10-fold increase in oyster biomass.

Maryland

Kennedy and Breisch (1981) provide a comprehensive review of oyster research and resource management. Tarnowski (1999) presents a chronology of factors affecting oyster resource management in Maryland. From a historical perspective, oyster resource management dates back more than a century and restoration efforts to about 1921, when the state funded projects to replace processed shell to rehabilitate oyster reefs. Maryland initiated an annual funding mechanism in 1927 to provide a more reliable means of financing the seed- and shell-planting program.

By 1932 about a million bushels of oyster shell (cultch) were planted on natural reefs. These early restoration efforts, however, did not succeed in improving oyster harvests. In 1935 the state planning commission noted that depletion of the oyster resource could be traced to overfishing, exploitation of seed stocks, and a failure to return adequate supplies of cultch to the bay, which consequently resulted in a decline in oyster harvests and the demise of the oyster canning industry. In response to declining harvests, the commission made several recommendations, including developing seed areas, transplanting seed to public reefs, planting cultch on suitable reefs, amending leasing laws, and increasing potential lease areas. The commission also recommended that every shell taken from Maryland waters be returned as cultch for restoration.

In 1942 the Tidewater Fisheries Commission undertook a large seed-growing and seed-transplanting operation based on a tax on oysters taken from the grounds. The Board of Natural Resources was created in 1943 and defined a program for oyster management, including seed and shell planting, area rotation and closures, encouraging private leasing, and a bushel tax to help fund the program. Shell taxes on shucked oysters were enacted in 1947 and 1951 to support shell-planting programs, and in 1953 laws were enacted that allowed the state to collect 50% of all shells produced. Even with this law in place, shell collections were not sufficient to provide adequate quantities of processed shell for cultch, and efforts were made to identify sources for dredged shell (Kennedy and Breisch, 1981).

Maryland has been operating a large-scale dredging and shell-planting project, as part of its oyster repletion program, since about 1960 (Kennedy and Breisch, 1981; Rothschild et al., 1994; Tarnowski, 1999). In the early 1960s, large deposits of fossil shell were dredged from non-producing grounds and deposited on public reefs to supplement the shell-planting program. During the 1960s, cultch plantings increased five-fold, which, combined with good spat sets, resulted in the highest harvests in decades (Tarnowski, 1999). During the 1960s and 1970s the state contracted for the dredging, washing, and replanting of about 5 million bushels a year, about 80% of the shell planted for seed production. These quantities far exceeded the amount of fresh shell that was made available during this period (Kennedy and Breisch, 1981). Possibly in response to the supplies of dredged shell, laws were amended that reduced the percentage of fresh shell that processors were required to make available to the state. The oyster fishery appeared to be moving in a positive direction during the 1970s when a series of natural events sent the industry into a tailspin from which it has yet to recover. Poor spat sets, due to prolonged low salinities, contributed to widespread production declines (Tarnowski, 1999). Since the shell-planting program expanded in the 1960s, approximately 180 million bushels of shells have been planted to restore reefs in Maryland waters (C. Judy, Maryland Department of Natural Resources, Annapolis, personal communication, 2002). After 40 years of dredging, sources of fossil shell are nearly exhausted.

Oyster management practices in Maryland were primarily directed toward maintaining the fishery instead of restoring the functionality of oyster reefs. Management activity was dedicated to transplanting seed to augment fisheries production by moving seed oysters from seed reefs where recruitment occurred to other public reefs where recruitment was limited. The political realities of the management community required the State to provide harvestable oysters to maintain the public oyster industry (Paynter, 1999). The Oyster Management Plan, developed in 1989 and revised in 1994, provided a more comprehensive management approach. The plan recommended developing and initiating short- and long-term management actions to help stabilize harvests, maintain spawning stocks, promote conservation goals, and develop seed stock sources.

Rothschild et al. (1994) provided a four-point strategy to revitalize Maryland's oyster fishery: fishery management, repletion, habitat replacement, and broodstock sanctuaries. The repletion strategy involved the deposition of mined fossil shell to provide a substrate to increase recruitment and subsequently transplanting recruited spat into areas to improve growth and survival. The habitat replacement strategy involved creating new substrate to enhance recruitment, growth, and survival with the objective of long-term oyster recovery. The broodstock sanctuary strategy would include harvest restrictions and would amplify the positive attributes of the fishery management, repletion, and habitat replacement strategies.

Maryland's oyster production during the mid-1990s continued to rely on seed movement programs that transplanted 1-year-old juvenile oysters from moderate-salinity areas (southern regions) to more brackish waters (northern region). This management practice was based on the observation that oysters growing in high- and moderate-salinity areas are not expected to survive their second summer and will not grow to market size in the presence of severe *P. marinus* infections. Since the parasite occurs at high prevalence and intensity among oysters growing in higher-salinity waters of the bay, the adverse impacts of disease were reduced by establishing seed reefs in areas where recruitment is high and survival is low and then moving seed to supplement areas where recruitment is low but survival is higher. Some projects have proven to be successful and some have not, and success has been variable from region to region (Paynter, 1999). Another key issue is the loss of brood stock and recruitment potential resulting from the transfers to the upper part of Chesapeake Bay, where the larval survival rate is drastically reduced by environmental conditions (temperature-salinity combination).

While the repletion program has contributed to maintaining, to some degree, harvests and the watermen's way of life, there are several important drawbacks to the program. Foremost, transplanting infected oysters provides a potential pathway to introduce pathogens to growing areas where they might not normally occur (Paynter, 1999).

In 1993, Maryland convened the Oyster Roundtable with the goal of developing sound, broadly supported recommendations for reviving oyster populations in the bay. Specific objectives included maximizing and enhancing ecological benefits, maximizing and enhancing economic benefits derived from harvest from public and private oyster grounds, and maximizing the ability of government to respond effectively to the magnitude of the problem. The Maryland Oyster Roundtable Action Plan developed action items concerning five general issues related to oyster production and ecology: diseases affecting oyster production, habitat and water quality, production and management, institutional barriers, and funding (Paynter, 1999). Subsequently, repletion programs developed annual work plans based on the Maryland Fall Dredge Survey, site surveys, and fisheries management criteria. The annual work plan follows the guidelines established by the action plan and is reviewed by the Oyster Roundtable Steering Committee.

Paynter (1999) provides a detailed summary of the Maryland Oyster Roundtable Action Plan, including addressing issues directly related to restoring oyster production, restoring oyster habitat, and the repletion program. Restoration activities included large-scale construction and seeding programs, restricting harvests, and monitoring. There was continued support for the repletion program, since the bulk of the oysters harvested resulted from those activities.

The concept of oyster recovery areas was developed to set aside areas in the bay where shellfish harvesting and planting are restricted and carefully controlled. These sanctuaries, where harvest is prohibited, were established to provide greater control over the potential movement of diseases, to maximize the reproductive potential of brood stocks, to provide an opportunity to evaluate different aquaculture methods, and to set aside areas where controlled research could be conducted (Paynter, 1999). The Maryland Department of Natural Resources is currently managing 24 sanctuaries throughout the bay, ranging in size from 5 to over 5,800 acres. It also restores managed areas called reserves, which are closed to harvests initially but may be opened for managed harvests when adequate stocks are present.

In 1994 the Chesapeake Bay Aquatic Reef Plan specified oyster restoration as a management practice. The Chesapeake Bay Program designated approximately 5,000 acres in Maryland to create new oyster habitat by 2000. The Oyster Recovery Partnership was established to accomplish these goals in Maryland.

Currently, the Maryland Department of Natural Resources Shellfish Division is responsible for maintenance and restoration of the state's oyster populations. Key elements in efforts to restore native oysters include habitat restoration, disease research, hatchery seed production, sanctuaries, and reserves. Maryland's restoration program has two components: the repletion program directed toward maintaining oyster harvests and a sanctuary component directed toward ecological investigations. Public reefs are restored by constructing man-made shell piles and rehabilitating natural underwater elevations by planting shell and seed stocks. Seed stocks are derived from natural spatfall and hatchery production. Currently about 150,000 to 500,000 seed are planted each year. About 2 million to 2.5 million bushels of dredged shell and about 200,000 bushels of processed oyster shell are planted to restore from 400 to 800 acres of public reefs each year. Shell- and seed-planting operations are rotated so that new acreage can be rehabilitated on a cyclical basis and to separate year classes. It is estimated that 80% of the oysters harvested from public reefs come from areas that the Department of Natural Resources planted with shell or seed (C. Judy, Maryland Department of Natural Resources, Annapolis, personal communication, 2002).

In 2000 and 2001 the Oyster Recovery Partnership planted or assisted in planting over 92 million disease-free, spat-on-shell in the bay. About 72 million spat were planted on managed harvest reserve reefs, and 20 million spat were planted on sanctuary reefs. Researchers are currently monitoring these reefs to evaluate the success of the plantings.

Evaluation of Oyster Resource Restoration Programs Before 1990

Haven et al. (1978) reported the catastrophic decline in oyster production in the Chesapeake Bay and later reported that the condition had not improved under current management practices (Haven et al., 1981). Likewise, Hargis and Haven (1999) reiterated that the reef restoration efforts were not enough to sustain commercial fisheries at historic levels or to maintain productive habitats to support the fishery itself.

In the 1960s, Maryland devoted substantial resources to an oyster repletion program, planting from 4 million to 6.5 million bushels of processed and dredged shell to enhance oyster production. The result of increased enhancement activity was evident when Maryland's oyster production increased from 1.5 million to 3 million bushels and the value increased from $7 million to $13 million in 1966 (Lipton et al., 1992). However, production declined by the end of the decade and continued on a downward trend through the 1970s, remaining over 2 million bushels until 1981.

During this period the repletion program was recognized as a critical element in resource management in Maryland, moving away from reliance on natural production to a "put-and-take" fishery. The importance of this period was the change in resource management strategy from fishery regulations (to control harvesting) to resource development to increase or sustain production. The discovery that planting seed was a relatively inexpensive option contributed to the shift away from reliance on natural oyster sets toward enhancing the population through hatchery production (Lipton et al., 1992).

Numerous researchers and resource managers (Haven et al., 1981; Kennedy and Breisch, 1981; Rothschild et al., 1994; Lenihan and Peterson, 1998; Hargis and Haven, 1999; Mann, 2000) have identified the problems associated with fishery-driven management practices, particularly overfishing on recently enhanced reefs. Hargis and Haven (1999) concluded that restoration of oyster reefs on public grounds followed by subsequent effective management, including closures, offers the best hope for restoration of self-renewing natural oyster populations, emphasizing that early public reef rehabilitation was rarely accomplished because enhanced reefs were often harvested with no long-term benefit to the resource because repleted public grounds were operated as "put-and-take" fisheries. They recommended that part of the overall management of restoration on public grounds should include sanctuaries where harvesting is restricted (allowable harvest quotas) or eliminated. This type of management is not directed toward providing short-term economic benefit to oyster harvesters. Direct benefit to oyster-dependent businesses will result from long-term resource recovery.

CURRENT OYSTER RESTORATION PROGRAMS

The Chesapeake Bay Program sponsored an Oyster Restoration Workshop in January 2000 to address issues related to current oyster restoration efforts that might lead to revising the Aquatic Reef Habitat Plan and the Oyster Management Plan (Chesapeake Bay Program, 2000). Several consensus statements were developed from the workshop, which were to be incorporated into the Chesapeake 2000 Agreement to act as a guideline for future restoration efforts. The long-term goal, as presented earlier, was to achieve a 10-fold increase in oysters in the bay by 2010, while a short-term goal was to develop and implement a strategy to achieve this goal. The strategy included the use of sanctuaries, aquaculture, and other management approaches to emphasize the ecological and economic benefits of oyster reef habitat. Oyster reef design and construction, disease management, and stocking were identified as critical elements in habitat restoration. Monitoring progress was also identified as important to achieving goals and answering questions relevant to management and improving restoration strategies (Chesapeake Bay Program, 2000).

It has been calculated that 15,000 acres in the bay need to be restored in order to reach the 1960s oyster population level and that 1,500 acres need to be restored in the next 10 years (150 acres per year) to reach the Chesapeake 2000 Agreement goal of a 10-fold increase in oyster abundance by 2010. Annual funding to complete these projects is projected to be about $3.1 million. More than 50,000 acres were designated under the Chesapeake Bay Program between 1996 and 2001. Within those designated areas (34 areas), 330 acres of oyster habitat have been constructed. The Virginia Oyster Heritage Program included construction of 1-acre, three-dimensional, broodstock sanctuary reefs; enhancement of 25 acres of two-dimensional public oyster grounds surrounding sanctuary reefs for sustainable commercial harvesting; monitoring spatfall, water quality, and habitat quality; and an educational component. The estimated cost of constructing the sanctuary reef (1 acre) and the public grounds (24 acres) was $350,000.

Evaluation of Contemporary Oyster Restoration Programs

Successful restoration should result in a combination of positive effects that are inextricably linked, and the synergy of these effects should be evaluated when determining the success of oyster restoration projects. Restoring oyster populations should lead to:

- increased oyster populations that ultimately form self-sustaining reef communities that contribute to species diversity, trophic dynamics, and community stability;
- functional reef communities that perform specific ecological services contributing to the overall water quality, nutrient cycling, hydrodynamics, and habitat aspects of the estuarine system; and
- increased harvests that result in revenues that provide economic benefits to all sectors of the oyster industry.

A broader view of successful restoration has been set forth by Pinit et al. (1999), where a fully functioning restored system is described as resilient and self-sustainable and able to produce a quantity and diversity of organisms of similar composition to natural systems. Successful restoration includes both functional (colonization of new recruits and diversity of flora and fauna) and structural components (water quality and hydrodynamics). Pinit et al. (1999) list the functional and structural characteristics that should be considered when measuring success of oyster reef restoration projects. They also list a number of reasons for the lack of success in many restoration projects, including unclear project objectives, inadequate design criteria, careless implementation, poor coordination, funding limitations, and lack of identified success criteria. Pinit et al. add that achieving success is not a pass/fail test; rather it is the measurement of gradual progress toward ecological recovery.

Wesson et al. (1999) provided an overview of past restoration efforts and preliminary results from contemporary oyster restoration programs. Progress in these programs demonstrated the advantages and disadvantages of various practices, including transplanting seed stocks, clearing shell and live oysters from selected restoration sites, direct application of processed shell (cultch) to extant (productive) oyster reefs, reef reconstruction using dredged shell (exhumation), construction of elevated reefs, and establishment of broodstock sanctuaries.

Resource managers in Virginia (VMRC, VIMS) have conducted investigations of several restoration projects to evaluate different methods for reef construction and to assess the value of reef construction parameters on recruitment and survival. Site selection based on competent data (Hargis and Haven, 1999; Coen and Luckenbach, 2000), locations related to existing oyster populations, reef elevation (Bartol and Mann, 1999; Hargis and Haven, 1999; Southworth and Mann, 1998), orientation of the reef to the prevailing circulation patterns (Hargis and Haven, 1999), cultch materials (processed oyster shell, dredged oyster shell, clam shell), structure (Hargis and Haven, 1999), substrate depth (Bartol and Mann, 1997), broodstock enhancement (Southworth and Mann, 1998), and costs were examined and evaluated. Recent replenishment projects have focused on construction of three-dimensional reefs in contrast to traditional projects that focused on spreading thin veneers of shell over coastal and estuarine bottoms (Coen and Luckenbach, 2000; Southworth and Mann, 1998). Three-dimensional reefs are built on the footprints of former reefs and consist of shell mounds that provide bottom elevation and protrude from the water at low tide (Bartol and Mann, 1997).

Wesson et al. (1999) summarized the early results of these projects, suggesting that restoration efforts were slowly progressing in a positive direction, and adding that oyster recovery will only be accomplished by the combination of committing to long-term management, protecting broodstock populations, and controlling harvest limits. Although many restoration plans provided reasonable goals, Wesson et al. (1999) described numerous factors that when combined make the recovery goals extremely difficult to achieve. The depleted state of extant oyster stocks, resulting from destruction or debilitation of estuarine and marine environments by man-made and natural changes, dictate that recovery will be slow and limited to areas where stocks remain in sufficient numbers to be reproductively active (Wesson et al., 1999). Limited reproductive potential in many areas, unpredictable consequences of disease, and high mortality are major obstacles to successful restoration. Southworth and Mann (1998) demonstrated the positive impact of transplanting brood stock when its addition was associated with substantially increased recruitment on the Great Wicomico reef. However, subsequent assessments showed that recruitment could not be sustained because the initial population suffered extensive mortality over the following years (Mann, 2000). Mann (2000) summarized restoration activities indicating that there was modest improvement of recruitment immediately following reef construction but that recruitment was not maintained in subsequent years. Declining recruitment was associated with population structures where mature and reproductively active oysters were poorly represented. Mann also reported that densities of spawning stocks must be increased and maintained in order to sustain recruitment and population stability. Management options include efforts to reduce natural mortality (site selection) and harvesting pressure (sanctuaries).

Berman et al. (2002) prepared an atlas of oyster reef restoration sites for the Virginia portion of the Chesapeake Bay. The atlas compiles a series of maps that summarize historical and current data relevant to oyster distribution and restoration efforts and provides details about the location, history, current status, and restoration potential for 30 individual projects. Restoration potential was categorized as modest at five sites because of low spat set, moderate or consistent disease risk, and high freshet risk. Restoration potential was categorized as limited at the remaining 25 sites because of low spat set, consistent disease risk, sedimentation, cultch availability, and user conflicts. Clearly, the risk of disease is the primary deterrent to successful restoration at most sites. Failure of oysters to reach marketable size at all sites strongly suggests that oyster survival is the problem and that disease is the causative agent. Oyster population dynamics, as described in dive surveys, on numerous restoration sites confirms that mortality among larger oysters as a result of disease continues to be the most serious obstacle for successful reef restoration (J. Wesson, VMRC, Newport News, personal communication, 2002).

Success or failure of specific restoration efforts have been correlated with salinity and water temperature and the concomitant intensity and prevalence of disease. Successful restorations projects occurred during the period from 1996 until 1998, correlating with significant declines of oyster diseases, high streamflows (lower salinity), and relatively cooler water temperatures (Burreson, 2000). Positive trends in restoration were reversed in 1999 when water temperatures warmed and salinity increased as a result of extended drought conditions. Severe epizootics occurred in most tributaries of the bay, resulting in significant oyster mortalities and a substantial setback for restoration projects (Burreson, 2000). The reoccurrence of disease under conditions of high salinity and warm water temperatures underscores the fact that oyster restoration must include disease management strategies. Although numerous management strategies include disease management, none have proven to be successful in the long term.

Experimental restoration efforts in Maryland have focused on experimental design to compare various population parameters (density, longevity, mortality), reef design and construction methods, management options (predredging, reserves), and the use of hatchery-produced seed stocks (K. Paynter, Department of Biology, University of Maryland, College Park, personal communication, 2002). These restoration efforts suggested that specific management options will increase the likelihood of success. Avoiding disease was determined to be critical for avoiding high disease-related mortalities; selecting restoration sites in low-salinity waters was the best practice for avoiding disease. Management strategies that included an aquaculture component showed positive results: relying on the use of hatchery-reared seed stocks to supplement natural recruitment resulted in higher oyster densities. Planting spat at high densities (2 million/acre) was shown to maximize ecological value. The benefits of using hatchery-reared and disease-resistant seed were also tested and showed some positive indication of disease resistance among specific disease-resistant strains. Paynter concluded from experimental restorations that specific pathogen-free spat, planted on clean shell survived for more than 4 years in low-salinity (<12 ppt) waters with minimal Dermo disease-associated impacts.

Experimental restorations also demonstrated factors that constrain successful restoration on a larger scale. There was little evidence of natural spat set, suggesting that reproductive output remained low. However, even in locations with high spawning densities (assuming high reproductive potential), there were severe limitations on setting potential (larval mortality, larval dispersal, lack of suitable substrate). Mortality associated with disease was recognized as significant (10 to 95%), unpredictable, and devastating on reefs in moderate- or high-salinity waters (K. Paynter, Department of Biology, University of Maryland, College Park, personal communication, 2002).

The Army Corps of Engineers (ACOE) is the largest federal stakeholder involved in restoration; both the Baltimore and Norfolk ACOE districts have active reef restoration projects. The Baltimore District has restored 194 acres of two-dimensional reefs (<3 feet elevation) and 6 acres of midrelief reefs and has planted about 57 million spat during these projects. The Norfolk District has restored about 240 acres of low-relief reefs and 11 acres of high-relief reefs. The ACOE has taken a proactive role in restoring oyster habitat and is seeking funding for long-term projects that incorporate the advantages of management and apply the best-available science. The ACOE's restoration strategy is based on building reefs to increase biomass and increase ecological functionality; constructing sanctuary reefs in retentive "trap" estuaries; and seeding reefs with selected, disease-resistant stocks to promote "self-recruitment." The use of hatchery-reared spat ("spat-on-shell") is an important component in seeding the reefs (D. Schulte, U.S. Army Corps of Engineers, Norfolk, personal communication, 2002).

The ACOE has developed a "Decision Document" to provide technical guidelines for oyster reef restoration projects. The project report describes activities that will contribute to the restoration of oyster biomass and populations in the Virginia portion of the Chesapeake Bay. Construction and related activities to be undertaken in the proposed project include creating new oyster habitat, planting disease-free spat and adult brood stocks on restored habitat, and relocating disease-resistant spat-on-shell to other portions of the bay (U.S. Army Corps of Engineers, 2003). The guidelines emphasize enhancing biogenic stability and ecological services, activities that are consistent with the mandate to restore habitat. From the ACOE's perspective, successful restoration will require a long-term strategy that is linked to commitment and funding. The ACOE is uncertain about whether this level of commitment can be sustained over the long term, and its approach does not specifically include restoration for the purposes of supporting a commercial fishery.

Although the answers to questions concerning the future of oyster restoration are not evident from preliminary results, experimental and pilot restoration projects do provide the basis for formulating future management strategies. Recent oyster restoration programs have taken advantage of earlier projects and the lessons learned by earlier researchers and have incorporated many of the biological and technical factors that were previously identified as necessary for success (O'Beirn et al., 1999). Moreover, many of the political and socioeconomic conflicts have been put aside in efforts to focus on specific management and restoration objectives. A group of oyster experts met in 1999 to develop recommendations to restore and protect the oyster resources of Chesapeake Bay (Chesapeake Research Consortium, 1999). They identified essential components of oyster restoration projects: construction of three-dimensional reefs, maintaining permanent sanctuary reefs, and selecting sites where natural spatfall will occur. The proposed goals were to restore 10% of the historic productive reef acreage, to restore a sustainable public fishery, to enhance natural recruitment, and to demonstrate the effectiveness of sanctuaries. The consensus of a group of oyster experts was that restoration efforts must move away from strictly fishery-driven objectives in order to focus on ecological objectives. The restoration philosophy should be to restore and manage oyster populations for their ecological value but in such a way that a sustainable fishery can exist (Chesapeake Research Consortium, 1999). A baywide oyster assessment is currently being conducted under the aegis of the Chesapeake Bay Program. The principal objectives are to develop quantitative projections of the efficacy of various management options, to develop management recommendations based on the most biologically effective combinations of options, and to develop concise recommendations for managing commercial oyster fisheries consistent with restoration goals.

When the primary objective for oyster resource restoration is to increase landings, evaluating success is straightforward. The economic return from increased landings and sales combined with the economic benefits to various industry sectors provides a measurable outcome for restoration programs. The amount of money spent on restoration programs can be compared directly with the revenues generated by the harvesting and sale of oysters. From 1993 through 2002, oyster harvests have not increased with increasing expenditures and efforts from Virginia's restoration programs. Coen and Luckenbach (2000) estimated that current returns (total harvest value) from Virginia's shell-planting program account for between 0.25 and 1 times the cost of the restoration program. Figure 6.1 shows the marked increase in funding since 1999, from about $1 million in 1999 to $4.5 million in 2002. Despite increased funding and expanded restoration efforts, reported oyster landings in 2002 are expected to be the lowest recorded (see Figure 6.1). State funding for restoration projects in Maryland is projected to exceed $4.5 million in 2003, compared to $0.6 million in 1995 (see Figure 6.2).

Fishery-Driven Restoration Versus Ecological Restoration

Mann (2000) delved into the question of whether fishery-driven restoration is a reasonable goal for ecological restorations and suggested that projections in which restored resources sustain historical harvests are unrealistic. Mann added that the direct harvest economic value of a fishery based on a restored resource will not reach historical levels unless there is an accompanying goal of long-term, self-sustaining community development. This argument prompted Mann to conclude that resource managers and relevant stakeholders and the ecology of Chesapeake Bay would be better served to view oyster restoration as the reestablishment of functional oyster reef communities, one of several cornerstones in the ecosystem.

Coen and Luckenbach (2000) proposed that the success of an oyster reef restoration effort would be judged by the ability of the habitat to

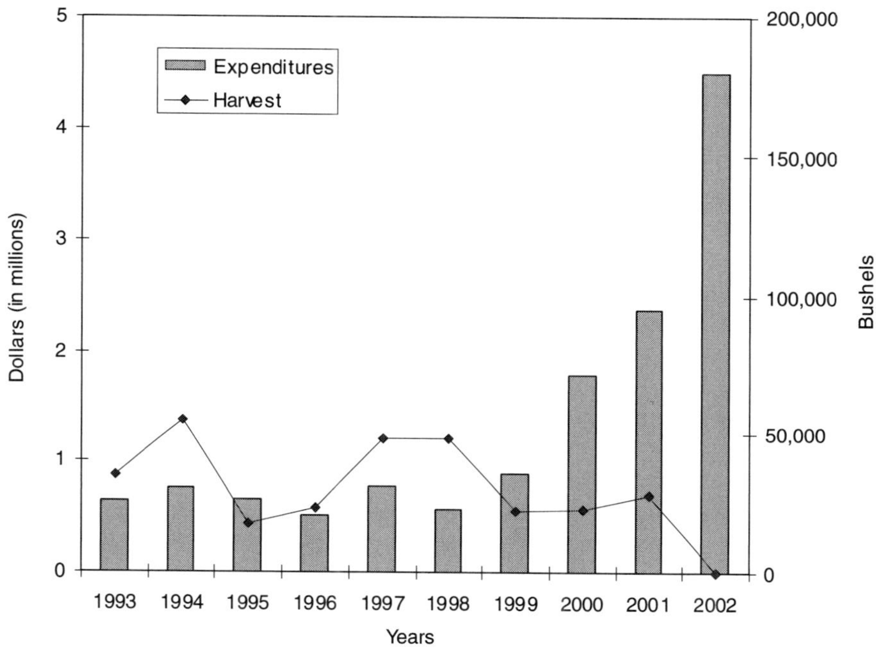

FIGURE 6.1 Virginia expenditures on oyster restoration and oyster landings.
SOURCE: Data from J. Wesson, VMRC, Newport News, personal communication, 2003.

support a self-sustaining oyster population. More specifically, the restored habitat should provide three-dimensional substrate in locations where recruitment and water quality support the growth and development of oysters. Similarly, Mann (2000) recognized several factors that will facilitate restoration, including developing high-density, three-dimensional reef structures in areas with favorable water quality; selecting sites where positive impacts are visible in a short time frame; and concentrating efforts on a river system basis rather than attempting wholesale restoration across the bay. Hargis and Haven (1999) recommended that restoration of the seed area in the James River estuary be a priority but did not discourage restoration efforts in other historical oyster-producing river estuaries.

Mann (2000) suggested that the problem for proponents of reef restoration is not so much demonstration of biological recruitment in the field as social and political recruitment of citizens to support such efforts on a long-term basis. Successful restoration efforts provide a vehicle to educate the public and foster vested interest groups. Likewise, Coen and Luckenbach (2000) suggested that the most critical element in establishing meaningful success criteria was achieving a proper balance of

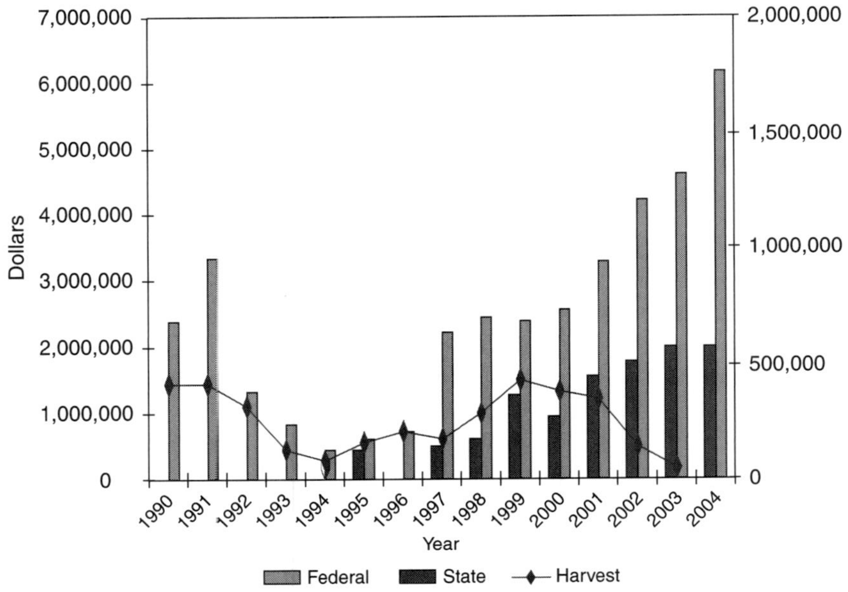

FIGURE 6.2 Annual funding of Maryland oyster projects versus harvest.
SOURCE: Data from C. Judy, Maryland Department of Natural Resources, Annapolis, personal communication, 2003.

sociopolitical constraints and ecological objectives. Balancing short-term fishery-driven interest with the need to establish long-term sustainable, ecologically functional oyster reefs poses a formidable resource management challenge.

Since oyster reefs and oyster populations are essential elements in the estuarine ecosystem, oyster restoration should be viewed as a component in a holistic approach to applied resource management. In a holistic approach, oyster restoration efforts must be combined with numerous multifaceted resource management and development options that will contribute to successful reestablishment of productive oyster grounds. In this approach the economic benefits derived from oyster harvests may be only a secondary benefit.

Alternative Hatchery-Based Management Strategies

The restoration of oyster reefs may include an aquaculture component when hatchery-reared larvae and spat are used to seed reefs and supplement natural recruitment. Stocking programs using hatchery-reared stocks may become important for "jump-starting" oyster populations on newly constructed or depopulated reefs (Allen and Hilbish, 2000). The use of hatchery-reared stocks for restoration will require careful selection of brood stocks that may be chosen for specific applications and genetic characteristics. Additionally, hatchery practices must be established to ensure that selective breeding programs are operated to maintain genetic variability among extant populations. Stocking programs that use hatchery-reared larvae and spat increase the risk of diluting natural population variation. Inevitably, the widespread use of selected hatchery-reared seed stocks will favor genetic transfer among extant stocks as interbreeding takes place among natural and/or hatchery-reared stocks. Changes in genetic diversity may have long-term impacts on postrestoration populations, if extant populations become increasingly inbred. Conversely, genetic improvements in hatchery-reared stocks may be swamped by inbreeding within post-restoration stocks, possibly with no recognizable benefits from selective breeding to extant populations (Ryman and Laikre, 1991). The actual genetic impact of selective breeding on restoration programs will depend on three parameters: the magnitude of augmentation, the effective size of the hatchery contribution, and the effective size of the recipient population. This information is essential to understanding the risks and potential consequences of using selective-bred, hatchery-reared seed stocks to augment natural populations (U.S. Department of Commerce, 2002).

Avoiding the presence of pathogens and alternatively using pathogen-free seed stocks have been identified as a critical element in restoration. Management could include the use of disease-resistant seed to infuse (interbreeding) selected genetic traits (alleles) and provide inheritable and sustainable disease resistance across an oyster population (Allen and Hilbish, 2000). Both Virginia and Maryland have established programs using hatchery-reared seed to evaluate disease resistant strains of *C. virginica*. The Aquaculture Genetics and Breeding Technology Center is testing disease-resistant strains to develop selected strains that are resistant to MSX and Dermo diseases and that can be ultimately used to produce disease-resistant seed for restoring oyster populations. Maryland and Virginia also participate in the Cooperative Regional Oyster Selective Breeding Project (CROSBreed) to develop dual-disease-resistant *C. virginica* capable of restoring dehabilitated populations. Results from the selective breeding project suggest that selected strains slow the development of lethal infections and demonstrate increased survival rates and longevity, as discussed in Chapter 4.

Genetic transfer from selected parental stocks to natural oyster populations (genetic rehabilitation) may also be an objective, particularly where disease is a major threat to population recovery. The desired outcome of interbreeding is hybridization favoring introgression of disease-resistant alleles into the natural population (Allen and Hilbish, 2000). To evaluate the effects of interbreeding and genetic transfer, disease-resistant seed should be stocked in closed/retentive systems (rivers, trap estuaries) where auto-recruitment rates are expected to be high. Interbred progeny can be monitored to determine introgression rates and production parameters (Allen and Hilbish, 2000).

When an intensive (highly controlled) aquaculture component is introduced into the management strategy, the question of cost emerges. Reef restoration programs, based on hatchery-reared seed, may be severely limited by problems of scale and economic limitations. It is widely recognized that the levels of hatchery production today are generally an order of magnitude or two too low to effectively provide the numbers of seed stocks needed to achieve restoration goals (Allen and Hilbish, 2000).

While restoration efforts have resulted in limited progress in establishing sustainable oyster populations, there remains an opinion among some researchers and resource managers that a more comprehensive management approach will ultimately lead to some level of oyster resource recovery. The approach would rely on applying a more stringent genetic improvement component based on newly emerging technologies, developing disease-resistant strains; selecting locations where environmental conditions are favorable for recruitment, growth, and survival; designing and constructing optimal reef habitat to encourage spat setting; avoiding disease, including growing oysters in areas or in a manner that reduces the chance of infection and not using infected seed; managing postrestoration populations for multiyear class distributions; and setting a long-term time frame (decade) for success (multigenerational approach to genetic introgression and auto-recruitment).

Draft Comprehensive Oyster Management Plan

The Comprehensive Oyster Management Plan (COMP) was developed by representatives from state and federal agencies, academia, environmental organizations, and the oyster industry through the Chesapeake Bay Program. The COMP provides both the general framework and specific guidance for implementing a strategic, coordinated, multipartner effort to restore and manage native oyster populations in the Chesapeake Bay (Chesapeake Bay Program, 2002). The main strategies presented in the COMP are managing around disease, establishing sanctuaries, rebuilding habitat, increasing hatchery production, managing harvest, improving coordination among partners, and developing a database to track projects. Sanctuaries are one of the main strategies for managing recovery by regulating the oyster fishery; special management areas (reserves) will be established to provide control over harvesting. The COMP also recognized the importance of using the Maryland Priority Restoration Areas or the Virginia Oyster Restoration Plan for identifying suitable sites for sanctuaries and other restoration activities.

The COMP recognizes the major impediments to rebuilding oyster resources (diseases and habitat condition) and acknowledges that restoration will require a multigeneration, long-term effort, without guarantees that the objectives will be met. The plan's objectives include:

- increase oyster populations to levels that restore important ecological functions, habitat, and self-sustaining regional populations;
- achieve a sustainable oyster fishery through a combination of harvest from public oyster grounds and private aquaculture;
- reduce the impacts of disease on oyster populations; and
- increase hatchery production and develop disease-resistant strains.

POTENTIAL CONSTRAINTS TO LONG-TERM RESTORATION PROJECTS

Funding

Estimated costs of achieving the Chesapeake 2000 Agreement's goal of a 10-fold increase is $100 million over the next 10 years. It is expected that the federal government will contribute about 50% of the funds to achieve this goal through projects supported by the ACOE, the Environmental Protection Agency, and National Oceanic and Atmospheric Administration (Chesapeake Bay Program, 2000). The states and participating partnerships will have to generate the remaining funds. Funding for oyster restoration projects in Virginia has increased from about $1 million in 1999 to $4.5 million in 2002, and the annual projected cost is about $3.1 million. Maryland currently spends about $1 million a year on its oyster repletion projects but has committed to spending $25 million over the next 10 years. Maryland's projected funding for 2003 includes $3.552 million for fishery restoration projects and $2.458 million for sanctuary restoration projects. Because substantial increases in state spending are necessary to support the projects proposed in the Chesapeake 2000 Agreement, there is concern among state resource managers that funding for restoration projects may be more difficult to obtain, especially during times when state budgets are facing increased demands and decreased revenues. In addition, if short-term outcomes fail to demonstrate success in terms important to stakeholders, political support for restoration efforts could diminish.

Time

It will take decades and possibly centuries to restore native oyster populations and oyster reefs. The time frames presented in the Chesapeake 2000 Agreement appear ambitious and probably naïve. There may be pressure to demonstrate success over the short term. Political support, partnerships, and funding may be closely linked to perceived success and may diminish when project goals are extended over decades. The question of who will benefit from the success of long-term projects may diminish current commitments and partition supporters. Clearly, the oyster industry seeks immediate relief and declares that the industry will not survive under a long-term recovery strategy.

Continuous Epizootics

The biggest challenge to oyster restoration in the Chesapeake Bay is overcoming mortalities associated with diseases. Changing seasonal, climatic, and environmental conditions make it difficult to manage oyster stocks in the presence of disease. Environmental conditions, especially temperature and salinity, affect the distribution and abundance (prevalence and intensity of infection) of parasites. Currently, *P. marinus* occurs among all productive oyster populations in the bay.

The approach includes managing for diseases by avoiding disease in project site selection, planning, prereef development preparation, and habitat rehabilitation. Disease management strategies may include efforts to clear reefs of infected oysters prior to replanting in an attempt to limit the use and distribution of infected stocks (seed and adults), avoiding waters and reefs where disease is likely to occur, and using specific pathogen-free seed. These strategies may reduce the impacts of disease and increase survival, but the challenge is to extend site-specific effects to a baywide scale (Chesapeake Bay Program, 2002).

Baywide Recovery (10-fold Increase in Biomass)

It has been estimated that 1,500 acres of productive oyster habitat need to be restored to achieve a 10-fold increase in oyster populations. The sanctuary concept is the structural basis for restoration and will require about 10% of the bay's historically productive grounds. It appears that initial success may be attained on a regional or tributary-specific basis where specific management strategies can be applied to overcome specific challenges—that is, seeking optimal salinity regimes to avoid diseases, achieving spawning stock biomass sufficient to sustain reproductive potential, and retaining larvae in trap estuaries. However, restocking may become an integral component in sustaining these populations, since recruitment may be unpredictable.

Optimal salinity zones (lower salinity where the impact of disease is reduced or higher salinity where recruitment is more successful) fluctuate, making management difficult.

In Virginia, reliance on broodstock sanctuaries may not be sufficient for establishing stable, multiyear class oyster populations. There are few locations with favorable salinity regimes, and fluctuation in the location of optimal salinity zones (lower salinity where the impact of disease is reduced or higher salinity where recruitment is more successful) makes it more difficult to choose stable sites for sanctuaries.

Self-Sustaining Oyster Populations

Poor recruitment and the absence of multiple-year classes are recognized as obstacles to establishing self-sustaining oyster populations in many once-productive parts of the bay. Eliminating harvests, enhancing substrates, and controlling habitat degradation are potential management strategies to increase oyster populations, population size distributions, reproductive potential, spawning, and recruitment. The draft COMP incorporates restoration components that establish management areas to protect extant brood stocks by prohibiting or restricting harvesting. Oyster sanctuaries can be established to protect adult oysters, thereby protecting spawning stocks and contributing directly to the reproductive potential of the population. The concept of managed areas can also be expanded to include special managed areas and harvest reserves where oyster resources can be designated for harvest based on specific harvest criteria. Managing both sanctuaries and reserves according to specific management practices is expected to enhance reproductive potential and increase recruitment, contributing to self-sustaining oyster populations.

However, the dilemma still exists when faced with the challenge of managing for sustainability under changing environmental conditions. During periods of lower salinity, oysters demonstrate a greater capacity to overcome the adverse effects of disease, but spat setting and recruitment are at best unpredictable and at worst nonexistent. During periods of higher salinity, spat setting and recruitment improve, but survival to reproductive age is markedly reduced. Restored oyster reefs may be subject to both extreme conditions for extended periods over several years, thus negating previous progress toward self-sustainability.

Hatchery Production

Restoration projects, especially in areas where recruitment is low, may become increasingly dependent on selected hatchery-reared seed stocks to maintain year classes and diverse population dynamics. This management strategy may rely on continuous seed input over an extended period. In situations where genetic introgression and auto recruitment are successful, long-term reliance on hatchery-reared seed stocks may become counterproductive by increasing the potential for adverse consequences associated with inbreeding and result in diminished genetic diversity among postrestoration populations.

Current hatchery production cannot substitute for natural recruitment, except in limited site-specific restoration efforts. Hatchery production will have to be substantially increased to meet the demand for a 10-fold increase in oyster populations. However, incorporating an intensive aquaculture component (hatchery and nursery systems) into the restoration strategy will require increased funding.

Sources of Shell for Reef Construction

Known sources of shell for cultch are limited and may or may not be sufficient for large, long-term reef restoration projects. The dominant source of shells for oyster restoration since 1960 has been dredged shells from buried shell deposits.

The sources of dredged oyster shell for Maryland's repletion program are dwindling, and permits to dredge shell deposits from the upper bay may become more difficult to acquire. Estimates of available shell from the 1960s suggested that sources for dredged shell would last for about 50 years; after 40 years of planting these sources have been largely exploited.

Wesson (VMRC, Newport News, personal communication, 2002) has estimated that 6 million to 8 million bushels of fossil shell could be produced annually in Virginia and, if properly managed, restoration efforts could continue for many years based on using dredged shell, processed shell, and alternative cultch materials. Permits have been issued to dredge shell deposits in Virginia. Resource managers are looking to alternative materials for cultch as well as material to construct reef cores. Alternative materials should be evaluated to determine properties that will be advantageous for reef construction.

SOCIAL AND CULTURAL ASPECTS OF RESTORATION

Restoration efforts have cultural-environmental meaning in addition to ecological and economic benefits. Some of this cultural-environmental meaning is linked to ecological benefits, since it contains the values and beliefs that individuals draw on to motivate themselves to be involved and supportive of oyster restoration.

What is the relevance of symbolic and cultural meanings of oyster restoration to discussions of whether current restoration or repletion efforts are ecologically or economically successful? The average environmentally concerned citizen in Maryland and Virginia cannot be expected to understand the intricacies of oyster management and restoration under the current ecological and economic conditions. Most have an understanding that oysters filter water and that historically the bay was full of oysters and the waters were cleaner. Most will have the environmental belief that the biggest threat to the bay ecosystem is poor water quality and that current restoration efforts are attempting, without much progress to date, to restore oyster populations to levels that will result in improved ecological conditions.

What is also important to consider when evaluating current restoration efforts, in addition to the ecological and economic impacts, are the cultural perceptions and attitudes surrounding oyster restoration as a cultural-environmental activity in and of itself. Restoration has strong public support and provides bay state citizens with hands-on opportunities to contribute something to help the bay through oyster gardening and monitoring programs. It is important to include in the evaluation of the current restoration programs an assessment of the cultural-environmental significance of the restoration efforts from a public perspective. This is particularly important given the environmental value and positive meaning applied to things native, pristine, and historic in the Chesapeake Bay. For example, the most widely disseminated and publicly influential annual status report on the bay's health is the Chesapeake Bay Foundation's State of the Bay Report, which uses an index to evaluate the current status of a range of bay resources and ecological parameters. The index scores 100 as the level of each resource/parameter before John Smith sailed up the bay. For the past 2 years oysters received a score of 2. For 2002 the average score for all resources/parameters was 27. That is a long way from pristine, but there is obvious environmental value in promoting the pristine as the ecological benchmark. To reemphasize: concepts of native, pristine, and historic carry strong cultural meaning to environmentally concerned and active bay state citizens, who will almost certainly, whether implicitly or explicitly, apply these cultural values to discussions of the current oyster restoration efforts. How they do so and the ecological and economic significance of these cultural factors are legitimate and important social science research questions (see the more extensive discussion of these issues in Chapter 5).

There is precedence from other restoration efforts for a contested view of "restoration" leading to strong resistance or support by multiple stakeholders to what could be argued as the ecologically correct and feasible restoration activities (Gobster and Hull, 2000). Social science and humanistic research on restoration has found that it means many things to many people and that very few people are against restoration in general. However, what seems to trigger strong reactions of approval, disapproval, and support are the specific activities and practices undertaken during restoration (Gobster and Hull, 2000). The significance of this latter point for oyster restoration, and specific examples exist from other ecosystems as well, is the possibility that the activity of introducing a nonnative versus restoration of the native species in the Chesapeake Bay could be a practice that touches strong cultural-environmental values.

A possible reduction of effort to restore the native oyster population (and perhaps its fishery), despite the acknowledged and recognized difficulties of this effort, could be met with resistance by people who value native species and pristine ecosystems and should be raised as a research and public policy concern. Restated, the bay is a "heritage seascape," with strong cultural beliefs and values with regard to protecting and maintaining that environmental heritage. This heritage in turn is linked to the concepts of pristine, native, and historic. The risks of such perceptions and oppositions arising may be increased given the recent public concern and media coverage of the presence of northern snakehead (*Channa sp.*) in Maryland in 2003.

SUMMARY

The potential for restoring the Eastern oyster, *C. virginica,* to self-sustaining populations is a critical issue in restoring the overall integrity and functionality of the Chesapeake Bay ecosystem. However, it is also important to understand that restoring productive oyster reef habitat is only one part of a complex ecosystem, and resource managers and researchers must guard against the sentiment that oyster restoration can single-handedly resolve all of the ecological and environmental problems facing the bay.

Successful restoration should result in a combination of positive effects that are inextricably linked and the synergy of these effects should be evaluated when determining the success of oyster restoration projects. Restoring oyster populations should:

- increase oyster populations that ultimately form self-sustaining reef communities that contribute to species diversity, trophic dynamics, and community stability;

- establish functional reef communities that perform specific ecological services contributing to the overall water quality, nutrient cycling, hydrodynamics, and habitat aspects of the estuarine system; and
- increase harvests that result in revenues that provide economic benefits to all sectors of the oyster industry.

Although restoration efforts have made limited progress in establishing sustainable oyster populations, there remains some optimism that a more comprehensive management approach will ultimately achieve recovery of the oyster resource. A comprehensive management approach relies on applying a more stringent genetic improvement component to develop disease-resistant strains based on newly emerging technologies; selecting locations where environmental conditions are favorable for recruitment, growth, and survival; designing and constructing optimal reef habitat; avoiding disease, including growing oysters in areas or in a manner that reduces the chance of infection and not using infected seed; managing multiyear class distributions for sustainability; and providing a long-term time frame for success.

Restoration efforts have cultural-environmental meaning in addition to ecological and economic benefits. Some of this cultural-environmental meaning is linked to values, beliefs, and perceptions that individuals draw on for protecting and maintaining their environmental heritage. This is particularly important given the environmental value and positive meaning applied to things native, pristine, and historic in the Chesapeake Bay.

7

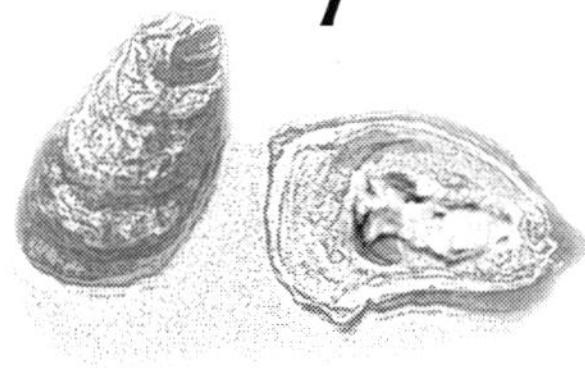

Oyster Aquaculture

PRESENT WORLDWIDE STATUS

Based on statistics from the Food and Agriculture Organization (FAO), worldwide oyster production reached a record high of 4.3 million metric tons in 2000; 4 million metric tons (93%) of those landings originated from aquaculture, of which 99.3% consisted of a single species, *Crassostrea gigas* (FAO, 2001). In contrast, wild harvest fisheries produced 157,409 tons, mainly *C. virginica* in North America (85%). These statistics indicate a significant shift toward intensive aquaculture practices. Oyster culture has been practiced since ancient times. For instance, the Romans built ponds to stockpile the harvest and collected spat on wooden branches (Clark, 1964; Héral and Deslous-Paoli, 1991). However, the true development of oyster culture in Europe was initiated during the 18th century to sustain the harvest after increased fishing effort depleted the natural beds and fishing regulations failed to halt the precipitous decline in oyster landings. The development of new techniques for using spat collectors to control spat supply was key to the development of aquaculture production during the 19th and 20th centuries. The most recent advance is the introduction of hatchery production in the 1980s (Jones and Jones, 1982; Chew, 1984). Production of oyster seed in hatcheries has allowed greater control of reproductive output and initiated the use of selectively bred oyster strains. Advances in containment system design have facilitated the expansion of aquaculture into offshore waters (Goulletquer and Héral, 1997).

OYSTER CULTURE IN CHESAPEAKE BAY

In Chesapeake Bay, vast stretches of prolific oyster grounds supported a large public fishery until the late 19th century, reducing the interest in developing aquaculture techniques. In 1884, 615,000 tons of oysters were produced in Chesapeake Bay, around 20% of the current worldwide production (Goulletquer et al., 1994a). Maryland and Virginia adopted various approaches and priorities when landings began to decrease. In the 1870s, Virginia oystermen established a system where they harvested seed from public grounds and transferred it to their leases to grow to market size. This may be considered a rudimentary oyster culture system, reliant on natural spatfall originating in public beds but based on private-sector initiative and investment. In Maryland waters only a small portion of the bottom is available for leasing, a total of 11,000 acres in 1892 and 12,000 acres in 1952 (MacKenzie, 1997). From the 1920s to the 1950s, annual harvests from leased grounds were about 100,000 bushels, an insignificant amount when compared to landings from public fisheries (F. Sieling, Maryland Department of Natural Resources, Annapolis, personal communication, 1994). In 1960, the state of Maryland started a repletion program based on shell deployment to maximize recruitment and support the oyster fishery (MacKenzie, 1997). Although this may also be considered a foray into oyster husbandry, it is based on the public fishery rather than the private efforts undertaken in Virginia. Most of the seed supplying both Virginia and Maryland has originated from public beds (e.g., James River). Most leaseholders on Virginia's Eastern Shore collected their own seed by spreading shell material in parallel rows on intertidal grounds and transplanting the seeded shell on tidal leases (Haven, 1972). Oyster culture in the Chesapeake Bay has been a technologically unsophisticated practice, relying mainly on natural spatfall in public beds and using extensive on-bottom culture. Aquaculture methods have not been employed to any appreciable extent in the Chesapeake Bay, not even methods to maximize spat recruitment using artificial spat collectors.

OYSTER CULTURE WORLDWIDE

Worldwide aquaculture practices for shellfish have been highly variable, depending on a range of internal and external constraints. The wide variety of options for culture practices reflects the physiological flexibility of oysters, such as tolerance of low oxygen conditions, relatively high turbidity, and various salinity and temperature regimes. Oysters filter feed on microscopic algae that are typically abundant in coastal waters. Because oysters do not need additional food to sustain growth, this species is relatively inexpensive and easy to culture. In some productive coastal areas it is possible to achieve high-density yields characteristic of artificial intensive culture systems. The wide range of techniques employed in oyster aquaculture can be characterized by the source of seed and the grow-out methods or facilities (see Table 7.1). Specific culture practices are highly dependent on internal and external determinants. Key internal determinants include:

- choice of culture site to maximize yield (survival and growth rates are dependent on water temperature, oxygen, salinity, turbidity, currents [flow], primary productivity [food supply], disease prevalence, and pollutant levels);
- protection from or removal of predators and fouling organisms;
- ease and maintenance of equipment used, including suitability for extreme weather conditions, such as ice or severe storms, and amount of labor required for maintenance and harvesting; and
- targeting production for either the half-shell or shucked market.

External constraints may be factors facilitating or impeding aquaculture; these include a long-term leasing system to guarantee use and investment, acceptance by the community with regard to aesthetics and enforcement, user conflicts with recreational activities, and impediments to navigation. These constraints underlie current regulations, such as no use of the water column in the Chesapeake Bay, which prevents deployment of suspended longlines. It should be noted that in several countries the development of integrated coastal zone management planning has facilitated the use of coastal resources through spatial allocation of re-

TABLE 7.1 Simplified Description of Various Methods for Oyster Culture Production

Seed Supply	Natural Spatfall		Hatchery Products	
			Natural strains	Selected strains
	Harvest (public beds)	Spat collectors (oyster shell to plastic PVC tubes and dishes)	Remote setting	

Increasing Technology →

Grow-Out Facilities	Intertidal		Subtidal (offshore)			
	On bottom (directly or in bag	Off bottom (tables, iron trestles, oyster bags, and racks)	On bottom	Tables, fixed rafts (not mobile)	Floating rafts, suspended bags	Longlines, cages, racks, lanterns

sources for various uses such as aquaculture and tourism. Zoning helps to prevent conflicts among different user groups. By considering these factors an aquaculture manager can select the optimal site and culture practices and therefore decide what level of investment will be necessary to achieve cost-effective oyster production.

Seed Supply

Traditionally, natural oyster spat has been collected by dredging and by hand picking in tidal and intertidal beds, respectively. Dredges, hand tongs, patent tongs, and rakes are common tools for this purpose. Artificial structures are often used to collect spat to enhance recruitment because they can be deployed at the optimal time and location. Cultch (the substrate for larval settlement) is cleaned to maximize spat survival. Moreover, since larval survival rates are drastically affected by temperature and salinity, environmental monitoring and consideration of interannual variability are used to decide when and where to deploy spat collectors (Lough, 1975; Goulletquer et al., 1994a, b). In most countries that rely on natural spatfall, larval abundance is monitored to determine the appropriate timing (Héral and Deslous-Paoli, 1991). Because oyster larvae only require a clean substrate that is not fouled or covered by silt, many types of materials can be used for spat-collecting operations: oyster shell, limed tiles, tiles, slate, wood, or iron. Plastic polyvinyl chloride (PVC) tubes and dishes with roughened surfaces have recently become favored in several countries because this material is lightweight (reduces field labor), maximizes spat-collecting area (modular deployment), and facilitates the removal of spat using automated equipment (Goulletquer and Héral, 1997).

The development of commercial methods for remote setting of hatchery-produced seed dates to the early 1980s on the U.S. West Coast (Jones and Jones, 1982). Hatchery-produced oyster larvae made oyster farmers independent of natural seed sources and paved the way for the development of broodstock management and genetic improvement programs. Hatchery and nursery facilities on the U.S. West Coast produce cultchless (single) oyster seed as well as the more traditional cultched spat and larvae for remote setting (see Figure 7.1). Brood stock from natural populations is gradually being replaced with selected strains or crossbred varieties, especially polyploid strains. The sterile triploid oysters offer the advantage of year-round harvest because their meats do not become depleted during the annual reproductive cycle (see section on triploidy below). West Coast hatcheries produce 37.5 billion eyed larvae annually, of which 12 billion are triploid (Nell, 2002). The bulk of these 37.5 billion larvae are produced by three large commercial hatcheries.

Oysters raised predominantly for shucked meat production are set onto oyster shells from the processing plants after the shell has been aged for at least a year. The shells are usually washed and placed into plastic mesh bags that are roughly a meter long and 20 to 25 cm in diameter. These bags, referred to as "cultch bags," are placed in large tanks, usually at or near the farm where the seed will ultimately be planted. The tanks are filled with seawater heated to 22°C and larvae from the hatcheries are added. The larvae complete metamorphosis in the tanks, attaching to the oyster shells in the bags. Generally, farmers aim for about 40 spat per shell, adding enough larvae to the tank to compensate for the roughly 30 to 40% mortality associated with metamorphosis. The cultch bags remain in the tanks for a few days and are then transported to protected nursery locations in adjacent bays. This process is referred to as "remote setting" and has become the norm on the West Coast for shucked meat production.

After several weeks in the summer (or months in the winter) when the young oysters have reached 1 to 2 cm, the shells are removed from the bags and planted on the bottom or longlined intertidally. Longlining is used in

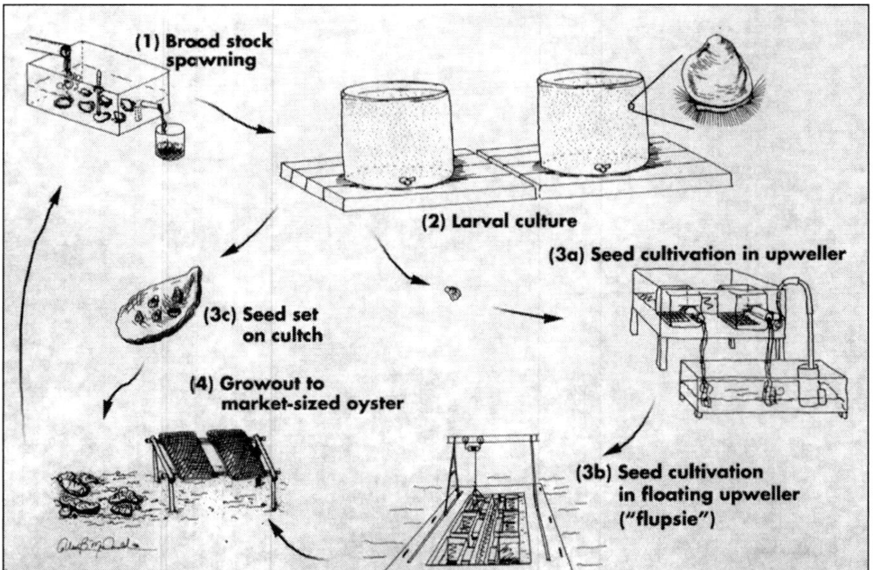

FIGURE 7.1 Diagrammatic of steps in oyster aquaculture for cultched (3c) or cultchless (3a and b) oyster production.
SOURCE: (Elston, 1999); reprinted with permission from the World Aquaculture Society.

areas where the bottom is too soft to support the oysters as they grow. It involves stringing the mother shell (with attached seed) into the braid of a poly rope and supporting it on 19-mm-diameter, 75-cm-long PVC pipe stakes. The stakes are stuck in the mud every half to three-quarter meters.

For cultchless oyster production, farms generally purchase seed from nursery facilities rather than purchasing larvae and setting it themselves. Larvae are set on microscopic shell fragments or induced to metamorphose without shell by the addition of epinephrine. The tiny young oysters require weeks to months in a variety of nursery systems before they can be planted on farms. The nursery systems require high maintenance and are expensive to operate. As a consequence, many farms opt to purchase seed from nursery facilities at sizes ranging from 2 to 20 mm. Growth to harvestable size is accomplished in a variety of systems such as rigid plastic mesh bags attached to ropes anchored on the bottom or secured to intertidal racks or suspended in trays or lantern nets suspended from floating longline or raft structures.

Grow-Out Facilities

Traditionally, oysters have been grown intertidally where they are directly exposed to air during each low tide. Intertidal methods include both bottom (beach) and near-bottom deployments. Oysters are either scattered directly on the bottom or placed in grow-out systems such as stacks, racks, bags, or intertidal longlines. Typically, the oyster grounds are leased, although there are some privately owned tidelands. Grow-out systems may be deployed directly on the bottom and anchored or supported by a variety of systems (iron trestles, posts and cables, and trellis) that suspend the crop a short distance above the bottom. The use of semirigid plastic mesh bags and cages has been highly successful in aquaculture systems around the world. Bags are made of extruded polyethylene material and are available in various forms and mesh sizes to match oyster size (from seed to market size).

In subtidal areas (areas not exposed at low tide), on-bottom culture is common because it requires limited investment for a boat, dredge, and oyster seed. More recently, this technique has been significantly improved by using ground enhancement and management, hydraulic dredges, global positioning systems, and automatic harvesters to improve efficiency. Seed is scattered directly on the bottom, reducing costs for both containment systems and labor. Fixed rafts or tables may be used in areas where tides are small and there are no conflicting uses of the tidewaters. Fast growth rates may be achieved using suspended down lines or trays deployed underneath the fixed structure.

In the Pacific Northwest, on-bottom culture was complicated by two species of indigenous burrowing shrimp that undermine the stability of oyster bottom culture beds. Since the late 1950s, the shrimp populations in Northwest coastal estuaries have burgeoned, causing severe damage to oyster beds. To maintain oyster beds, the industry had to reduce the impacts of the burrowing shrimp. The only effective treatment to date has been the use of the pesticide carbaryl. Recently, the industry has intensified efforts to find nonchemical alternative controls and is developing an integrated pest management plan for burrowing shrimp to phase out the use of carbaryl.

Floating rafts and surface or subsurface longlines are the most recent deployment methods for culturing oysters. Suspended longlines are flexible enough to handle various grow-out and harvest methods and can be used in high exposure areas. When deployed offshore, oysters are farther away from pollutant sources and increased water flow through the containment system helps maximize growth. Japanese lanterns, trays, and tray stacks can be hung vertically from rafts and longlines. Although market size can be reached rapidly, a large initial investment is required for longlines and specific boats that must be geared with hydraulic arms tailored to handle the lines.

In the Chesapeake Bay there are few intertidal areas, limited to Virginia's Eastern Shore, and the tidal range is small (1 to 3 feet; see Table 2.1). Most of the available areas for oyster culture are distributed subtidally in both Maryland and Virginia. The subtidal culture practices include on bottom, submerged on-bottom cages, fixed tables and rafts, floating rafts, and longlines. Unless policies are developed to manage concurrent uses of bay waters, spatial conflicts are likely to arise from the use of raft and floating systems that impede navigation and will consequently limit the amount of area available for oyster culture. Container-based oyster culture, whether on bottom or suspended from the sea surface, is more capital intensive and typically limits the farmer to the higher-value half-shell market. Shucked oyster meat prices have been too low generally to make container-based culture economically feasible. On-bottom culture may be the most feasible large-scale method for the Chesapeake Bay because it requires lower infrastructure investment and produces product for the existing shucked market. If *C. ariakensis* is more susceptible to predation by blue crabs or other predators than *C. virginica,* direct on-bottom culture without containment may not be practical. On a smaller scale, oyster culture operations can use oyster bags and immersed systems such as a flexible belt, chub ladder, and Taylor float (Luckenbach et al., 1999).

TRIPLOIDY AND REVERSION

As mentioned above, hatchery technology has made it possible to produce nonreproductive triploid oysters. Triploidy is the state in an individual cell or an organism of having three sets of chromosomes in each cell nucleus, rather than the two sets (diploidy) typically found in most animal cells. Triploidy was first artificially induced in the Eastern oyster, *C. virginica,* more than 20 years ago by chemical inhibition of meiosis in fertilized eggs (Stanley et al., 1981). Since then triploidy has been induced in several species of oysters, although, to date, it is commercially exploited mostly with the Pacific oyster, *C. gigas,* (Nell, 2002). More recently, a method has been developed for producing triploids by mating tetraploid and diploid oysters, described in detail below. Currently, hatcheries on the U.S. West Coast produce about 37.5 billion Pacific oyster "eyed larvae" each year, of which about 12 billion are triploid (Nell, 2002). The commercial value of triploid Pacific oysters comes from reduced sexual maturation, which results in retention of better meat quality through the spawning season and superior growth, at least in productive waters (Davis, 1988; Garnier-Gere et al., 2002).

Triploids may also reduce the risk for reproduction and spread of a nonnative species introduced for aquaculture, owing to their near-complete sterility (Allen, 1993). For this reason, limited trial introductions of triploid nonnative oysters into the Chesapeake Bay have been carried out since 1997, when Pacific oysters were first introduced by scientists from the Virginia Institute of Marine Sciences (VIMS). After triploid Pacific oysters performed poorly in these early trials, attention shifted to the Suminoe oyster, *C. ariakensis,* as a potential candidate for introduction. Triploidy had already been chemically induced in the Suminoe oyster for commercial testing on the West Coast (Langdon and Robinson, 1996). To evaluate the growth and survival of *C. ariakensis* in the Chesapeake Bay, VIMS researchers initially deployed individually certified, chemically induced triploids at six locations (Calvo et al., 2001). In 2001, VIMS researchers, having reared a small number of tetraploid *C. ariakensis* brood stock, conducted a second trial at 13 sites with 60,000 mated triploids.

Three properties of triploids need to be considered to assess risks associated with an introduction of triploid Suminoe or other nonnative oysters into the Chesapeake Bay: the *fidelity* with which triploids are produced, the *stability* of the triploid state, and the *sterility* of triploid adults. Although there is literature on the fidelity of triploid induction for oysters and other bivalves, there are fewer published data on stability and even fewer on sterility.

Fidelity of triploid production differs between chemical and mated methods of induction. Initially, triploidy was routinely induced in Pacific oysters by inhibiting the formation of the second polar body with cytochalasin B (Allen et al., 1989), so that eggs retained two sets of maternal chromosomes in addition to the one set of paternal chromosomes. Because eggs develop at slightly different rates, however, chemical inhibition of meiosis is usually not 100% effective. Experienced commercial hatcheries routinely obtain ~80% triploids after chemical inhibition, although percentages as high as 90 to 100% are possible (Downing and Allen, 1987).

In 2002 the Virginia Seafood Council (VSC) proposed introducing a million chemical triploids at 39 locations in the lower Chesapeake Bay, estimating that 99% of the treated seed would be triploid. Such a high percentage of triploids appears to be optimistic based on the reported rates of triploid induction in Pacific oysters by commercial hatcheries (Guo et al., 1996). Even at 99% triploidy, an introduction of this size could potentially introduce 10,000 reproductively competent, nonnative oysters to bay waters, as critics and an ad hoc review panel of the Chesapeake Bay Program argued. However, the issue of the fidelity of chemical triploids has been resolved by the availability of tetraploid *C. ariakensis* stocks developed by VIMS researchers. Indeed, the 2002 VSC application, which was withdrawn, was subsequently modified for use of mated triploids, resubmitted, and approved in 2003 by the Virginia Marine Resources Commission and the U.S. Army Corps of Engineers.

Because chemical induction of triploidy is much less than 100% and because cytocholasin B is toxic to both humans and oyster larvae, concerted efforts were made in the early 1990s to induce tetraploidy (four chromosome sets) in Pacific oysters (Guo et al., 1994; Guo and Allen, 1994a). In principle, 100% triploidy could be induced simply by fertilizing eggs from normal diploid females, which carry one set of chromosomes (haploid), with diploid sperm from tetraploid males. Guo and Allen (1994a) obtained the first successful induction of viable tetraploid Pacific oysters by inhibiting the first meiotic division of eggs from triploid females. Guo et al. (1996) subsequently demonstrated production of triploid larvae and spat from crosses of tetraploid males with diploid females. Triploids induced in this manner are called mated triploids to distinguish them from chemically induced triploids.

What, then, is the fidelity of mated triploids? Theoretically, 100% of the progeny of a tetraploid by diploid cross (4n×2n) should be triploid. In practice, however, the percentage of triploidy may fall short of this theoretical expectation, as shown by unpublished studies on the Pacific oyster (S. K. Allen, Jr., Virginia Institute of Marine Science, Gloucester Point, personal communication, 2003). Examination of 2- and 4-day-old larvae produced from 178 4n×2n commercial crosses of Pacific oysters in 1999 showed that 93% of the matings were 100% triploid, and of the remaining 7% that were not entirely triploid, 2% had more than 99% triploid larvae, but 5% had less than 95% triploid larvae. The latter were attributed to errors in the commercial spawning trials, perhaps the inadvertent contamination of a 4n×2n cross with sperm from a 2n male. The number of diploid progeny in many spawns (about 1 in 1,000) approaches the limit of detection by flow cytometry, the standard method for distinguishing triploid and diploid cells. Analysis of 2,148 mated Pacific oyster triploid larvae produced in 2000 revealed 2,133 triploids, 13 individuals that appeared to be mosaics of triploid and diploid cells, and only 2 (0.09%) individuals that were diploid. If these results are typical of mated triploids, it can be inferred that a batch of 1 million mated triploids would contain about 900 normal reproductive diploids. This assumes that the relatively larger risk of hatchery error can be managed and reduced through careful certification of hatchery protocols and products. The risk of introducing a small percentage of diploids from true 4n×2n crosses can be managed by considering the density of triploid planting and the distance over which broadcast spawning might be affective (e.g., Levitan et al., 1992).

A second question is whether tissues in triploid individuals remain triploid. Some triploid Pacific and Suminoe oysters have shown signs of reversion to the diploid state. Diploid somatic (i.e., nonreproductive) cells have been found in oysters that were previously confirmed as triploid. Both Pacific and Suminoe oysters have been observed to develop into a mosaic of diploid and triploid cells (S. K. Allen, Jr., Virginia Institute of Marine Science, Gloucester Point, personal communication, 1999, 2000). The percentage of diploid cells is higher for chemically-induced triploids and varies among individuals, as a function of tissue type, and possibly among species. The risk that reversion presents to any proposed introduction of triploid nonnative oysters is that the germinal tissue of triploids may revert to the diploid state, making the oyster reproductively competent. However, the prevalence of mosaics is low in chemically induced *C. ariakensis* triploids that have been reared in the Chesapeake Bay, averaging 2.5% (range from 1 to 5%) over the first 6 to 7 months and reaching as high as 10% by 18 months. The prevalence of mosaics in mated triploids is reported to be lower and more stable over time, averaging only 0.6% even in post-harvest-size animals (S. K. Allen, Jr., Virginia Institute of Marine Science, Gloucester Point, personal communication, 2003). The percentage of diploid cells in these mosaics is usually less than 10% but can range as high as 25 to 70% in some individuals at 9 months of age. The percentage of diploid cells increases with age, so that an individual with 25 to 30% diploid cells at 9 months of age can have up to 70 to 84% diploid cells by 18 months. Thus, based on these limited preliminary studies, probably much less than 1% of a mated triploid population would be able to produce normal gametes if they remained in the field for more than 3 or 4 years.

There are few data on the ability of mosaics to produce normal haploid gametes, and detection methods can only resolve the status of male oysters. Reversion of reproductive tissue has been observed in a *C. gigas* male that produced haploid sperm at an age of 4 years (S. K. Allen, Jr., Virginia Institute of Marine Science, Gloucester Point, personal communication, 2003). Another element of the risk of using triploids, then, is the escape of individuals from containment and from harvest, leaving them with sufficient time for reversion of gonadal tissue to the diploid state and recovery of reproductive potential. There are insufficient data to quantify this element of risk at present, but it appears to be quite low and manageable through constraints on methods for containment and harvest of triploids introduced into the Chesapeake Bay.

Finally, one must consider the sterility of triploids even without reversion. The reproductive potential of triploid Pacific oysters has been evaluated in detail (Allen and Downing, 1990; Guo and Allen, 1994b). These authors showed that triploid Pacific oysters are not completely sterile and that progeny, mostly 2n or 3n, are obtained in all laboratory crosses. Fecundity of triploid females was reduced on average to 1.2 million eggs, about 2% of the fecundity of diploid females, but with a considerable range, from 19,000 to 21.5 million eggs per female. The relative fecundity of triploid males could not be ascertained but appeared to be lower than that of females. Fertilization of triploid gametes was only slightly lower than that of diploids, suggesting normal gamete interactions between the eggs produced by triploids and sperm. Survival of progeny from crosses of triploid males with triploid females to the spat stage was less than 1 in 100,000 compared to 1 in 5 survival in diploid crosses. Guo and Allen (1994b) estimate the reproductive potential of triploid relative to diploid Pacific oysters as the product of the relative fecundity and relative survival of progeny, $0.02 \times 0.0004 = 0.0008\%$. However, most progeny of triploid crosses (90%) are themselves triploid; the probability of producing diploid progeny is very low but not zero. The relative reproductive potential of triploid females crossed with diploid males is 0.0045%, which suggests that inadvertent introductions of diploid males with triploid females might raise the reproductive potential of nonnative triploid introductions by an order of magnitude. Similar data on the reproductive potential of triploid Suminoe oysters have not yet been published.

SUMMARY

Triploidy, in which cells have three sets of chromosomes rather than the more typical two sets (diploidy), reduces the risk of reproduction and spread of a nonnative species because they are almost completely sterile. Triploidy induced by chemical means is effective in producing about 80%

triploid larvae in commercial hatcheries; triploidy induced by mating tetraploids with diploid oysters approaches 100% effectiveness, but the small percentage of diploids occurring in mated triploid offspring (0.09%) becomes a significant number in commercial-scale operations involving millions of offspring. Triploids may also revert to the diploid state as they age, but probably less than 1% of a mated triploid cohort might eventually produce normal gametes if they remain in the field more than 3 or 4 years. Triploids are not always completely sterile. Careful examination of *C. gigas* triploids indicates that the fecundity of triploid females averages 2% of diploids, but survival of progeny of triploid × triploid and triploid × diploid crosses is extremely small.

8

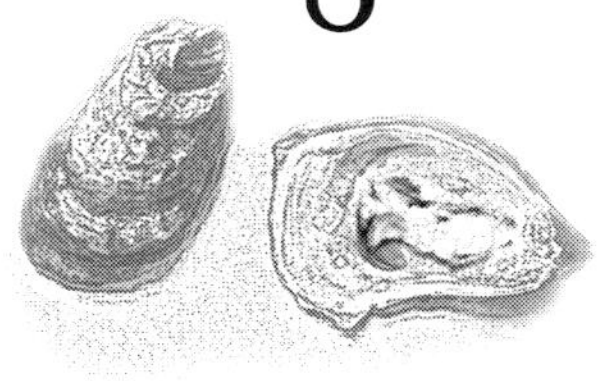

Regulatory Framework for Managing Proposed Introductions

INTRODUCTION

This chapter summarizes and analyzes the regulatory framework applicable to intentional introductions to U.S. waters of nonnative marine species such as an introduction to Chesapeake Bay of *Crassostrea ariakensis*. The analysis focuses especially on the regulatory framework applicable to the Virginia Seafood Council (VSC) proposal to introduce *C. ariakensis* to Virginia's waters as described in a public notice of the Norfolk District of the U.S. Army Corps of Engineers (see Appendix E). The VSC proposal raised concerns that existing federal regulations were inadequate for addressing a nonnative introduction that potentially could affect marine resources in many coastal states. The committee was asked to investigate the adequacy of this framework "to monitor and oversee" the introduction of *C. ariakensis* to Chesapeake Bay. There are four primary levels of regulation relevant to such introductions: state, federal, interjurisdictional (consisting of multistate and state-federal regulatory institutions), and international agreements applicable to introductions in U.S. waters that are described in the following sections.

STATE

Through the public trust doctrine, coastal states have a proprietary as well as a regulatory role with respect to oyster cultivation and harvesting. Tidal and submerged lands that were not conveyed prior to statehood are owned in trust by the states. This trust means that navigable waters, lands beneath those waters, and living resources within those waters are owned by the state for the benefit of its residents (Macinko, 1993; McCay, 1998). Recognition of the public trust doctrine can be traced to *Martin v. Waddell's Lessee*, 41 U.S. 367 (1842), a case involving oysters in New Jersey. In *Illinois Central R.R. Co. v. Illinois*, 146 U.S. (1892), the U.S. Supreme Court found that:

> The State can no more abdicate its trust over property in which the whole people are interested, like navigable waters and the soils under them, so as to leave them entirely under the use and control of private parties than it can abdicate its police powers in the administration of government and the preservation of the peace.

Although this and related decisions have limited the conveyance of submerged lands, state courts differ in their application of the public trust doctrine and thus in the conditions under which submerged lands can be conveyed or leased and introduction, cultivation, and harvest of living resources allowed (Ajuzie and Altobello, 1997; Power, 1970).

Virginia

Virginia has enacted a state statute that explicitly refers to the introduction of some nonnative aquatic and marine species. Virginia Code Section 28.2-825 provides:

> A. It shall be unlawful for any person to import any fish, shellfish or crustacea into the Commonwealth with the intent of placing such fish, shellfish or crustacea into the waters of the Commonwealth unless one of the following conditions exists:
> 1. The fish, shellfish or crustacea are coming from within the continental United States from a state or waters which are on the Marine Resources Commission's list of approved states and waters, and are species which are on the Marine Resources Commission's list of approved species; or
> 2. The person has notified the Commissioner of Marine Resources of such intent and has received written permission from the Commissioner of Marine Resources.

Virginia's administrative regulations also address the introduction of species into the state's waters. Along with incorporating the provisions of Section 29.3-825 of the Virginia Code, state regulations provide that certain listed species can lawfully be placed into waters of the Commonwealth. See Chapter 754, Section 10, of the Virginia administrative regulations. The Suminoe oyster, *C. ariakensis*, is not included in the approved list. A nonbinding 2002 resolution (Virginia House Joint Resolution No. 164, see Box 8.1 and Appendix D) of the Virginia legislature supports introduction of *C. ariakensis*.

On February 25, 2003, under these laws, the Virginia Marine Resources Commission (VMRC) approved a VSC proposal that 1 million triploid *C. ariakensis* be deployed at 10 locations in Virginia's coastal waters, as discussed more fully below in this chapter's federal section.

Virginia's regulatory scheme contains elements of the "clean list" approach supported in a 2001 report to the Pew Oceans Commission (Goldburg et al., 2001), as compared with the federal Lacey Act's "dirty list" approach discussed below. A clean list bans the introduction of species other than those listed. Furthermore, those seeking to list a species must show that the introduction will not cause unacceptable negative impacts. Any permits issued can impose bonding and insurance requirements in case unforeseen consequences occur. A clean list approach is a cautious approach to nonnative introductions. By contrast, a dirty list bans only species that have been deemed injurious (see discussion of the Lacey Act under the federal heading).

Since most of Virginia's oyster production historically has come from beds leased from the state by oyster growers, any lease provisions regarding the use of nonnative oysters are also a significant form of state regulatory control. Furthermore, such leasing must be carried out in accordance

BOX 8.1
State Law Documents (see Appendix D)

- U.S. Army Corps of Engineers State Program Regional Permit 97-RP-19
- Virginia Marine Resources Commission (VMRC) General Permit #3
- VMRC, "Pertaining to Importation of Fish, Shellfish or Crustacea," Chapter 4VAC 20-745-10 et seq.
- Virginia Coastal Zone Management Program Information and Federal Consistency
- Maryland Coastal Zone Management Program Information and Federal Consistency
- 1995 Session, Virginia General Assembly, House Joint Resolution No. 450, "Requesting the Virginia Institute of Marine Science to Develop a Strategic Plan for Molluscan Shellfish Research and Begin the Process of Seeking Necessary Approvals for In-Water Testing of Nonnative Oyster Species"
- 2002 Session, Virginia General Assembly, House Joint Resolution No. 164, Amendment in the Nature of a Substitute (proposed March 9, 2002), "Proclaiming Support for the Revitalization of the Virginia Oyster Industry"

with Virginia's public trust doctrine whose rules are determined primarily by Virginia's courts. Finally, some lessee operations may also be regulated by Virginia local governments.

Virginia's Coastal Management Program (CMP), adopted pursuant to the federal Coastal Zone Management Act (CZMA) discussed below in this chapter's interjurisdictional section, focuses on "protect[ing] and restor[ing] coastal resources, habitats, and species of the Commonwealth." The Virginia CMP does not mention nonnative species in any way. The CMP is a program designed pursuant to the CZMA to help facilitate Virginia's duties as a coastal state to manage its coastal region (Virginia Department of Environmental Quality, 2001). Virginia originally developed the CMP in the mid-1980s, and the latest version was printed in 1994. As a federally approved CMP, the CZMA obligates federal agencies such as the Corps of Engineers to act consistently with its "enforceable policies."

Maryland

Different from Virginia, most of Maryland's oyster production has come from licensed fishermen harvesting public oyster beds rather than from leased beds. In Maryland, Department of Natural Resources (DNR) regulations control the introduction of nonnative aquatic species, utilizing the equivalent of a "clean list" approach. No one may import or possess shellfish, including oysters taken from waters outside Maryland without a DNR permit. Such a permit can only be issued if proof is presented that the nonnative shellfish will not be harmful to native shellfish. A recently enacted Maryland law (2002 House Bill 353) calls for study of the nonnative *C. ariakensis* oyster as well as the native *C. virginica*. The law authorizes in-water experiments with nonnatives so long as biosecurity measures, including the International Council for the Exploration of the Sea (ICES) protocols discussed below, and this committee's recommendations are followed.

DNR permits also are required to engage in aquaculture in Maryland. A permit will not be issued if the activity will adversely affect wild stocks of fish or result in the release of nonnative or genetically altered species or contamination. Thus, for oyster aquaculture production from beds leased in accordance with Maryland's public trust doctrine, these rules are an important additional state regulatory control. Maryland Code § 4-11A-12 limits lessees to the use of *C. virginica* in their oyster aquaculture operations. Some lessee operations may also be regulated by Maryland local governments. Under Maryland Code § 4-11A-05, the leases do not include the water column above the leased bed. Maryland, like Virginia, has a federally approved CMP that does not mention nonnative species.

North Carolina

In North Carolina, nonnative introductions in state waters are prohibited without a permit from the director of the North Carolina Division of Marine Fisheries. Under these laws, sterile *C. gigas* and *C. ariakensis* oysters were deployed in double-hulled cages in three locations in North Carolina coastal waters for 18 months during 1999 and 2000 in growth, taste, and mortality experiments (Peterson et al., 1999). Proposed follow-up experiments funded by North Carolina's tobacco-settlement-based Golden Leaf Foundation involve deploying sterile *C. gigas* and *C. ariakensis* in bags and on racks at 12 coastal locations. Like Virginia's and Maryland's CMPs, North Carolina's federally approved CMP does not mention nonnative species. Unlike Virginia and Maryland, due to its location, North Carolina does not participate in the Chesapeake Bay Program, and its procedures for nonnative introductions are discussed below. However, all three states are members of the Atlantic States Marine Fisheries Commission (ASMFC) discussed below, and North Carolina has applied ASMFC's 1989 policy on nonnative introductions (Peterson et al., 1999).

Other States and Countries

In addition to Virginia, Maryland, and North Carolina, New Zealand and Hawaii and several other states use a "clean list" approach to regulate fish and wildlife introductions. Information on New Zealand's approach is available in Bean (1996), Clout (1999), and Clout and Lowe (2000). Detailed information on Hawaii's as well as Florida's approaches is provided by the U.S. Congress, Office of Technology Assessment (1993). That report cites Florida's laws and their implementation as models and also includes proposed model state laws. A Florida shellfish aquaculture submerged-lands lease, which prohibits the use of nonnative species, and includes 6 inches of water column, is included in McCoy (2000) as Appendix I.

Relevant laws and regulations as of 1993 of Chesapeake Bay Program states and those nearby are summarized in the program's 1993 nonindigenous species policy document (see Appendix E and Chesapeake Bay Program, 1993). When contacted by the ASMFC, some Atlantic coastal state officials opposed and some supported the VSC's first introduction proposal. All wanted an opportunity to comment on future introduction proposals.

Washington state's "dirty list" law is described in Dentler (1993). West Coast state laws are discussed in Nadol (1999). State approaches also can be gleaned from the plans they have prepared under the federal Aquatic Invasive Species Act (e.g., the Massachusetts Aquatic Invasive Species Management Plan of July 2002).

FEDERAL

Executive Order 13112

This presidential executive order, entitled "Invasive Species" (see Box 8.2 and Appendix E for federal law documents), is aimed at preventing the introduction of invasive species into the United States as well as providing for control and minimization of the impacts caused by the introduction of such species. The order establishes an Invasive Species Council, which is intended to provide national leadership regarding invasive species. As required by the order, the Council has issued a National Invasive Species Management Plan to address goals and objectives and provide specific measures for federal agencies, including the Corps, to carry out under the plan.

Section 2(a)(2)(vii) of the executive order instructs federal agencies such as the Corps of Engineers to "not authorize, fund, or carry out actions that they believe are likely to cause or promote the introduction or spread of invasive species in the U.S. or elsewhere unless . . . the agency has determined and made public its determination that the benefits of such actions clearly outweigh the potential harm caused by invasive species, and that all feasible and prudent measures to minimize risk of harm will be taken in conjunction with the actions." Under the order, "invasive species" means "an alien species whose introduction does or is likely to cause economic or environmental harm or harm to human health."

BOX 8.2
Federal Law Documents (see Appendix E)

- 1999 Presidential Executive Order 13112 on Invasive Species
- 50 Code of Federal Regulations Part 16 (regarding a finding under the Lacey Act)
- Fish and Wildlife Service Lacey Act Evaluation Criteria, revised March 23, 2001
- Fish and Wildlife Service Regulations on Injurious Wildlife, 50 C.F.R. Part 16, Injurious Wildlife (Importation or Shipment of Injurious Wildlife, Permits, Additional Exemptions)
- USFW Fall 2002 Lacey Act Listing of the Snakehead as Injurious
- Clean Water Act Section 117—Chesapeake Bay (33 U.S.C. § 1267)
- Chesapeake Bay Program Materials: 1987 Chesapeake Bay Agreement; Chesapeake Bay Agreement: 1992 Amendments; Chesapeake Bay Policy for the Introduction of Non-Indigenous Aquatic Species, December 1993; Chesapeake 2000
- U.S. Army Corps of Engineers, Nationwide Permit 4-67, Federal Register 2020, 2063, 2064, and 2078 (January 15, 2002)—including a Summary of NWP 4, as well as Provisions Relating to Shellfish Beds and Shellfish Seeding Activities; U.S. Army Corps of Engineers, Nationwide Permit 4-67, Federal Register 6692 (February 13, 2002), Correction of the Original Notice of Nationwide Permit 4
- U.S. Army Corps of Engineers Joint Federal/State Public Notice, May 2, 2002

Lacey Act

The Lacey Act is administered by the U.S. Fish and Wildlife Service (USFWS). Utilizing a "black list" approach, the act prohibits importation into the country of certain "wild animal" species that are listed as "injurious to human beings, to the interests of agriculture, horticulture, forestry, or to wildlife or the wildlife resources of the United States" (18 U.S.C. § 42(a)(1)). The act also sanctions interstate movements of state-listed species within the United States (16 U.S.C. § 3372). In addition, a regulation (50 C.F.R. § 16.13 (a)(1)) issued under the act bans the release "into the wild" of live or dead "fish, mollusks, crustacean, or any progeny or eggs thereof" except "by the State wildlife conservation agency having jurisdiction over the area or by persons having prior written permission from such agency." Thus, for introductions of those species, including oysters, a "clean list" approach applies across the United States, administered initially at the state level and enforced at the federal level under the Lacey Act. The list of species considered to be "injurious" found at 50 *Code of Federal Regulations*, Part 16, is included in Appendix E together with the fall 2002 USFWS listing of the snakehead fish as "injurious." *C. ariakensis* is not currently listed as an "injurious species." Thus, VMRC approval of an introduction could meet the state approval requirement of 50 C.F.R. § 16.13 (a) (1) quoted above.

National Invasive Species Act

The act (16 U.S.C § 4701) is focused on unintentional introductions of nonindigenous species via ballast water introductions. Section 4701 outlines the following five objectives:

1. to prevent further unintentional introductions of nonindigenous aquatic species;
2. to coordinate federally funded research, control efforts, and information dissemination;
3. to develop and carry out environmentally sound control methods to prevent, monitor, and control unintentional introductions;
4. to understand and minimize ecological damage; and
5. to establish a program of research and technology development to assist state governments.

Section 4725 provides that federal agency aquatic nuisance programs under the act shall be carried out consistently with other applicable federal, state, and local environmental laws and that nothing in the act is intended to supercede state and local aquatic nuisance species controls.

A September 2002 bill that the 107th Congress did not enact would have supported federal and state regulation of imports of live aquatic organisms with screening guidelines developed by the federal Invasive Species Council and grants to the states. The act's reauthorization is pending before the 108th Congress.

Federal Animal Protection Laws

The Animal and Plant Health Inspection Service (APHIS) of the U.S. Department of Agriculture has authority under the Farm Securities and Rural Investment Act and the Animal Health Protection Act of 2002 to promulgate regulations to prevent the introduction of exotic diseases. APHIS has not yet released import regulations to prevent introduction of exotic diseases of farm-raised aquatic species. In the future, as funds become available, APHIS is planning to develop and implement a policy consistent with international agreements and domestic programs that would include epidemiology, surveillance, and disease-free certification components.

Rivers and Harbors Act § 10 and Clean Water Act § 404 and Related Statutes

As the Corps of Engineers public notice (May 2, 2002, see Appendix E) illustrates, activities involving *C. ariakensis* using in-water structures or fill can require Corps permit approval under Section 10 (33 U.S.C. § 403) and Section 404 (§ 1344). Once the Corps has permit jurisdiction due to such in-water activities, it reviews the entire project of which the in-water activities are a component and issues a permit if it finds the entire project to be in the "public interest." See, for example, *North Carolina v. Hudson,* 731 F. Supp. 1261 (E.D.N.C. 1990).

The advance approval for noncommercial riparian shellfish growing in Virginia provided by the Corps nationwide and regional aquaculture permits and VMRC General Permit No. 3 is not available where nonnative species are involved. Corps permits issued under Section 404 can be vetoed by the U.S. Environmental Protection Agency (EPA) under Section 404(c) if EPA finds that "unacceptable" adverse environmental effects would result from the permitted introduction activities. Corps permit issuance could be challenged in federal court as arbitrary and capricious based on the evidentiary record by interested individuals or private, non-profit, or public-sector entities opposed to the introduction. See, for example, *North Carolina v. Hudson,* 731 F. Supp. 1261 (E.D.N.C. 1990). Corps permit denials, stringent conditions imposed on issued permits, and EPA vetoes of Corps permits similarly can be challenged by disappointed permit applicants. See, for example, *James City County v. Environmental Protection Agency*, 12 F. 3d 1330 (4th Cir. 1993).

Corps permit issuance is statutorily conditioned on the proposed activities' consistency with relevant state CMPs under CZMA Section 307 discussed below, compliance with state water quality standards under Clean Water Act Section 401, and Corps compliance with the Endangered Species Act (ESA), the Sustainable Fisheries Act's essential fish habitat (EFH) provisions, the National Environmental Policy Act's (NEPA) environmental impact statement (EIS) procedures, and the Regulatory Flexibility Act's regulatory impact review and initial regulatory flexibility analysis requirements. See for example, *Environmental Defense Fund v. Corps of Engineers*, 348 F. Supp. 916 (N.D. Miss. 1972; EIS for waterway project not defective based on invasive species analysis). NEPA regulations include a methodology for dealing with scientific uncertainty in the EIS process. The information and methodologies utilized in the preparation of this report could prove useful in future federal and state EIS processes involving nonnative introductions.

Under Section 7 of the ESA, the Corps would review the proposed introduction for possible jeopardy to ESA-listed species and their designated critical habitat. Depending on the species, consultations with USFWS or the National Marine Fisheries Service and a biological opinion could follow. If the biological opinion found jeopardy, it would also identify reasonable and prudent alternatives to the introduction as proposed for the Corps and the applicant to consider. In its May 2, 2002, public notice included in Appendix E, the Corps concluded that the VSC's proposed introductions of *C. ariakensis* would not jeopardize endangered or threatened species or adversely modify their critical habitat in Chesapeake Bay.

The Corps' May 2, 2002, public notice succinctly describes the applicability of the Sustainable Fisheries Act's EFH provisions:

> The Magnuson-Stevens Fishery Conservation and Management Act, as amended by the Sustainable Fisheries Act of 1996 (Public Law 104-267), requires all Federal agencies to consult with the National Marine Fisheries Service on all actions, or proposed actions, permitted, funded, or undertaken by the agency, that may adversely affect Essential Fish Habitat (EFH). The Chesapeake Bay, its tributaries, and the Atlantic Ocean all contain Essential Fish Habitat (EFH) for species managed under a Federal Fishery Management Plan. The following is a list of species that may be affected: scalloped hammerhead, sandbar, dusky, and sand tiger shark, red drum, cobia, Spanish and king mackerel, black sea bass, scup, summer, winter, and windowpane flounder, butterfish, bluefish, Atlantic sea herring, and red hake. The habitat which the structures may affect consists of shallow waters and mudflats.

However, because *C. virginica* is restricted to state waters not covered by a federal fisheries management plan, the EFH review process does not include the potential effects of *C. ariakensis* introductions on essential habitat for *C. virginica*. More generally, the EFH regulations (50 C.F.R. § 600.815 (a)(3)) authorize regional fishery management councils such as the Mid-Atlantic States Fisheries Management Council to list in their fishery management plans introduction of "exotic species" as an activity that "may adversely affect EFH." Furthermore, the councils and the Secretary of Commerce make recommendations to state agencies such as the VMRC and Maryland DNR as well as federal agencies such as the Corps about the adverse habitat impacts of proposed introductions (16 U.S.C. § 1855(b)). Under the Sustainable Fisheries Act Section 306(b) and associated laws for the ASMFC, the councils can even take over management of ocean fisheries in state waters if state management is conflicting with council or commission management.

Under this regulatory framework and utilizing its May 2002 public notice, in April 2003 the Corps indicated it would approve with conditions the VSC's proposal to deploy 1 million triploid *C. ariakensis*. The 15 conditions described in the Corps' April 14, 2003, Statement of Findings and Final Environmental Assessment were derived from recommendations provided in a February 2003 letter from this committee to the VMRC and a February 2003 Chesapeake Bay Program ad hoc panel report (see Appendix G) discussed below in this chapter's interjurisdictional section. A key difference between the Corps and VMRC approvals was that, unless subsequently extended through a new public notice issued by the Corps, all *C. ariakensis* deployed would have to be removed by June 30, 2004. A permit reflecting these conditions became effective upon signature by VSC, the Corps, and the Virginia Institute of Marine Science (regarding its monitoring roles under permit conditions 8, 10, and 11). Also, under an informal March 2003 federal-state interagency agreement, prior to any subsequent commercial use of nonnative oysters, triploid or diploid, a full EIS, including risk assessments and alternatives analyses, would be prepared, a process estimated to take at least 2 years.

Other Issues Related to the Clean Water Act

While the Corps of Engineers regulates discharges of dredge and fill material into U.S. waters under Clean Water Act Section 404, other point source discharges of pollution are regulated by EPA and the states pursuant to the act's National Pollution Discharge Elimination System (NPDES) permits. EPA's position as stated to the committee at its first meeting has been that it does not consider a nonnative organism a pollutant for NPDES permitting purposes. With respect to nonnative species introduced through ballast water, that position has been successfully challenged in a federal district court, and an appeal before the Ninth Circuit Court of Appeals is pending. That court recently ruled that particulate matter such as feces and other emissions from cultured nonnative mussels were not a point source discharge of pollutants requiring an NPDES permit. See *Association to Protect Hammersley v. Taylor Resources*, 299 F. 3d 1007 (9th Cir. 2002). However, for pen-reared salmon operations in Maine, a federal magistrate judge has recommended that the federal district court rule that escaping non-North American origin salmon, excess fish feed, antibiotics, and uneaten chemicals are pollutants requiring an NPDES permit. See *United States Public Interest Research Group v. Heritage Salmon, Inc.*, Civil No. 00-150-B-C, Recommended Decision of Magistrate Judge Margaret Kravchuk (D. Maine, Feb. 19, 2002).

If introduction of a nonnative oyster or emissions from oysters were held to be a point source polluting discharge, for a Chesapeake Bay introduction, an NPDES permit would be required from the Maryland or Virginia water quality agency that administers the NPDES permit system for the state's Chesapeake Bay waters where the introduction would take place (Craig, 2002). If such a permit were issued, it might be challenged by a disappointed sister Chesapeake Bay or Atlantic coast state under the law of interstate water pollution, or by disappointed individual or private, nonprofit, or public-sector entities under the common law of public and private nuisances.

To date, no state has invoked the law of interstate water pollution against another state for authorizing the introduction of a nonnative species to waters they share. The legal issues such a claim would present are summarized here. The law of interstate water pollution is found in several recent U.S. Supreme Court opinions integrating recent federal water pollution legislation such as the Clean Water Act (CWA) with older common law principles.

Prior to enactment of the CWA, the Supreme Court had held that a sister state could sue under a federal common law of nuisance to abate pollution resulting from operations in another state (*Illinois v. Milwaukee*, 406 U.S. 91 (1972)). However, the court has since held that the CWA's passage has eliminated most legal bases for a federal court to impose more stringent pollution limitations than those imposed under the CWA regulatory regime (*City of Milwaukee v. Illinois*, 451 U.S. 304 (1981)). Only causes of action based on maritime tort may have survived. See *Middlesex County Sewerage Authority v. National Sea Clammers Association*, 616 F. 2d 1222 (3rd Cir. 1980), revised on other grounds, 453 U.S. 1 (1981).

CWA Sections 401 and 402 provide sister states with notice and an opportunity to be heard before NPDES permits are issued for discharges into an interstate waterway from a neighboring state. However, the sister state does not have the authority to block issuance of the permit (*Arkansas v. Oklahoma*, 503 U.S. 91 (1992)). The sister state's only recourse is to apply to the EPA administrator, who has the discretion under CWA Section 402(d)(2) to disapprove the permit upon concluding that the discharges will have an undue impact on interstate waters (*International Paper Co. v. Ouellette*, 479 U.S. 481 (1987)). While sister states have only these limited CWA procedural rights, the court held that residents of a sister state can still challenge the discharge under the source state's law of nuisance. Residents of the source state also may challenge the discharge as a nuisance even though it has been approved through the issuance of CWA and other regulatory permits. See, for example, *Boomer v. Atlantic Cement Co.*, 257 N.E. 2d 870 (New York Court of Appeals, 1970).

A private nuisance is defined by the courts as an unreasonable interference with a neighboring private property owner's use and enjoyment of their property. Thus, a nonnative introduction challenged as a private nuisance would have to be shown to be causing unreasonable harm to the plaintiff's private property (e.g., Chesapeake Bay bottom leased from the state, due to the introduction's negative impact on the plaintiff's use and enjoyment of that property). A public nuisance is defined more broadly by the courts as any act that unreasonably causes damage to the public in the exercise of rights common to all. Any affected member of the public has standing to sue the alleged unreasonable activity without necessarily showing any injury to a private property interest of theirs. See for example, *Town of Preble v. Song Mountain, Inc.*, 308 N.Y.S. 2d 1001 (Supreme Court of New York, 1970). Within the law of private nuisance, an aquaculture operation using nonnative species might be treated as an "abnormally dangerous activity" for which the operator is strictly liable for any injuries caused to persons or their property without any showing of negligence on the part of the operator. See for example, *Wood v. Picillo*, 443 A.2d 1244 (R.I. 1982; chemical company strictly liable for percolation of hazardous waste on nearby residents' property). No court has yet applied these doctrines to alleged public and private injuries from a nonnative introduction. However, the potential for litigation challenging a nonnative introduction under those theories has been discussed in the literature (Biber, 1999; McCoy, 2000; Terpstra, 1998).

Finally, a resident of the source state might claim in state courts that the state's approval of a nonnative introduction violated the state's public trust doctrine responsibilities discussed in the state section of this chapter and Chapter 10. One theory would be that the state has violated its trust responsibilities for living resources in tide and submerged lands by approving introduction of a nonnative that may contribute to the decline of native species (Johnson, 1989).

Under the CWA, EPA-approved state water quality standards must protect existing uses such as shellfish beds, including oyster beds (40 C.F.R. §. 131.12). Under CWA Sections 303 and 401, these water quality standards are to be achieved and maintained through state and federal permit processes. Reductions in the size of Chesapeake Bay oyster beds can reduce the area protected by these water quality mandates.

Code of Conduct for Responsible Aquaculture Development in the U.S. Exclusive Economic Zone

Prepared under the National Aquaculture Act of 1980 by the National Marine Fisheries Service (NMFS), the code utilizes a precautionary approach combined with adaptive management to promote sustainable aquaculture development in U.S. waters between 3 and 200 nautical miles offshore. According to the code, participants should conserve biodiversity and carefully regulate nonindigenous and genetically altered species (NMFS, 2002). Also, a system of monitoring should be enforced not only by federal and state authorities but also voluntarily through self-regulation.

A precautionary approach to nonnative introductions also is suggested by several of the international agreements discussed below to which the United States is a party or signatory.

INTERJURISDICTIONAL

Coastal Zone Management Act

As previously discussed, the purpose of the Coastal Zone Management Act (CZMA; 16 U.S.C. § 1451 et seq.) is to encourage the states to protect, preserve, develop, and restore natural coastal resources. Although the CZMA is a federal law, participation by the states is voluntary. All U.S. coastal states have implemented the CZMA by adopting CMPs for federal approval under the statute. The CZMA was designed:

> to encourage and assist the states to exercise effectively their responsibilities in the coastal zone through the development and implementation of management programs to achieve wise use of the land and water resources of the coastal zone, giving full consideration to ecological, cultural, historic, and esthetic values as well as the needs for compatible economic development. (16 U.S.C. § 1452(2))

The act does not address nonnative species (nor do any of the state CMPs discussed above). But the act does not prohibit states from incorporating invasive species issues and information into their CMPs with federal approval.

Interstate Consistency: Two Examples

"Federal consistency" is defined by the U.S. Department of Commerce as "the term used to describe the mechanism by which a state can review federal activities, including federally licensed or permitted activities, to determine whether they are consistent with the state's coastal management program."

- *In the Consistency Appeal of Vieques Marine Laboratories from an Objection by the Puerto Rico Planning Board,* U.S. Department of Commerce, Office of the Secretary, 1996
  - ♦ In this case a nonprofit corporation chartered by the commonwealth of Puerto Rico wanted to operate a nonnative shrimp farm off the coast of Puerto Rico. The shrimp farm would involve the use of floating cages anchored on the bottom of a pristine bay, covering an area of no more than 5 acres (or 2% of the bay's total area). This required the use of mooring buoys and floating docks.
  - ♦ Puerto Rico's Planning Board (PRPB) objected to the application for a permit submitted by the nonprofit group to the Corps. The PRPB based its objection on the grounds that the project was not consistent with Puerto Rico's CMP provisions protecting coastal water quality.
  - ♦ Since the PRPB entered a consistency objection, that objection precluded the Corps from issuing any federal permit unless the Secretary found either "that the activity is consistent with the purposes and objectives of the [federal] CZMA (Ground I) or is [a] necessary [project] in the interest of national security (Ground II)."
  - ♦ There are three elements involved in Ground I (see 15 C.F.R. § 930.121). Those three elements are:
    - ■ the activity furthers the national interest as articulated in Section 302 or Section 303 of the act, in a significant or substantial manner;
    - ■ the national interest furthered by the activity outweighs the activity's adverse coastal effects, when those effects are considered separately or cumulatively.
    - ■ there is no reasonable alternative available that would permit the activity to be conducted in a manner consistent with the enforceable policies of the management program. When determining whether a reasonable alternative is available, the Secretary may consider but is not limited to considering previous appeal decisions, alternatives described in objection letters, and alternatives and other new information described during the appeal.

  - The proposed shrimp farming project failed the second element and thus was not consistent with the objectives or purposes of the CZMA and thus could not be approved by the Corps.
- *In the Consistency Appeal of the Virginia Electric and Power Company from an Objection by the North Carolina Department of Environment, Health, and Natural Resources*, U.S. Department of Commerce, Office of the Secretary, 1994
  - The city of Virginia Beach is the largest city in Virginia, with over 400,000 residents, and must purchase its water from sources outside the city limits.
  - The Virginia Electric and Power Company (VEPCO) on behalf of the city of Virginia Beach, appealed a decision by the Secretary of Commerce regarding the state of North Carolina's objection to VEPCO withdrawing water from Lake Gaston for the city's water supply needs. The Secretary, in this consistency appeal, overruled North Carolina's objections and allowed the city of Virginia Beach to acquire the requisite federal permits to construct a pipeline for the removal and use of water from Lake Gaston.
  - The CZMA requires that an applicant for a federal license of an activity that will affect a water use or natural resource of the coastal zone must certify to the permitting agency (here it is the Federal Energy Regulatory Commission) that the proposed activity (building of the pipeline to withdraw water from Lake Gaston) complies with the Virginia CMP.
  - A relevant issue on appeal in this case is whether the CZMA authorizes a state to review for consistency with its CMP an activity that is occurring in another state. The Secretary of Commerce held that the CZMA does authorize such interstate consistency review. Although the Secretary noted that the CZMA does not give a state the authority to control activities that occur in another state, it does allow states that have federally approved CMPs "the right to seek conditions on or prohibit the issuance of federal permits and licenses that would 'affect' their state."
  - The Secretary held that North Carolina, under the CZMA, has a right to review for consistency with its own federally approved CMP, the proposed activity of Virginia in the withdrawal and use of water from Lake Gaston if that withdrawal and use could affect "any land or water use or natural resource in North Carolina's coastal zone."

The Department of Commerce subsequently adopted interstate consistency regulations (U.S. Department of Commerce, 1994). Thus, even if a C.

ariakensis introduction in one state's waters received all other necessary state, federal, interjurisdictional, and international approvals, a neighboring coastal state could raise CZMA interstate consistency objections to the federally issued permits based on alleged impacts to its coastal resources contrary to enforceable policies in its CMP. The state's objections would in turn be subject to override by the Secretary of Commerce as described above.

Chesapeake Bay Program

Clean Water Act § 117 provides for the establishment of a Chesapeake Bay Program (CBP) Office within the EPA (see Appendix E). It also discusses management strategies for the CBP as well as funding and grants to support the program. Parties to the original 1987 Chesapeake Bay Agreement are the United States EPA, representing the federal government; the District of Columbia; Maryland; Pennsylvania; Virginia; and the Chesapeake Bay Commission. The commission is a 21-member tristate commission created in 1980 to advise the members of the Maryland, Virginia, and Pennsylvania legislatures on bay issues. The members, mostly legislators from the three states, are responsible for coordinating bay policy issues across state lines and developing shared solutions. The commission supported this National Research Council study by letter dated January 16, 2002, enclosing a supporting resolution.

The program's Chesapeake 2000 Agreement specifically addresses goals for oysters in the bay. The text states that "[b]y 2010, achieve, at a minimum, a ten-fold increase in native oysters in the Chesapeake Bay, based upon a 1994 baseline."

The 1993 Chesapeake Bay Policy for the Introduction of Non-Indigenous Aquatic Species is a basinwide regional policy that considers only first-time introductions of nonindigenous, nonnaturalized aquatic species (see Appendix G). Under the policy, an introduction is considered "first time if 1) the species is not indigenous or naturalized, or 2) the jurisdiction has not previously promulgated rules, regulations or otherwise issued a permit allowing the introduction of that aquatic species into an unconfined system, excluding permits issued for research."

The 1993 policy has four specific goals:

1. to provide technical reviews of proposed nonindigenous species introductions to identify potential nuisance species;
2. to provide the permitting decisionmakers with the best-available information and assessment regarding a nonindigenous species' potential for becoming a nuisance in the ecosystem or to human activities;
3. to create a mechanism for sharing information among all bay jurisdictions, including Delaware, the District of Columbia, Maryland, New

York, Pennsylvania, Virginia, and West Virginia, regarding species being considered by other bay jurisdictions; and

4. to not unduly lengthen or burden the existing permitting process within the signatory jurisdiction.

Pursuant to the 1993 policy, some experiments with *C. ariakensis* had been approved, but a May 31, 2002, ad hoc panel report (included in Appendix G) recommended against the proposed introductions of triploid *C. ariakensis* described in the Corps of Engineers, May 2, 2002, public notice. The VSC subsequently withdrew this permit request. No federal or state law requires such private- or public-sector compliance with recommendations generated by CBP processes. A CBP ad hoc panel also recommended against the VSC's subsequent proposal to the VMRC as submitted (the CBP panel's February 20, 2003, report is included in Appendix G).

Atlantic States Marine Fisheries Commission

The Atlantic States Marine Fisheries Commission (ASMFC) is a coalition of the 15 Atlantic states joined together under an interstate compact approved by Congress (P.L. 77-539 and 81-721) to manage their shared coastal fishery resources. The governing legislation is the Atlantic Striped Bass Conservation Act, Atlantic Coastal Fisheries Cooperative Management Act (16 U.S.C. § 1501 et seq.), and the Interjurisdictional Fisheries Act (P.L. 99-659). Under 16 U.S. Code Section 5101, "fish" are defined broadly to include all forms of marine animal life other than marine mammals and birds; a "coastal fishery resource" subject to management by ASMFC includes any fish that "is broadly distributed across waters under the jurisdiction of two or more states." A 1989 ASMFC report contains a plan to control interjurisdictional transfers and introductions of shellfish (U.S. Atlantic States Marine Fisheries Commission, 1989). The U.S. Atlantic States Marine Fisheries Commission (2002) lists potential risks and benefits of a *C. ariakensis* introduction and includes position papers from states, federal agencies, and CBP committees. The ASMFC is developing aquaculture guidelines.

Potomac River Fisheries Commission

The Potomac River Fisheries Commission operates under the Potomac River Compact of 1958, authorized by Congress and ratified by Virginia and Maryland. The commission is a semiautonomous agency, but its work and policies are coordinated closely with the Fisheries Service of the Maryland DNR and the Marine Resources Commission of Virginia. Fishery agencies of both states provide law enforcement on the Potomac River for the commission. The Potomac River Fisheries Commission works to conserve and improve seafood resources of the Potomac River. The commission also regulates and licenses fisheries and the dredging of soft-shell clams in the River. From the sale of crab, oyster, fish, and clam licenses and an oyster inspection tax, the commission receives proceeds. Annually, each state appropriates $150,000 to the commission's work. Eight members constitute the commission. Four are from Maryland, four from Virginia. A 1992 Virginia law (Virginia Code § 28.2-1004) authorizes the commission to engage in a "pilot program for experimental oyster hatchery seed planting" with private- and public-sector agencies. The law's effectiveness is conditioned on the enactment of a similar law by Maryland, which does not appear to have happened yet. The commission's role if any in discussions of *C. ariakensis* to date is unclear.

INTERNATIONAL

Five international agreements to which the United States is a party or signatory establish a risk-adverse general framework for nonnative introductions (see Box 8.3). Some of these agreements adopt the international environmental protection concept known as the "precautionary approach."

International Council for the Exploration of the Sea Convention

This convention is in force, and the United States is a party. The 1994 International Council for the Exploration of the Sea (ICES) Code of Prac-

BOX 8.3
International Law Documents (see Appendix F)

- International Council for the Exploration of the Sea (ICES), 24 U.S.T. 1080; T.I.A.S. 7628 (September 12, 1964)
- ICES Protocol, 27 U.S.T. 1022; T.I.A.S. 8238 (August 13, 1970)
- ICES Code of Practice on the Introductions and Transfers of Marine Organisms, 1994 (http://www.ices.dk/iceswork/wgdetailacme.asp?wg=WGITMO)
- Code of Conduct for Responsible Fisheries, Food and Agricultural Organization of the United Nations, 1995
- United Nations Convention of the Law of the Sea, Part XII, Section 1, Art. 196 (http://www.un.org/Depts/los/convention_agreements/texts/unclos/closindx.htm)
- Convention on Biological Diversity (1992), United Nations Environment Programme, Alien Species: Guiding Principles for the Prevention, Introduction, and Mitigation of Impacts; Articles 5-9, specifically Art 8(h)

tice on Introductions is included in Appendix F. The code includes widely utilized risk assessment protocols for both "transfers" of native species in their current range and "introductions" of nonnative species.

With respect to introductions as summarized in Peterson et al., (1999), the ICES code consists of guidelines to follow prior to the introduction as well as steps to follow after deciding to proceed with an introduction. Under the guidelines, the initial step is to describe the purpose of the introduction. Next a thorough review of the biology and life history of the organism proposed for introduction must be provided to the ICES. If a decision is made to proceed with an introduction, the guidelines suggest use of quarantined brood stock that has been approved by the receiving country and is raised under quarantine long enough to establish its health. Then, if no diseases or parasites are recognized, only the F1 or subsequent generation of the brood stock can be transplanted to the natural environment. The code then calls for complete destruction of the initial parental brood stock and sterilization of all hatchery effluents. The quarantined F1 generation may then be placed into open water on a limited scale for assessment of interactions with native species. For large-scale introductions, a continuing study of the introduced species in its new environment should be made and impacts reported.

Convention on Biological Diversity

This convention is in force, and the United States is a signatory but has not yet ratified the convention or its biosafety protocol and thus is not a party to either. Under the convention, draft guiding principles regarding "alien" species introductions have been developed. Guiding principle 10 regarding intentional introductions provides:

> No intentional introduction should take place without proper authorization from the relevant national authority or agency. A risk assessment, including environmental impact assessment, should be carried out as part of the evaluation process before coming to a decision on whether or not to authorize a proposed introduction. States should authorize the introduction of only those alien species that, based on this prior assessment, are unlikely to cause unacceptable harm to ecosystems, habitats or species, both within that State and in neighboring States. The burden of proof that a proposed introduction is unlikely to cause such harm should be with the proposer of the introduction. Further, the anticipated benefits of such an introduction should strongly outweigh any actual and potential adverse effects and related costs. Authorization of an introduction may, where appropriate, be accompanied by conditions (e.g., preparation of a mitigation plan, monitoring procedures, or containment requirements). The precautionary ap-

proach should be applied throughout all the above-mentioned measures.

Guiding principle 1 defines the precautionary approach:

> Given the unpredictability of the impacts on biological diversity of alien species, efforts to identify and prevent unintentional introductions as well as decisions concerning intentional introductions should be based on the precautionary approach. Lack of scientific certainty about the environmental, social and economic risk posed by a potentially invasive alien species or by a potential pathway should not be used as a reason for not taking preventative action against the introduction of potentially invasive alien species. Likewise, lack of certainty about the long-term implication of an invasion should not be used as a reason for postponing eradication, containment or control measures.

1995 Food and Agriculture Organization Code of Conduct for Responsible Fisheries

Article 9 of the Food and Agriculture Organization of the United Nations (FAO) Fisheries Code regarding aquaculture is included in Appendix F. Article 7.5 of the code urges nations to "apply the precautionary approach widely to conservation, management, and exploitation of living aquatic resources in order to protect them and preserve the aquatic environment." Application of the precautionary approach to nonnative species introductions is elaborated in FAO 1996.

Ramsar Convention

The goal of the Ramsar Convention is the preservation of wetlands of international significance. The Ramsar Web site contains the full text of the convention and amendments (*http://www.ramsar.org/index_very_key_docs.htm*). As of November 12, 2002, there were 133 parties to the convention. The United States joined the convention as a party in 1987. Under the convention, the Chesapeake Bay is designated as a wetland of international significance, which under Article 3 should be used wisely. In 1987 a conference of the convention parties defined "wise use" as "sustainable use . . . compatible with the maintenance of the natural properties of the ecosystem."

United Nations Convention on the Law of the Sea

The United States is a signatory to the United Nations Convention on the Law of the Sea (UNCLOS) but the Senate has not yet voted to accede to the convention and thus the United States is not a party. The Bush administration and the Pew Oceans Commission support U.S. accession, and the U.S. Commission on Ocean Policy will also recommend that the United States become a party.

UNCLOS Article 196 regarding introductions is included in Appendix F. It obligates nations "to prevent . . . the intentional or accidental introduction of species . . . which may cause significant and harmful changes" to the marine environment.

In addition, the United States is a party to two international agreements that are particularly relevant to imports of *C. ariakensis* into the United States: the World Trade Organization (WTO) Agreement on the Application of Sanitary and Phytosanitary Measures (SPS) and the Office International des Epizooties (OIE) agreement. Those two agreements are summarized in Biosecurity Australia (2002) as follows:

World Trade Organization SPS Agreement

The SPS agreement applies to measures designed to protect human, animal, and plant life and health from pests and diseases, or a country from pests, and which may directly or indirectly affect international trade. It also recognizes the right of WTO member countries such as the United States to determine the level of protection they deem appropriate and to take the necessary measures to achieve that protection. Sanitary (human and animal health) and phytosanitary (plant health) measures apply to trade in or movement of animal- and plant-based products within or between countries.

In the SPS agreement, sanitary and phytosanitary measures are defined as any measures applied:

- to protect animal or plant life or health within the territory of the member from risks arising from the entry, establishment, or spread of pests, diseases, disease-carrying organisms, or disease-causing organisms;
- to protect human or animal life or health within the territory of the member from risks arising from additives, contaminants, toxins, or disease-causing organisms in foods, beverages, or feedstuffs;
- to protect human life or health within the territory of the member from risks arising from diseases carried by animals, plants, or products thereof, or from the entry, establishment, or spread of pests; and
- to prevent or limit other damage within the territory of the member from the entry, establishment, or spread of pests.

Key provisions of the SPS agreement include:

- an importing country has the sovereign right to adopt measures to achieve the level of protection it deems appropriate (its appropri-

ate level of protection, or ALOP) to protect human or animal life or health within its territory, but such a level of protection must be consistently applied in different situations and
- an SPS measure must be based on scientific principles and not be maintained without sufficient evidence.

Office International des Epizooties Agreement

The OIE, the world organization for animal health, is an intergovernmental organization created by a 1924 international agreement. The objectives of the OIE are:

- to keep member countries informed of the occurrence and course of significant animal diseases throughout the world and of means of controlling these diseases;
- to coordinate, at the international level, studies devoted to the surveillance and control of significant animal diseases; and
- to harmonize health standards covering trade in animals and animal products.

The OIE currently comprises more than 160 member countries, including the United States, and operates under the authority of an international committee formed by permanent delegates designated by the governments of all member countries.

The standards referenced in the SPS agreement include the following OIE codes and manuals:

- the OIE International Animal Health Code, prepared by the International Animal Health Code Commission, contains standards, guidelines, and recommendations designed to prevent the introduction of pests and diseases into the importing country during trade in animals, animal genetic material, and animal products;
- the Manual of Standards for Diagnostic Tests and Vaccines, prepared by the Standards Commission, lists laboratory diagnostic techniques and requirements for production and control of biological products (mainly vaccines); and
- an Aquatic Animal Health Code and a Diagnostic Manual for Aquatic Animal Diseases, prepared by the Fish Diseases Commission. These are sister publications to the OIE code and manual above.

The OIE has developed guidelines for risk analysis which recognize that the importation of animals and animal products may involve a degree of risk to the importing country. The OIE supports risk analysis because it provides importing countries with an objective method of assessing risks associated with importation and of determining how those risks may be managed. It notes that analysis should be transparent so that the exporting country is provided with a clear and documented decision on the measures imposed on imports or the reasons for refusing to allow importation.

SUMMARY

The four-level regulatory framework reviewed in this section can be characterized as a patchwork with significant gaps, especially in regions outside the Chesapeake Bay. The patchwork characterization refers to four significant gaps: the absence of federal regulatory approval for species not listed as injurious, lack of federal jurisdiction in circumstances where the Corps does not have permitting authority, the absence of regional-level review processes (i.e., multistate and federal) outside of the Chesapeake Bay region, and the limit of CBP review to "first time" introductions.

Within the Chesapeake Bay, the Chesapeake Bay Program's 1993 Policy for the Introduction of Non-Indigenous Aquatic Species presents a working prototype for regional decision making. The policy would be strengthened if it included review of proposed uses of nonnative species beyond "first-time" introductions, since the first introduction may be shortlived and subsequent introductions could be more significant in terms of magnitude and geographic extent. Outside the Chesapeake Bay, equivalent review processes for proposed nonnative introductions could be established using the CPB policy as a model. Existing multistate entities that could implement such review processes include the ASMFC and its sister interstate-compact-based regional fisheries commissions, which together cover U.S. coastal waters outside Alaska and Hawaii, and the Magnuson Fisheries Conservation and Management Act's regional federal fisheries management councils, which together cover the entire U.S. Exclusive Economic Zone extending from 3 to 200 nautical miles offshore (Hildreth, 1991). As described above, ASMFC has played a constructive role in regional deliberations about *C. ariakensis*. Additional regional coordination on nonnative introductions could be provided through the coastal zone management interstate consistency process described in this chapter's interjurisdictional section. This would necessitate amending current federally approved state coastal zone management programs to include compatible enforceable policies regarding nonnative introductions.

While the 1993 policy and recommendations made pursuant to it are not legally binding, the general respect for CBP decisions, including those made under the 1993 policy, is impressive and demonstrates an alternative approach to command and control regulation of nonnative introductions. Other features of the CBP approach worth incorporating into any adjustments made to the current framework include the CBP policy's use of best-available scientific information, timely responses to requests for decisions pursuant to the policy, and minimization of regulatory overlaps. While CBP decisions pursuant to the 1993 policy generally include specific recommendations regarding monitoring and biosecurity, the nonbinding nature of those recommendations indicate that even in the Chesapeake Bay the current regulatory framework does not provide adequate monitoring and oversight. Furthermore, it appears that the CBP does not itself have either the budget, personnel, or mandate to engage in the necessary monitoring and oversight.

The 1993 policy is consistent with a precautionary approach to nonnative introductions (e.g., in its requirement that environmental and economic evaluations be conducted in order to ensure that risks associated with first-time introductions are acceptably low). Also, the 1993 policy illustrates a "clean list" approach to introductions, an approach that the committee generally recommends for all levels of decision making about nonnative introductions as contrasted with the "dirty list" approach. Under the 1993 policy and many state laws, introductions of nonnative species are prohibited unless specifically approved. Utilizing a "clean list" is a key step in implementing a precautionary approach. In contrast, under a "dirty list" approach, only introductions of species that have been specifically blacklisted by legislation or regulation are prohibited. However, merely requiring approval before any nonnative species is introduced is insufficient unless the person or entity proposing the introduction is also required to provide environmental, economic, and social evaluations of the risks and tradeoffs involved. Furthermore, important uncertainties in information should be weighed against approval of a proposed introduction. Interested parties throughout the potentially affected region should be provided meaningful opportunities to participate in the decision-making process. Relevant legislation and regulations governing the decision-making process should be amended accordingly.

These steps toward implementing a risk-adverse approach to introductions are reflected in several of the international documents reviewed by the committee to which the United States is a signatory or a full party. These steps are particularly important given the relatively weak capabilities of the current framework for sufficient monitoring, evaluation, and adaptive management. Finally, if a proposed nonnative species is approved for introduction, control measures will be required to reduce remaining risks and, where feasible, compliance could be guaranteed by bonding or insurance requirements.

Many interested parties find it surprising that under the current framework, depending on the methods used for intentionally introducing a nonnative species, including *C. ariakensis*, no federal regulatory approval is required. Corps approval is required only if the introduction involves in-water structures or fill; even in the latter situation, Corps general and regional permits may eliminate the need for a complete regulatory review of the proposed introduction. Further study is required to determine whether nonnative introductions should be comprehensively regulated at the federal level through, for example, statutory amendments instituting a "clean list" approach under the Lacey Act. The pending reauthorization by the 108th Congress of the federal Invasive Species Act could provide a forum for discussion of this and related changes in federal law to provide a better-coordinated and better-focused approach to intentional introductions at the federal level.

These discussions appropriately could be extended to U.S. implementation of the various international agreements regarding intentional introductions policy and procedures, including the Convention on Biological Diversity, its Cartegena Protocol on Biosafety, and draft Guiding Principles on Intentional Introductions; the ICES Convention and its code of practice regarding the transfers of natives and introductions of nonnatives; NMFS, FAO, and other aquaculture codes; UNCLOS Article 196, that requires nations that are party to the convention to control intentional introductions of nonnative species that may cause significant and harmful changes to the marine environment; and relevant WTO and OIE agreements regarding imports of nonnative species to the United States.

9

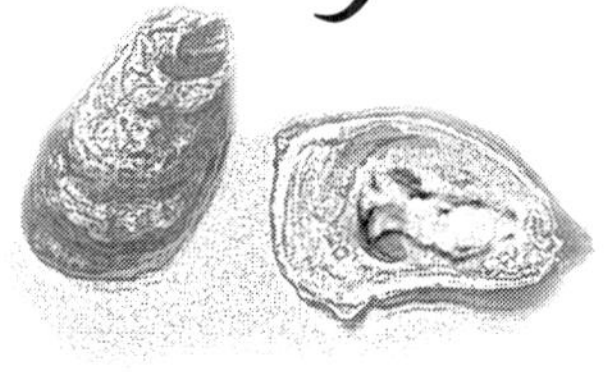

Elements of Risk Assessment for the Introduction of *Crassostrea ariakensis* in the Chesapeake Bay

BACKGROUND ON RISK ASSESSMENT

The committee was asked to assess whether the breadth and quality of existing information on oysters and other introduced species are sufficient to support risk assessment of three management options: no use of nonnative oysters, open-water aquaculture of triploid nonnative oysters, and introduction of diploid nonnative oysters. We must begin by identifying an appropriate scope for the risk assessment. Structuring a conceptual or numerical risk assessment highlights what is known and what is not known about the modeled system, breaks complex issues into more understandable problems, and helps to identify critical assumptions. Some previous National Research Council (NRC) studies (e.g., NRC, 1983, 2002b) define risk assessment as the identification and characterization of hazards and the determination of the likelihood that hazards will result in harms. In these studies, risk management is defined as a decision-making process that takes into account the probability distribution of harms given exposure to the hazard and the associated conditional costs and benefits. That is, risk assessment and risk management are reduced to characterizing the branches and conditional probabilities of a decision tree and selecting a strategy based on preferences regarding the conditional net benefits. While this may be workable for well-understood systems with well-defined objectives, complex systems require a more general framework for risk assessment and a closer integration of risk assessment with risk management (NRC, 1993, 1996a, 1996b, 2002b). This closer integration is necessary because the risks that we choose to manage determine the risks that need to be assessed. Following Pratt et al. (1995), we define risk assessment as a decision-making technique that incorporates objective and subjective estimates of the probability of uncertain factors and identifies preferred actions with respect to multiple objectives (see Box 9.1).

BOX 9.1
Glossary of Risk Analysis Terms

Bayes theorem: a statistical rule for combining relative frequencies and subjective probabilities or prior information about the likelihood of certain future conditions.
Conditional costs and benefits: outcomes that can be expected given the occurrence of specific management actions or particular stochastic events.
Decision: an active or passive choice.
Harm: costs incurred as a consequence of specific hazards; a conditional payoff. For example, some Chesapeake Bay stakeholders would construe the establishment of a self-sustaining population of nonnative oysters as a harm related to exposure to the hazard of reversion.
Hazard: an action or event that has the potential to result in an undesired outcome. In the context of this study, a hazard occasioned by open-water aquaculture of triploid nonnative oysters is that reproductively competent oysters are released and the undesired outcome is that the nonnative oysters become invasive.
Likelihood: the probability that an outcome will occur given exposure to a hazard.
Multiple criteria decision analysis: procedures for evaluating alternative outcomes with respect to multiple objectives.
Objectives: goals of stakeholders.
Payoffs: conditional outcomes evaluated in terms of management objectives.
Probability and probability distributions: statistical descriptions of relative frequencies, often expressed in terms of expected value, variance, skewness, and kurtosis.
Relative frequency: the frequency of particular outcomes relative to the frequency of observed outcomes
Risk: the possibility of undesired outcomes being realized as a result of management action or natural variability. In the context of this study, a risk associated with open-water aquaculture of triploid nonnative oysters could be defined as the joint probability that some oysters might revert and that their progeny might become established in the Chesapeake Bay.
Risk analysis: synonymous with risk assessment.
Risk assessment: a decision-making technique that incorporates relative frequencies and subjective probabilities of uncertain factors and identifies preferred actions with respect to one or more objectives.
Risk management: the avoidance, mitigation, reduction, shifting, pooling, or buffering of risk.
Risk preferences: faced with a choice between an action that leads to a guaranteed payoff and an action that leads to an equal but uncertain payoff, the selection of a particular payoff by a decision maker. Most decision makers in most circumstances will select the certain payoff; they are said to be risk averse.
Stakeholder: any person, group, or organization interested in the management decision.
Subjective probabilities: estimates of the likelihood of events and their distribution based on expert judgment with limited historical or experimental observations.

The structure and dynamics of the Chesapeake Bay ecological and socioeconomic systems are complex, not well understood, and subject to environmental, social, and political influences beyond the scope of management control. Consequently, decision makers are faced with uncertainty—uncertainty about the structure and dynamics of integrated physical, biological, economic, and sociocultural systems, uncertainty about how the systems will respond to the actions taken, and uncertainty about the merits of alternative outcomes. Decision making under these conditions entails risk to ecological systems, risk to socioeconomic systems and institutions, and risk that implementation of management actions will lead to unanticipated or undesired consequences. Actions taken to minimize one aspect of risk often increase the level of risk in other dimensions. When the consequences of management actions are uncertain, good decision making involves balancing risks and benefits.

In order to balance risks and payoffs, it is important to understand the characteristics of risk preferences and approaches to solving multiple criteria decision problems. Although individual risk preferences differ, most decision makers are averse to increased levels of risk unless the risk is offset by increased conditional gains. That is, given a choice between a risk-free payoff and an equal but uncertain payoff, most decision makers will select the risk-free payoff. Risks may be symmetric (equal likelihood of being above or below an expected value) or asymmetric (unequal likelihood of being above or below an expected value). Risk preferences are typically asymmetric (we prefer favorable outcomes) and discordant (we disagree on what is favorable). For example, stakeholders who favor the establishment of nonnative oysters might prefer management actions that increase the likelihood of reproduction, while those who oppose establishment of nonnative oysters might prefer management actions that decrease the likelihood of reproduction, decrease the likelihood of reproduction by reverted oysters, or decrease the likelihood of successful establishment of nonnative populations.

Multiple objectives may be balanced through political processes or formally examined using multiple criteria decision analysis methods (e.g., Keeney and Raiffa, 1976; Saaty, 1990). These methods have been used to address a variety of fishery management issues (e.g., Hilborn and Walters, 1977; Bain, 1987; Walker et al., 1983; Healey, 1984; Mackett, 1985; Merritt and Criddle, 1993). Solutions that emerge from the application of multiple criteria decision analysis often favor compromises that minimize maximum losses or maximize minimum benefits. Multiple criteria decision analyses that incorporate multiple stakeholders with overlapping objectives often select management options that enjoy broad support and limited objection. Stakeholders with conflicting objectives may prefer similar options for dissimilar reasons. For example, stakeholders opposed to the establishment of nonnative oysters might strongly prefer that triploid aquaculture field trials be based on genetic triploids (mated tetraploid-diploid crosses) because of the reduced likelihood of reproduction. Stakeholders interested in commercial-scale aquaculture of triploid nonnatives might favor field trials based on genetic triploids to reduce liability of diploid escape.

Risks that cannot be avoided can often be mitigated, reduced, shifted, pooled, or buffered. A policy mandating the early harvest of triploid nonnatives might mitigate the risk of open-water triploid aquaculture by minimizing the number of oysters that would revert from a triploid to diploid condition and reducing the likelihood that reverted oysters would spawn before they are harvested. Risk reduction might entail actions that partially reduce the level of risk. For example, triploids created from tetraploid-diploid crosses have a lower reversion rate than chemically induced triploids. Risk shifting entails a transfer of risk from one individual or stakeholder class to another. For example, oyster farmers may be able to shift the risk and liability associated with broodstock maintenance and larval production to state hatchery facilities. Risk pooling distributes individual risk across a class of stakeholders. Traditional insurance programs are a mechanism for pooling low-frequency risks with severe negative consequences. Buffering consists of actions that increase the resilience of systems to adverse events. For example, oyster farmers may be able to buffer production risk by distributing their oyster beds across a broad geographic region or across salinity gradients.

The three management options entail differing arrays of risks and are subject to the diverse objectives of multiple stakeholders. Development of a risk assessment framework for the decision would involve characterizing and developing probability distribution functions for the various risks, evaluating those risks according to the diverse objectives, and balancing those objectives in a multiple criteria decision analysis. At this stage there is insufficient information for a formal risk assessment of management options concerning the introduction of a nonnative oyster into the bay. Moreover, it is not possible to complete a risk assessment because the objectives and goals for the Chesapeake Bay and its dependent communities are not well defined or fully agreed upon. The objectives of state, federal, and local governing agencies are unclear, conflicting, or both. In addition, the objectives of the watermen, environmental groups, aquaculturalists, and other users are also unclear, conflicting, or both. Until these diverse objectives are sorted out, meaningful risk assessment cannot be undertaken. Nevertheless this chapter reviews areas of risk and identifies specific risk factors, indicates relative degrees of risk in the ecological and socioeconomic realms, and evaluates the adequacy of the information available.

Ecological risks associated with the options include the environmental and ecological consequences of continued low native oyster population levels, the risk that nonnative oysters would become established and pervasive or fail to become established or pervasive, the risk of further collapse or impaired recovery of native oyster populations, and the risk of adverse consequences for other stocks of fish, shellfish, and vascular plants. Ecological risks could arise through competitive interactions for food and space through technological interactions (e.g., joint harvesting of scarce stocks of native oysters admixed with abundant stocks of nonnative oysters) or through other interactions (e.g., enhancement of predator, parasite, or disease organisms, or fertilization competition between native and nonnative oyster gametes). An example of a qualitative risk assessment for shellfish farming in Tasmania considered the spread of predators or pests, habitat disturbance, and effects on food resources for other filter feeders (Crawford, et al., 2003). The risk assessment model described by Dew et al. (2003) provides a useful example of a simple model of the likelihood of self-sustaining populations resulting from commercial production of supposed triploid nonnative oysters. Examples of more comprehensive approaches to ecological risk assessment would include models coupling hydrodynamics and larval transport, models assessing age- or size-dependent predator-prey and competitive interactions, and models that incorporate variability in environmental conditions (e.g., salinity, temperature, substrate) to oyster growth, survivorship, and reproduction.

The decision to introduce or not introduce *C. ariakensis* can be expected to generate differing arrays of risk to economic and sociocultural institutions and systems. Economic and sociocultural risks would differ under each of the three management options and would impact, for example, the availability of oysters for public- and leased-bottom fisheries, the opportunity and incentive for consolidating and vertical integrating of harvesting and processing operations, the sustainability of public- and leased-bottom fisheries, household production and the structure of fishery-dependent communities, the net use and option benefits of recreational and amenity services, and non-use benefits. The economic and socioeconomic risks associated with the management options and the adequacy of available data to assess the magnitude and significance of risk to social, cultural, and economic systems and institutions are also examined below.

Implementation risk includes the risk of political objection, the consequences of management actions that may differ from the intent of those actions, the possibility that actions taken by one regulatory entity might adversely affect the efficacy of actions taken by another regulatory entity, and the likelihood that the selected management option might spur unauthorized introductions. Endangered Species Act Section 7 consultations are an example of an assessment of the risk that implementation of a proposed action could jeopardize the recovery of an endangered species or adversely modify its critical habitat. The adequacy of information available for assessing implementation risks is the final consideration discussed below.

RISK FACTORS

Ecological Risk

Disease

The possibility that a new disease-causing organism might be introduced along with *C. ariakensis* has been one of the major concerns of all agencies and individuals involved in the deliberations about a possible introduction. This fear is not unwarranted since a number of disease outbreaks in oysters have been linked to the introduction and transfer of molluscs for commercial culture (Rosenfield and Kern, 1979; Andrews, 1980). Adherence to the International Council for the Exploration of the Sea protocols, discussed in more detail below, would significantly reduce the risk of bringing in a new disease-causing organism as a consequence of introducing a nonnative oyster. Some of the linkages are more robust than others, and some researchers are more confident of a causal relationship than others. Manifestation of disease following an introduction could result from the concomitant accidental introduction of an exotic pathogen or through the susceptibility of the introduced oyster to an endemic pathogen (Sindermann, 1990). The method mostly commonly envisioned is that the introduced oyster would bring a new pathogen, which would infect the native oyster (or other species). This might occur even though the introduced oyster displayed no disease symptoms. It might be extremely difficult even to detect the pathogen because so few oysters are infected because so few pathogens are present in any individual, or both. Since the native oyster would never have been exposed to the parasite, it would likely be highly susceptible, experience high infection rates, and suffer heavy mortalities. Less frequently considered but equally plausible is the possibility that the introduced oyster would be exposed to an enzootic or "resident" pathogen, which might or might not cause problems in the native oyster (or other) species. In this case it is the introduced oyster that would develop disease. The causative pathogen may never have been recognized simply because it never caused a problem in native species. It must also be stressed that pathogens can be transported by means that are not related to a nonnative introduction for fishery or aquaculture purposes, including water currents, ballast water, other (vector) animals, or discards from restaurants and processing plants. Case studies illustrating various scenarios are presented here.

Haplosporidium nelsoni (MSX Disease)

When *H. nelsoni* was first identified in 1958 as the cause of massive oyster mortalities in Delaware Bay, it was a new parasite to the investigators who saw it in tissue sections of the affected oysters. Mortalities of the scale caused by the parasite in Delaware Bay and later in Chesapeake Bay had never before been recorded in those estuaries, and it was assumed that the parasite was new to the region (Ford and Tripp, 1996). Later, two separate investigations (Katkansky and Warner, 1970; Kern, 1976) reported the finding of a parasite that was morphologically identical to *H. nelsoni* in the tissues of the Pacific oyster, *C. gigas*, in California and Korea. Friedman and Hedrick (1991) and Friedman (1996) found what appeared to be the same organism in *C. gigas* from Japan and California. The observed prevalence of *H. nelsoni* in these samples of *C. gigas* was low (<2%), and no commercially noticeable mortalities of *C. gigas* were reported. Burreson et al. (2000) determined that the small subunit ribosomal DNA sequence of *H. nelsoni* is identical to that of the parasite in *C. gigas* and concluded that *C. gigas* was the source of *H. nelsoni*, which was highly pathogenic in *C. virginica* even though it caused little damage in the original host. Further, they strongly suggested that the parasite was introduced in shipments of *C. gigas* for commercial trials. However, they noted a lack of known introductions of *C. gigas* into the mid-Atlantic in the years immediately preceding the initial outbreaks of *H. nelsoni* and acknowledged the possibility of other mechanisms of introduction. Subsequent citations of Burreson et al. (2000), including many of the position papers and documents prepared by agencies concerned with the possible introduction of *C. ariakensis*, state, *without qualification*, that *H. nelsoni* was introduced in "unauthorized" oyster shipments. There is little argument that *H. nelsoni* came from the Pacific where it infects *C. gigas*, but the pathway of its introduction is simply not known. Introduced *C. gigas* might well have been the source, but other possibilities must be considered. Particularly noteworthy is the great increase in ship transit between Pacific and Atlantic ports that occurred during and after World War II. Shipping could have introduced *H. nelsoni* via infected *C. gigas* attached to ships hulls or via release of *H. nelsoni* spores in the discharge of ballast water. The spore is a thick-walled stage in the life cycle of *H. nelsoni*; its role in transmission is not known, but the spore in other species is typically a transmission stage that can remain "dormant" for long periods and is highly tolerant of environmental extremes. MSX must be considered a disease in which the causative agent was surely introduced but for which the means of introduction is unknown.

Perkinsus marinus (Dermo Disease)

Although *P. marinus* was first described and associated with oyster mortalities in the Gulf of Mexico in the late 1940s, it had probably infected oysters in the southern United States for a long time (Mackin and Hopkins, 1962). The parasite was identified in tissue slides of oysters collected in the 1930s, and mortalities similar to those caused by *P. marinus* were reported in documents dating back to the early part of the 20th century. Further, this parasite was identified in tissues of oysters in the southeastern United States, including the Virginia portion of Chesapeake Bay, as soon as investigators looked in the late 1940s and early 1950s. It was not detected in the northeastern United States, however, and numerous studies documented that the parasite multiplied and killed most readily at elevated water temperatures. However, coincident with large-scale commercial transplanting of infected oysters from Virginia into the New Jersey portion of Delaware Bay for several years in the mid-1950s, the parasite was found in nearby native Delaware Bay oysters, although it caused no mortalities (Ford, 1996). When *H. nelsoni* began killing oysters in the bay in 1957, an embargo was placed on all imports and exports, and after several years *P. marinus* was rarely detected in Delaware Bay oysters.

It was argued that without repeated introductions of the parasite, Delaware Bay water temperatures were simply too low for the parasite to maintain a self-sustaining population. Nevertheless, in 1990, a severe outbreak of *P. marinus* infections and consequent oyster mortalities began in the bay, without the concomitant transfer of oysters from outside the estuary. That outbreak and other outbreaks over a 500-km range north to Cape Cod, Massachusetts, which followed over the next 2 years, occurred during a period of unusually warm water temperatures. Although this range extension was not associated with contemporaneous transplantation of oysters, historical transfers of oysters from the south to overfished regions of the north would probably have introduced *P. marinus* repeatedly over many earlier decades. In this new, colder region the parasite may have persisted at low and undetectable levels until the temperature became warm enough to stimulate an epizootic, as occurred in Delaware Bay. This account of Dermo disease illustrates what appears to be an example of a pathogen that can be present but suppressed by unfavorable environmental conditions (low temperature), then stimulated to become epizootic when conditions become favorable.

Marteilia refringens (Marteiliosis, Aber Disease)

M. refringens infects and causes severe disease in the European oyster, *Ostrea edulis*. Mortalities ascribed to the pathogen were first noted in Brittany, northern France, in 1967 and 1968 and over the next several years, caused mortalities in most Breton culture areas (Balouet et al., 1979). It is now found south along the French coast and into Spain, Portugal, and the Mediterranean. Like *H. nelsoni*, the mechanism of transmission and origin of *M. refringens* are unknown, although recent studies indicate that a copepod may be a second host (Audemard et al., 2002). *M. refringens* was a "new" parasite when first associated with the *O. edulis* mortalities; neither it nor any similar parasite had been seen before. Some authors (Andrews, 1980; Farley, 1992) have pointed to the propinquity of the marteiliosis outbreak and the inception of *C. gigas* introductions into France in the mid-1960s (Grizel and Héral, 1991) as a cautionary example of a disease agent brought with an introduced host. Others are less sure of the connection (Grizel and Héral, 1991). The timing of the introduction of *C. gigas*, which was found to have very low prevalence of a Marteilia-like parasite (Cahour, 1979) and the outbreak of Marteiliosis in *O. edulis*, may be simply a coincidence. If it does represent a cause-effect situation, the linkage is less strong than in some other instances. Interestingly, the mussels *Mytilus edulis* and *M. galloprovincialis*, which inhabit the same European waters as *O. edulis*, are infected by very similar parasites (Berthe et al., 2000; Le Roux et al., 2001; Longshaw et al., 2001). An alternative explanation for the appearance of *M. refringens* is that *M. maurini* "jumped" to a new host, *O. edulis*, concomitant with a gene mutation.

Bonamia ostreae (Bonamiosis)

B. ostreae also infects and causes mortality in *O. edulis*. It came to the attention of molluscan disease specialists when it was implicated as the cause of epizootic mortalities of *O. edulis* in Brittany, France. The mortalities were first noted at one site in mid-1979 and later during the same year in other Breton oyster farms (Balouet et al., 1983). Within a year or two *B. ostreae* was found to be causing oyster deaths in Ireland, England, the Netherlands, and Spain. In conjunction with *M. refringens*, *B. ostreae* continues to depress European oyster production.

The probable movement of *B. ostreae* through oyster shipments has an interesting history. Although it was described simply as a "microcell" at the time, the same parasite was found in the mid-1960s in *O. edulis* reared at the Milford Laboratory in Connecticut (Farley et al., 1988). Some of the Milford progeny were shipped to California where the microcell was later observed in dead and dying oysters (Katkansky et al., 1969). Seed oysters from at least one California hatchery were then transferred to Washington state, Maine, Spain, and France (Katkansky and Warner, 1974; Elston et al., 1987; Figueras, 1991; Friedman and Perkins, 1994). *B. ostreae* has since been detected in *O. edulis* grown in California, Washington, and Maine (Elston et al., 1986; Friedman and Perkins, 1994; Zabaleta and Barber, 1996). Since the parasite is directly transmissible between oysters, a case can be made that the source of *B. ostreae* responsible for the European outbreak was the "large amounts of *O. edulis* seed transferred to France in the years prior to the detection of the disease there" (Elston et al., 1986). The finding of *B. ostreae* in *O. edulis* shipped from the Milford Laboratory to California, where the oysters experienced heavy mortality, and the presence of the same parasite in oysters from the Milford Laboratory held in quarantine at the Oxford, Maryland, National Marine Fisheries Laboratory, suggests that the parasite may have been introduced from the East to the West coasts of North America (Farley et al., 1988). The original shipments of *O. edulis* to the Milford Laboratory came from the Netherlands, where oyster mortalities caused by *B. ostreae* were reported only after the French outbreaks. The evidence available thus suggests that the parasite was probably not native to Europe at the time those shipments were made (see Table 3.2). *B. ostreae* is thus an example of a disease agent that is known to be transmitted directly between oysters, with a relatively well-documented linkage with the movement of the host oyster for commercial purposes.

Vibrio tapetis (Brown Ring Disease)

Brown Ring Disease is a disease that has caused mortalities of the Manila clam, *Ruditapes philippinarum,* in Western Europe (Paillard et al., 1994). It is caused by a marine bacterium, *Vibrio tapetis,* and is characterized by the deposition of a ring of organic material (periostracum) on the inner edge of each valve. Manila clams were introduced to France from the U.S. West Coast for aquaculture because they grow faster than the native clam, *R. decussatus.* For several years Manila clams did extremely well in culture and became established in the wild, but in 1987 heavy mortalities associated with the brown ring symptom were reported in culture parks in northern Brittany. The disease spread north and south with Manila clam culture in Europe and is now found in England, Spain, and Portugal as well as France, and it affects naturalized populations of *R. philippinarum* as well as those under culture. It is not found in the northwestern United States or in the native range of the Manila clam along the Asian Pacific coast. The pathogen *Vibrio tapetis* can be found in environmental samples and the native clam, *R. decussatus,* is highly resistant to the disease (Maes and Paillard, 1992). Brown Ring Disease appears to be an example of an introduced host being highly susceptible to a resident pathogen, which has little effect on the native clam.

Quahog Parasite X (QPX Disease)

QPX is a newly recognized disease agent in the hard clam, *Mercenaria mercenaria* (Whyte et al., 1994). Although it has yet to be given a scientific name, QPX is a member of the phylum Labyrinthulomycota, a group of microorganisms that live in marine and estuarine environments on micro and macro algae and detritus. Sometimes they are pathogenic, and they have been associated with mortalities of molluscs in captivity or under culture (Ford, 2001). QPX has been found in clams from Virginia to the Canadian maritime provinces, but it appears to be most serious in the northern culture areas. Recently, it has become evident that serious outbreaks in clams being cultured in New Jersey and Virginia occurred in stocks that had come from South Carolina and Florida and not in those produced locally. This observation is supported by results of an experiment in which brood stock originating in five states from Florida to Massachusetts were spawned in a single hatchery and their offspring grown in side-by-side plots in Virginia and New Jersey (Ragone-Calvo and Burreson, 2002). In both grow-out sites, clams from Florida and South Carolina acquired numerous and heavy QPX infections and suffered high mortalities. Those from Massachusetts and New Jersey had few infections and low mortality. Clams from Virginia exhibited intermediate traits. QPX outbreaks appear to be an example of a disease caused by an enzootic parasite, which may evoke no detectable problems in a resident host but is highly pathogenic to nonlocal stocks of the same species.

Considerations for Disease Risk Assessment

Implications of Detecting Pathogens in Disease Surveys

One of the questions raised about the potential introduction of *C. ariakensis* is the lack of knowledge of parasites and diseases of the species in its native range. A number of samples have been examined at the Virginia Institute of Marine Science (VIMS) and more are planned (see Chapter 4). So far the only identified pathogen has been a haplosporidian found in one of 155 oysters (0.6%) in a sample from northern China. This very low prevalence is not unusual for parasites that are present in "resistant" hosts but can cause epizootic mortalities when they infect susceptible hosts (e.g., *H. nelsoni* in *C. gigas* versus *C. virginica*). It is worth noting that the prevalence of *H. nelsoni* in various samples of *C. gigas* (considered "resistant" to *H. nelsoni*) from Korea, Japan, and California averaged <1%

(Kern, 1976; Kang, 1980; Friedman and Hedrick, 1991; Friedman, 1996; Burreson et al., 2000). After it appeared on the East Coast of the United States, infection prevalence in *C. virginica* was commonly 50 to 90% (Ford and Tripp, 1996). Thus, low prevalence in one host does not mean that, if introduced, a parasite would be harmless in a new environment.

Effectiveness of Diagnostic Methods

The standard diagnostic method for the detection of infectious agents in molluscs is the examination of fixed and stained tissue sections by light microscopy. The method is good for relatively large parasites such as worms or most protozoans, although the amount of tissue examined in the section is only a very small fraction of the total available and very light infections can easily be missed. Nevertheless, this is the technique that detected the low prevalence of a haplosporidian in *C. ariakensis* in China, the marteiliodes-like parasites in the same species from Hong Kong, as well as the low prevalences of *H. nelsoni* in *C. gigas*. In the last case, sample sizes of hundred to thousands of specimens were examined to document prevalences of <1%.

Tissue section histology is much less effective for the smaller organisms like bacteria, unless they form aggregates, such as is the case for the bacterium that causes "nocardiosis" in *C. gigas* (Elston et al., 1987; Friedman and Hedrick, 1991). Viruses are particularly difficult to detect using light microscopy unless infections cause obvious cellular damage in the host. Tissues displaying damage not attributable to an obvious cause can be examined by electron microscopy, which can detect viruses but is extremely labor intensive and expensive. Molecular techniques can detect bacteria and viruses not easily found using other methods, but they can also suffer from a need to limit the amount of tissue examined to a relatively small fraction of the total animal. Further, they require that the organism already have been isolated and characterized in order to develop the appropriate assay tools. Thus, such methods are not particularly helpful in screening for "unknown" pathogens. A DNA-based assay is available for the herpes virus that is associated with larval and juvenile bivalve mortalities in hatcheries and nurseries and is being employed at VIMS to examine *C. ariakensis* from China.

The International Council for the Exploration of the Sea Code of Practice

The greatest risk of *C. ariakensis* introducing a disease agent would occur if seed or adult oysters were to be placed in Chesapeake Bay directly from another location (e.g., "rogue" introductions from China or the U.S. West Coast). If, on the other hand, the International Council for the Exploration of the Sea (ICES) protocols (see Appendix F and *http://www.ices.dk/products/misc.asp*) are followed, the chances of introducing a pathogen, such as the parasites that have been cointroduced with oysters in the past, would be greatly minimized. The guidelines direct that the exotic oysters would be used only as brood stock in a quarantined hatchery and would be destroyed after spawning. Only the offspring, produced in the hatchery, would be placed in Chesapeake Bay and strictly monitored for evidence of disease. Hatchery protocols should further ensure that gametes and embryos are well cleansed to remove any parasites potentially included in the gamete mix. Although washing would never be 100% effective, the operation would provide an added measure of safety to the already strict ICES protocol. Neither the ICES protocol nor any washing procedure would, however, prevent the transmission of pathogens that infect the gametes themselves and are passed directly (vertically) to the offspring. Most such pathogens are viruses. Further, ICES protocols cannot prevent disease outbreaks caused by pathogens that are transported by means other than the commercial species under consideration for introduction. Among the case studies presented earlier, only the movement of *B. ostreae* would almost certainly have been prevented by proper use of ICES protocols. On the other hand, the ICES protocols would not have prevented the introduction of *H. nelsoni* into the United States or the northward-range expansion of *P. marinus* if these events resulted from commercial shipping, transport by water currents, or overboard disposal of infected animals by restaurants, processors, or private individuals. Nor would ICES guidelines have prevented disease outbreaks caused by "resident" parasites such as QPX or the bacterial agent of Brown Ring Disease.

A papovavirus is known to infect eggs of *C. virginica*, causing a condition known as viral gametocytic hypertrophy or VGH (Farley, 1978). The virus particles replicate inside the gametes, eventually resulting in large "inclusion bodies" that are easily visible under the microscope. It is not known whether the virus is transmitted vertically, but the prevalence of VGH is extremely low and causes no mortality. Viral diseases of oyster larvae and juveniles (e.g., herpes virus) have been described in hatcheries and nurseries in a number of countries. Although herpes virus has been found in the gonads of adult oysters (Arzul et al., 2002), its presence inside eggs, which would be evidence of true vertical transmission potential, has not been demonstrated (T. Rheault, Moonstone Oysters, Pt. Judith, RI, personal communication, 2002). It should be noted that herpes virus was found in *C. virginica* on the East Coast as early as 1972 in oysters that had been subjected to elevated temperatures in a power plant discharge pipe in Maine (Farley et al., 1972).

No bacteria have been reported to infect gametes of molluscs. On the other hand, protozoans belonging to microsporidia have been found infecting the eggs of several species of oysters, including *C. gigas* and *O. edulis* (Bower et al., 1994), and a *Marteilioides* has been found in the eggs of *C. gigas* (Comps et al., 1986). Vertical transmission of any of these parasites has never been reported; however, it should be noted that microsporidia can be vertically transmitted via the eggs of a number of insect and small crustacean (amphipods and copepods) species (Raina et al., 1995; Dunn et al., 2001; Bigliardi and Carapelli, 2002).

Crassostrea ariakensis as a Reservoir for Enzootic Pathogens

One concern raised about the possible introduction of *C. ariakensis* is that it may act as a reservoir for *H. nelsoni* and *P. marinus*, even though it does not succumb to the parasites. Although the possibility cannot be dismissed, the reservoirs for both of these disease agents (*C. virginica* and perhaps other molluscs in the case of *P. marinus* and "unknown" in the case of *H. nelsoni*, whose mode of transmission is unknown) are, and historically have been, plentiful enough to result in extremely high infection levels in oysters wherever environmental conditions are permissive. If *C. ariakensis* did act as a reservoir, its low parasite burdens would make it insignificant compared to the reservoirs already present.

Crassostrea ariakensis as Host for a "Native" Pathogen

The possibility that *C. ariakensis* might acquire a resident pathogen should not be ignored in the debate about its possible introduction. For instance, *C. ariakensis* may be highly susceptible to *B. ostreae*. While being held in a hatchery in southwestern France, a cohort of *C. ariakensis* suffered high mortalities associated with infections of a parasite identical morphologically to *B. ostreae* (Cochennec et al., 1998). The hatchery received raw seawater from a locality where *B. ostreae* was present in the native *O. edulis*, and it is probable that the *C. ariakensis* became infected and died with heavy *B. ostreae* infections (see Chapter 4). In this connection it should be recalled that *B. ostrea* is present, albeit at low infection levels, in *O. edulis* in Maine (Friedman and Perkins, 1994; Zabaleta and Barber, 1996).

The Risk of Disease

C. ariakensis has performed extraordinarily well in field trials conducted in the lower bay (Calvo et al., 1999). It grows rapidly and does not succumb to MSX or Dermo diseases (see Chapter 4). If ICES guidelines are strictly followed in an introduction process, the risk of introducing a new pathogen will be almost completely eliminated. However, the "risk of disease" associated with a possible introduction is not zero. Despite its admirable performance so far, *C. ariakensis* should not be considered disease resistant. Its susceptibility to the bonamia-like parasite described above is a case in point. Even *C. gigas*, known for being resistant to the parasites that have devastated both *C. virginica* and *O. edulis*, suffers mortalities in Japan, the West Coast of the United States, and more recently in France from causes not well understood (Koganezawa, 1974; Cheney et al., 2000; Lacoste et al., 2001). Finally, large-scale aquaculture in which juveniles are reared at high density with limited water exchange typically favors disease outbreaks in hatcheries and nurseries regardless of the species involved.

Ecological Risks Directly Associated with *C. ariakensis*

Since the potential ecological risks and benefits of nonnative species are difficult to quantify, it is not surprising that scientists differ on the value of deliberate introductions of species (Ewel et al., 1999). For example, some believe that the need for restoring ecosystem function is so great that concerns about possible harmful effects of deliberate introductions are not warranted. Others, in contrast, place primary emphasis on the biological, economic, and social costs of the introductions. All recognize that once an aquatic species is introduced it is virtually impossible to control its spread. Marine ecosystems have few biogeographic barriers, and the dispersal capabilities of nonnative species do not necessarily coincide with political and economic boundaries. A species that is desirable in one location may be regarded as a nuisance or as undesirable in another. There is also increasing evidence that the human alterations of ecosystems often influence the probability that an introduced species will become invasive and that time lags of several decades or longer often exist between the initial introduction of an organism and when that species becomes a nuisance.

The major ecological concerns centered on the proposed introduction of *C. ariakensis* deal with illegal or rogue plantings or placing reproductively viable diploid oysters (even with adherence to ICES guidelines) into the bay. If reproductively viable diploid organisms are introduced, the primary issues are fourfold: Where will *C. arakensis* grow in the bay and how might the oyster affect other resident species, especially the native Eastern oyster? Will *C. ariakensis* provide similar ecosystem services to the bay as the native oyster? Will *C. ariakensis* become a "nuisance species," which would result in negative impacts on the bay's ecosystem? What are the chances of the nonnative oyster dispersing to regions outside the bay? If an illegal introduction of *C. ariakensis* occurs, there is increased concern that disease agents or other species (e.g., "hitchhikers"), which may be attached to the oysters, would be introduced into the bay. The primary ecological risk associated with deploying triploid oysters in the bay is the probability that a self-sustaining population of nonnative oysters will be established there, because of the direct introduction of a small percentage of diploid individuals among mated triploids or the reversion of triploids to diploids. If this should happen, the ecological risks would be similar to those of introducing reproductively viable oysters into the ecosystem. Assessing the relative risk of the aforementioned issues is severely constrained by the lack of fundamental ecological information on *C. ariakensis*. Little is known about the ecology of the oyster in its native range and how it will interact with other species if it is introduced into the Chesapeake Bay.

Is C. ariakensis *capable of establishing reproductively viable populations in the bay? If so, will it compete with* C. virginica *or other resident species in the bay?*

C. ariakensis is well adapted to living in estuarine habitats, which characteristically experience a wide range of temperature and salinity variation and contain high levels of suspended material concentrations in the water column (Chapter 4). Field trials using triploid oysters conducted at several sites in lower Chesapeake Bay indicate that *C. ariakensis* grows well under a relatively wide range of salinity conditions (Calvo et al., 2001). Also, preliminary laboratory studies indicate that *C. ariakensis* is capable of reproducing over a similar salinity range as *C. virginica* (M. Luckenbach, Virginia Institute of Marine Science, Gloucester Point, personal communication, 2002). Coupled with its relatively rapid growth and resistance to MSX and Dermo diseases, it is very likely that *C. ariakensis* is capable of establishing wherever *C. virginica* was established historically in the Chesapeake Bay, with the exception of areas where sedimentation now prevents or inhibits larval settlement.

From the limited information available, it appears that environmental conditions tolerated by *C. ariakensis* broadly overlap those favored by *C. virginica*; there is little evidence suggesting the two species would occupy different types of habitats within the bay ecosystem. While there is conflicting information about the reef-building characteristics of *C. ariakensis*, a number of records indicate the presence of *C. ariakensis* reefs in certain coastal areas in China (Chapter 4). Given their functional and ecological similarities, it seems likely that both oyster species will utilize similar food and spatial resources. The intensity of competition between the species and whether *C. ariakensis* will outcompete *C. virginica*, however, is more difficult to predict. Often when ecologically similar marine species are competing for space, those exhibiting faster growth rates competitively dominate slower-growing species (Branch, 1984); however, there are exceptions to this pattern (e.g., Lang, 1973; Lang and Chornesky, 1990). M. Luckenbach (Virginia Institute of Marine Science, Gloucester Point, personal communication, 2002) has performed a series of interspecific competition experiments in the laboratory with juvenile life stages of both species. He found that *C. virginica* was the more effective spatial competitor, despite the fact that field trials indicate *C. ariakensis* grew much faster than *C. virginica*. Clearly, additional field and laboratory work is needed to more fully understand the nature of interactions between the two species and how variations in environmental factors and the presence of the oyster disease organisms may influence competitive interactions for space. In addition, little is known about the reproductive cycle of *C. ariakensis* in the bay and whether the species would spawn sooner and set more heavily than *C. virginica*. For species that are competing for similar settlement substrates, the outcome of competition is often controlled by the order in which they colonize a habitat (e.g., Osman, 1977; Sutherland and Karlson, 1977). However, if *C. ariakensis* became very abundant, it could act as an important competitor with other fouling species (e.g., barnacles, ascidians, bryozoans) resident to the bay. Also, if *C. ariakensis* does not form three-dimensional reeflike structures and the population expands by growing horizontally over the seafloor, it could have negative impacts on other resident species. For example, as zebra mussels expanded their populations in the Great Lakes by first colonizing hard substrates (e.g., rocks, pilings, intake pipes), they subsequently began forming dense aggregations across sedimentary habitats (Berkman et al., 2000). If *C. ariakensis* displayed a similar type of habitat shift following its introduction, it could begin outcompeting and/or smothering infaunal bivalve species (e.g., *Macoma balthica, Mya arenaria*) that are important food items for blue crabs.

Both oysters are also very likely to compete for food resources. Again, the degree of competition is difficult to discern given the general lack of knowledge about the physiological and feeding ecology of *C. ariakensis*. R. Newell (Center for Environmental Science, University of Maryland, Cambridge, personal communication, 2002) has conducted some preliminary laboratory experiments comparing the clearance rates of both species, finding that both exhibit similar feeding rates. This result is somewhat surprising given *C. ariakensis* grows much faster than *C. virginica*. Bayne (2002), for example, demonstrated that the competitive advantage the nonnative Pacific oyster, *C. gigas,* has over the native Australian oyster, *Saccostrea glomerata* was due to faster rates of feeding, particularly at high food concentrations, which resulted in greater metabolic efficiencies for both feeding and growth. Possibly, *C. ariakensis* has a higher metabolic or assimilation efficiency than *C. virginica,* which translates into a faster growth rate.

The lack of ecological information on *C. ariakensis* also greatly limits the ability to predict how it would impact other species in the bay. If *C. ariakensis* occupies a niche similar to that of the native oyster, it seems unlikely that *C. ariakensis* would outcompete other resident species. However, if *C. ariakensis* populations became very abundant, they could compete with other fouling organisms in the bay. Also, if the Suminoe oyster does not form reef structures, it could colonize habitats currently occupied by other sedentary species.

Another serious threat is the impact that a rogue introduction of the oyster would have on the bay ecosystem, since it is highly likely that other species attached to the oyster shell would also be introduced along with the oyster. There are many examples of the impact that "hitchhiking" species have had on coastal ecosystems (see Chapter 3).

Will C. ariakensis *provide ecosystem services similar to those provided by the native oyster?*

As mentioned in Chapter 4, very dense and spatially extensive populations of suspension-feeding bivalves can provide a number of estuarine ecosystem services. These include bio-filtering of water and removal of suspended materials and improving water clarity. Reef-building suspension-feeding species can provide important habitats for economically important species such as striped bass and blue crabs. The reefs also act to enhance biodiversity relative to surrounding soft sediment habitats. It seems likely that *C. ariakensis* is capable of providing similar types of ecosystem services as the Eastern oyster if sufficient population densities existed in the bay. While both oyster species are functionally similar, the degree to which *C. ariakensis* could influence the bay ecosystem in a manner similar to *C. virginica* would depend on the amount of oyster biomass in the bay and the role of oysters in the ecosystem.

Most of the ecosystem services provided by *C. ariakensis* will consist of the benefits provided through water filtration activities. The degree and types of ecological services provided by aquaculture of triploid Suminoe oysters would depend on the spatial scale in an individual commercial operation and on how the oysters are grown to marketable size. In situations of intense aquaculture grow-out with high concentrations of oysters, one should see positive benefits on water clarity. However, aquaculture of triploid nonnative oysters is likely to be limited in spatial extent initially, owing to high production and biosecurity costs, and thus unlikely to contribute substantially to total oyster filtration capacity of the bay. Nevertheless, if aquaculture of triploid *C. ariakensis* proves successful and more grow-out areas in the bay are established, presumably the ecosystem services provided by biofiltration would increase proportionately (e.g., Bayne and Warwick, 1998). Water clarity could be magnified locally if intensive aquaculture operations are located in some of the smaller "trap" or retentive estuaries and tributaries within the bay that have more restricted water movement. Finally, over the longer term, were on-bottom aquaculture of nonnatives permitted on a large scale—entailing a much greater risk of diploid establishment—more profound impacts on filtration capacity and water quality could be expected to follow. Indeed, there is a counterexample; intensive on-bottom aquaculture in France exceeded the carrying capacity of the areas (e.g., Grant et al., 1993; Raillard and Menesguen, 1994), decreasing oyster growth rate and depleting the food available to other filter-feeding organisms in the area. This is also likely if aquaculture operations were located in some of the smaller "trap" or retentive estuaries and tributaries within the bay that have more restricted water movement. If aquaculture of triploid *C. ariakensis* proves successful and more and more grow-out areas in the bay are used, presumably the ecosystem services provided by biofiltration would increase proportionately (e.g., Bayne and Warwick, 1998). The reef-type ecosystem services (e.g., habitat for other species) provided by *C. ariakensis* would be based on the length of time the oysters were grown in the water and whether they are grown on the seafloor or suspended in the water column. In both grow-out scenarios these services would be limited both spatially and temporally when compared to naturally occurring oyster reef habitat. Presumably, mobile organisms would be attracted to the oysters or the structures in which the oysters were being reared. There could be negative impacts on the bay ecosystem if culturing operations proliferated in the bay without following best management practices to minimize impacts on other parts of the ecosystem. For example, Everett et al. (1995) concluded that oyster stake culture methods adversely impacted eelgrass through increased sedimentation and physical disturbance associated with planting and harvesting. Limiting stake or other culture structure density in areas with submerged aquatic vegetation (SAV) could mitigate negative effects while still allowing oysters to provide the valuable ecological services noted above and in Chapter 4. Peterson and Heck (2001) and Heck and Orth (1980) demonstrated that benthic mussels (*Modilous americanus*) in Saint Josephs Bay, Florida, when cultured at appropriate densities, provide a variety of ecological functions that enhanced seagrass (*Thalassia testudinum*) productivity. Similar effects have been reported by clam farmers on the Eastern Shore of the Chesapeake and by shellfish farmers in the Pacific Northwest. As with bottom culture, aquaculture activities relying on suspended grow-out techniques, sited inappropriately or cultured at too high a density, may have negative impacts. Excessive accumulation of biodeposits and shell rubble may affect the benthic habitat beneath these operations (ICES, 1988). Enhanced amounts of anoxic sediments have occurred in several shallow bays in Japan as a result of oyster culture where large amounts of pseudofeces are produced (Nose, 1985). Conversely, when cultured at lower densities, suspended shellfish culture was shown to have little impact on the benthic environment in Tasmania (Crawford et al., 2003). On one culture site in Tasmania, dense beds of eelgrass were observed under suspended oyster trays as well as outside the boundary of the farm. Accumulations of shell rubble may alter benthic species composition and provide substrate for oyster settlement.

If a "wild" fishery for *C. ariakensis* was established in the bay through the introduction of reproductively active diploid oysters, the degree of ecosystem services is again dependent on the amount of oyster biomass and the extent of population distribution in the bay. As mentioned previously, it may have taken the Eastern oyster hundreds of years, with minimal fishing pressure, to form the extensive reefs in the bay's tributaries (Hargis, 1999). If *C. ariakensis* was not harvested for a number of decades, sufficient quantities of oysters might develop that would mimic ecosystem services provided by the Eastern oyster prior to its decline through overfishing, disease, and habitat degradation. The extent and time frame of this possible event are also highly dependent on how well *C. ariakensis* could adapt to the bay's environmental conditions and how quickly wild populations of oysters would become established and proliferate into sufficiently dense reefs to have an effect like that of the historical native oyster population.

Will C. ariakensis *become an invasive or nuisance species that may negatively impact the bay ecosystem?*

As mentioned in Chapter 3, it is very difficult to predict which species will become an invasive or nuisance species and which will not. While some attributes are hypothesized for successful aquatic invaders (see Table 9.1), few generalizations have been confirmed, and each has exceptions (e.g., Simberloff, 1989; Lodge, 1993; Ricciardi and Rasmussen, 1998). These attributes can, however, be used to provide a general guideline to facilitate the identification of potential invasive species. For example, species possessing wide environmental tolerance limits and natural mechanisms for rapid dispersal (e.g., zebra mussels, *Dreissena polymorpha*) are likely to colonize a large geographic range. Other studies have shown than species possessing broad ranges of distribution are often a good predictor of invasive ability (e.g., the Asian clam, *Corbicula*

TABLE 9.1 Some Hypothesized Attributes of Aquatic Invasive Species

1. Broad environmental tolerance
2. Rapid growth
3. Early age (size) of reproductive maturity
4. High reproductive capacity
5. Possessing multiple mechanisms of dispersal
6. Release from natural predators, parasites, and diseases
7. Short generation time
8. Broad diets
9. Gregariousness
10. Abundant and broadly distributed in native range

SOURCE: Based, in part, on Groves and Burdon (1986), Ehrlich (1986), Morton (1987, 1997), and Lodge (1993).

fluminea [Morton, 1997]; water hyacinth, *Eichornia crassipes* [Groves and Burdon, 1986]).

Many of the attributes that make *C. ariakensis* an attractive species for the establishment of a fishery in Chesapeake Bay are the same characteristics that have been attributed to aquatic nuisance species. For example, *C. ariakensis* has a very broad native distributional range (Chapter 4) and an equally wide range of tolerance to environmental conditions such as variations in salinity and temperature. It lives in estuarine habitats, both intertidally and subtidally, and is tolerant of turbid and eutrophic water conditions. *C. ariakensis* also grows relatively fast, has a high reproductive capacity, and is able to reproduce within several months following larval settlement. While nothing is known about what species may prey on *C. ariakensis*, it does appear immune to the oyster diseases that now plague the Eastern oyster in the bay. A number of oyster predators that reside in the bay (e.g., blue crabs, flatworms, cownose rays) might act to control the growth of the oyster population. As mentioned in Chapter 4, the shell of *C. ariakensis* does not appear to be a strong as *C. virginica*, which may make it more vulnerable to shellfish predators (particularly crab predators). It should be noted that some mollusc species have the ability to rapidly respond to the effects of crab predators by increasing the thickness of their shells (e.g., Trussell and Smith, 2000).

Several studies have noted that one of the most consistent attributes of a invasive species is the use of dispersal mechanisms that involve human activity (e.g., Ehrlich, 1986; Carlton and Geller, 1993; Morton, 1997; Ricciardi and Rasmussen, 1998). *C. ariakensis* possesses a planktonic larvae that could easily be transported in ship ballast water, and adults can attach to the hulls of ships. In addition to these unintentionally transported vectors, it seems highly likely that adult oysters will be intentionally transported by human activity as part of normal aquaculture or fishery practices. The relative importance of these vectors is generally dependent on whether the species is capable of establishing reproductively viable populations.

If the nonnative oyster became invasive and the population was not kept in check by harvesting or by native predators, it is conceivable that *C. ariakensis* could reach sufficient densities to shift the bay ecosystem back toward benthic dominance rather than pelagic dominance. Of course, the same thing could happen if the native oyster rebounded, but this is considered unlikely due to disease and harvest pressure. Reducing standing stocks of phytoplankton might facilitate improvement of water quality and reduce populations of gelatinous zooplankton (Chapter 4). An increase in SAV could have beneficial secondary effects on associated invertebrates and waterfowl. Altering the bay from pelagic to benthic dominance may also result in shifts in species composition and abundance at higher trophic levels. For example, pelagic finfish (e.g., menhaden, striped bass) populations may be reduced, while species that directly or indirectly rely on benthic productivity (e.g., sheepshead, bluefish) may be positively affected (see Baird and Ulanowicz, 1989). Rapid population expansion of the nonnative species may also displace native oysters and other fouling species. Rapid population expansion, however, could enhance rates of denitrification and alter water clarity in the bay. Lastly, the nonnative oyster could become a major fouling species, thereby increasing the economic costs associated with maintenance of water input pipes, boat hulls, and so forth.

What are the chances of the nonnative oyster dispersing to areas outside the bay?

If reproductively viable populations of *C. ariakensis* are established in the bay, it is highly likely that individuals will eventually spread outside the bay. As mentioned previously, the species is capable of dispersing through a variety of unintentional and intentional mechanisms (e.g., larval transport by water currents, transport of larvae and adults by ship traffic, human movement of adults) that will act to amplify its spread to regions outside the bay.

Rates of dispersal of the species outside the bay are difficult to predict. If dispersal is primarily through transport of larvae in the water column, movements will be dependent on the prevailing water circulation patterns, the degree of water column stratification, and flushing time and larval behavior (e.g. Dekshenieks et al., 1996). Once outside the bay, oyster larvae would be transported by prevailing long-shore current systems. There is limited evidence that larval swimming behavior of *C. ariakensis* may differ from *C. virginica* (M. Luckenbach, Virginia Institute of Marine Science, Gloucester Point, personal communication, 2002).

While the ecological ramifications of this behavior are unclear, it may result in differences in larval dispersal ability compared to the Eastern oyster. Additional work to couple hydrodynamics with larval behavior and transport, both inside and outside the bay, is needed before any attempts at estimating potential rates of dispersal can be made. Studies of larval settlement patterns, postsettlement mortality rates, and growth in natural conditions are also needed.

Dispersal of *C. ariakensis* outside Chesapeake Bay by other vectors is highly likely and may occur over much shorter timescales than larval transport via water currents. For example, larvae could be entrained in ship ballast water and/or attach to the hulls of ships. As vessels move from port to port along the eastern seaboard, larvae may be released with ballast water exchange or from adults attached to the bottoms of the vessels. In addition, intentional movement of the species along the eastern seaboard by humans is likely, especially if the species proves to be an economically attractive one relative to the native oyster.

Risk to Social, Economic, and Cultural Systems

Human Health

Assuming that monitoring of water quality and shellfish sanitation practices are followed, there is no known reason to expect the human health risks of consuming triploid or diploid *C. ariakensis* harvested from the Chesapeake Bay to be any different from those of consuming *C. virginica* harvested from the bay.

Economic Effects

Price

Oyster harvests from the Chesapeake Bay have declined to less than 3% of the total U.S. live, fresh, and frozen supply. Therefore, a doubling or even tripling of Chesapeake Bay oyster harvests over several years is likely to have only minimal impact on U.S. oyster prices. This is likely to be the case with the introduction of hatchery-based triploid *C. ariakensis* or with cautious introduction of diploid *C. ariakensis*. Nevertheless, changes in local harvests may influence price in local markets. The market for oysters is dynamic. Prices and sales volume vary across species, production region, seasonal and intergenerational changes, consumers' preferences, product-form innovations, and marketing efforts. It is unlikely that triploid aquaculture or even the introduction of diploid *C. ariakensis* will result in sufficient increases in production volume to contribute to an observable impact on prices in the shortterm. Media attention associated with the introduction of *C. ariakensis* could lead to increased or decreased consumer demand, depending on the message. However, it must be cautioned that in the event introduction of *C. ariakensis* or recovery of *C. virginica* leads to explosive growth in harvest, the price effect is likely to be negative. It is unlikely that price volatility will change significantly in the short run under any of the proposed options.

Oyster Harvests from Public Bottoms

Assuming there is no directed release to public oyster beds in the short run, the segment of the oyster industry dependent on public bottoms is unlikely to experience significant direct benefits or costs in the near term. This is because restoration activities associated with the no-action option are not expected to result in significant near-term increases in the stock of native oysters, and nonnative triploid-based aquaculture is not expected to result in sufficient production to affect regional and national oyster prices. It is possible that introduction of diploid nonnative oysters could result in rapid colonization of the Chesapeake Bay and provide a basis for a public-bottom fishery, but even if nonnative oyster populations expanded rapidly, it would take several years for significant numbers of adult oysters to recruit to the fishery. It is possible that the traditional public-bottom fishery could be adversely affected to a significant degree if the introduction of nonnative oysters led to the accidental introduction of new diseases or parasites.

Oyster Harvests from Private Bottoms/Aquaculture

Assuming disease, parasites, or other uncontrollable effects are managed to eliminate their likelihood, the introduction of *C. ariakensis* is most likely to have a positive influence on harvests in this sector. Oystermen with private leased bottoms, the majority of which are located in Virginia, are most likely to benefit from any introduction of hatchery-based *C. ariakensis*.

Assuming that sanctioned introductions adhere to ICES protocols and that rogue introductions do not occur, the inception of triploid-based *C. ariakensis* aquaculture will probably have a positive influence on harvests by those watermen with leased bottoms who can adapt to somewhat more intensive aquaculture-based management of their sites.

Processing Sector

The processing sector earns net revenue primarily by adding value to live oysters through processing, distributing, and marketing. With only minimal supply from the Chesapeake Bay, regional processors rely on oysters from outside the region. This clearly puts them at a competitive disadvantage relative to processors in regions where the supply is more abundant, such as in the Gulf of Mexico or Washington state. Any increase in the bay's supply, whether native or nonnative, is likely to have a positive effect on this sector as long as the price of the Chesapeake oysters is competitive with oysters from other regions.

Recreational and Amenity Services

Interdiction of nonnative oyster culture, inception of open-water aquaculture of triploid oysters, or outplanting of nonnative diploids could be expected to generate differing arrays of risk to recreational and amenity services. However, the information available to quantify these risks is very limited. In Chapter 5, it was suggested that the key linkages between the three management options and the magnitude of use and option benefits are probably through their effects on water quality, substrate characteristics, and the composition of benthic vertebrates, invertebrates, vascular plants, and algae. In addition to being affected by the same factors that influence use and option value, non-use benefits are likely to be influenced by individual preferences regarding the benefits or costs of the introduction of an alien species.

Public Institutions

Substantial change in public policy is usually accompanied by some degree of institutional risk. Shifts in public policy related to intensive aquaculture of triploid *C. ariakensis* or to the managed introduction of diploid *C. ariakensis* would involve institutional changes associated with implementing new management strategies. Changes associated with both propositions may result in profound differences in management paradigms with concomitant institutional risks. Risks may be associated with encountering divergent public opinions, the need for new and more complex regulatory mechanisms, the implementation of new management policies, and changes in institutional infrastructure.

Current oyster resource management policy is a product of extensive negotiation among stakeholders and representatives of state and federal agencies and is based on specific common objectives, including restoring natural oyster populations, restoring ecological services associated with functioning oyster reefs, and sustaining a traditional commercial oyster fishery. This common management policy is the basis for a multifaceted and multilevel approach to managing natural oyster resources. The scope of the public policy is demonstrated by the numerous partners that share this common management approach. However, commitment to the common management policy is largely dependent on the confidence that each party has in the commitment of other parties. In addition, commitment to the common management policy is dependent on real and perceived changes in the probable success of restoration efforts and the likelihood that nonnative oysters will be introduced.

Shifts away from this common management approach become more complicated because of the scope of existing management policy. For example, Maryland, Virginia, and the Chesapeake Bay Program have implemented long-range plans, such as the Oyster Recovery Partnership, the Oyster Heritage Program, and the Oyster 2000 Agreement. These partnership programs are the product of difficult political negotiations filled with compromise and are dependent on coordinated support from multiple funding sources. Changes in any element of these agreements may jeopardize support for current programs.

Institutional change, translating to institutional risk, can come as shifts in policy, politics, agency infrastructure, employment, strategic plans, and funding. Explicit institutional risks include reallocation of funding, loss of influence and/or power, fear of underfunded mandates, and added responsibilities and redirection or undermining of the institution's long-term goals. The least complex challenge may involve changes in management strategies within the same agency in response to such functions as permitting, compliance monitoring, and law enforcement as an agency shifts from managing a wild fishery to managing aquaculture-based production. A more complex challenge (i.e., in response to the managed introduction of nonnative oysters) results when the activity shift involves multilevel involvement by numerous other stakeholders, typically state and federal agencies. Shifts in activity-based responsibilities among agencies bring about institutional challenges that relate to risks, competition for funding, political support, and public interest. Multitiered institutional structures with overlapping regulatory responsibilities and diverse objectives contribute to the potential that one action may produce various outcomes, each with its own perceived risk and reaction.

There is a risk that institutional structures, such as the Virginia Marine Resources Commission (VMRC) and the Maryland Department of Natural Resources, will react differently to the management alternatives associated with the introduction of nonnative oysters. For example, the VMRC may be better situated to optimize outcomes related to aquaculture on privately held leases. Hypothetically, this scenario may prompt Maryland or another state to take action that is considered to be controversial and as such an institutional and public-interest risk. Leasing submerged public lands to private entrepreneurs may be considered bad public policy, creating a dilemma for resource managers.

There is also an inherent risk in the practice of relying on interested parties to provide oversight of the design and monitoring of protocols intended to minimize the unintentional release of sexually competent nonnative oysters and to prevent the cointroduction of disease or disease organisms. It seems appropriate to ask: "*Sed quis custodiet ipsos custodes?*" (Who is to guard the guards themselves?) when those charged with assessing and reducing the risk of introduction are simultaneously charged with devising a strategy for renewed economic opportunity in the fishery.

The institutional risk can be positively or negatively correlated with public-interest risk. For example, agency actions to support a specific management strategy may be strongly discouraged by public opinion when the management strategy is generally held to be inconsistent with prevailing environmental ethics. A potential loss of public support is an institutional risk, because shifts in public opinion influence political support and future funding.

One issue that has been discussed by numerous resource managers is related to the long-term commitment to restoring native oyster populations. Resource managers have acknowledged that there is an institutional risk involved in moving toward management options based on the purposeful introduction of nonnative oysters. The perceived risk is associated with the shift in funding to implement alternative management options and away from current programs to restore native stocks. Substantially more institutional risk is associated with the managed introduction of reproductively competent *C. ariakensis*, which may compete directly with current and future programs for funding. Stakeholders have reiterated the institutional risk associated with nonnative introductions, recognizing the potential consequence that support for future efforts to restore native oysters may be lost.

The degree of institutional risk cannot be ascertained without knowledge of how each institutional entity will react to the three management options. However, in an era of governmental budget constraints and mandates for increased efficiency, institutional risks are often magnified as agencies compete for funding. Additionally, the likelihood is that the institutional risk is high as agencies develop management structures and practices to deal with the introduction of nonnative oysters, because the overall challenge of managing the Chesapeake Bay's natural resources is a high-profile activity.

Management Efficacy

The three alternatives place some common and some unique burdens on management. The status quo involves the costs of monitoring and enforcement of seasons, catch limits, closed areas, and shellfish testing. If diploid nonnative oysters were intentionally introduced and became established with sufficient numbers to support a commercial fishery, it is likely that similar costs would be involved in monitoring and enforcement of seasons, catch limits, closed areas, and shellfish testing. In addition to the costs associated with monitoring and enforcement in an ongoing native oyster fishery, aquaculture of triploid nonnative oysters would involve monitoring and enforcement costs intended to reduce the risk that a diploid population of nonnative oysters could become established in the Chesapeake Bay. In addition, both in the case of intentional diploid introductions and the case of triploid-based aquaculture, monitoring and enforcement costs will increase to cover measures to reduce the risk of accidental cointroductions of exotic disease organisms and nuisance species. Measures that might be taken to reduce these risks include, for example, review and regulation of aquaculture operations plans; mandatory bonding of aquaculture facilities; random sampling of aquaculture oysters during the growing period to determine reversion rates, maturation, and the cause of unusual mortalities; genotyping of aquaculture brood stock; and sampling of adjacent grounds to detect the establishment of escaped nonnative oysters. Impatience with efforts to restore native oyster stocks coupled with the perception that *C. ariakensis* is a promising replacement increases the risk that unsanctioned introductions will occur under the no-introduction and triploid aquaculture options. The risks occasioned by unsanctioned—rogue—introductions are discussed below.

Community Structures and Social and Cultural Systems

In considering the social and cultural risks associated with the introduction of *C. ariakensis* into the Chesapeake Bay, it is important to consider both the community level and a broader baywide level. At the community level, the focus of a risk assessment is the possible impacts on watermen's livelihood, beliefs, and values upon which the identity of watermen communities are based. At the baywide level, the focus shifts to consumers, bay advocates, and the general public.

Community-Level Risk Factors

Continued decline in the long-term productivity and harvests of oysters will almost certainly increase pressure on the social and economic fabric of watermen communities and watermen living outside these communities. As noted in Chapter 5, oyster harvests have declined significantly over the past few decades and, due to recent drought conditions, the oyster harvest for 2002 is projected to be one of the lowest on record.

Because of differences in salinity levels and the associated prevalence of Dermo and MSX diseases, Virginia-based watermen have suffered greater economic losses than their Maryland counterparts in the past, but now both industries have been severely impacted. Public testimonies to the committee by leadership of watermen associations for both Maryland and Virginia corroborate harvest- and household-level information that the oyster fishery is becoming economically unsustainable.

Continued decline or low levels of productivity and harvest of the bay oysters could increase pressure on both blue crab and oyster fisheries. With continued low availability of native oysters, watermen may decide to continue commercial crabbing longer into the fall, during the period when the oyster and crab seasons overlap, and thus may increase pressure on blue crab stocks. Increased effort in fall crabbing may have a particularly strong adverse biological effect on blue crab reproduction because it is during this period that inseminated mature female crabs migrate south toward the mouth of the bay in order to spawn in warmer, high-salinity waters.

Continued low harvests of native oysters could also result in the opening up of more bottom for power dredging of oysters, as has recently occurred in Maryland. The risks and benefits of opening up more bottom for power dredging are uncertain and controversial. The risk is that dredging, which is one of the most efficient types of gear currently used by watermen to catch oysters, will lead to increased depletion of existing oyster beds and bars. Power dredging has been promoted by watermen who believe this activity cleans the beds of silt and turns over buried shell, with benefits for future spat sets and harvests, but this claim is controversial and has yet to be substantiated.

An additional risk of continued low levels of harvests is the decline in the number of young watermen willing to enter the profession. Without a viable and sustainable fall/winter oyster fishery, it may not be economically viable for watermen to make the investment in gear, repairs, and associated fees required to remain profitable in either the blue crab or oyster fishery. Decline of the oyster fishery is an important factor contributing to the continued decline in the number of commercial watermen and an increase in the average age of those who remain on the water. The prospect of augmentation of current and future oyster harvests is a critical factor in watermen's support for the common management plan.

The introduction of nonnative oysters, on the other hand, brings a risk of differential economic returns to watermen, depending on their state of residence and on whether the introduction is based on triploids used in aquaculture or reproductively competent diploids for wild harvest. As noted in Chapter 5, existing legislative interest and support, industry configurations (leased bottom, emphasis on aquaculture of oysters), and ongoing scientific activities make it almost certain that Virginia would have an advantage over Maryland in aquaculture of triploid oysters. Compared to *C. virginica,* the biological advantages of triploid *C. ariakensis* (resistance to MSX and Dermo diseases and enhanced growth rate) raise concerns about the ability of Maryland watermen, who are reliant on harvesting native oysters from public bottoms, to compete in local and national markets. If Maryland watermen are not able to compete effectively, many of the same risks and benefits associated with interdiction of nonnative oysters may occur. Moreover, there may be decreasing support for current restoration efforts with the increasing economic success of triploid aquaculture in Virginia. While Maryland policy and support for oyster aquaculture may improve in the immediate future, partly in response to any successful triploid *C. ariakensis* aquaculture in Virginia, many commercial watermen may not be able to withstand short-term declines in harvests and income in order to participate in longer-term oyster aquaculture development.

Open-water aquaculture of triploid oysters could result in a more vertically integrated oyster industry like the bay region's poultry-growing industry. Rather than extensive production by small-scale harvesters, the result may be fewer and larger growers and processors. Such a development would bring many changes to watermen's livelihood and their communities, along with a change in public valuation of watermen's livelihood and communities as part of Chesapeake Bay cultural and environmental heritage. However, without information about the costs of operation, the relationship between operational costs and scale, or about market opportunities available to niche producers or larger integrated operators, it is difficult at this time to evaluate the likelihood of triploid aquaculture resulting in increased vertical integration.

Inception of triploid nonnative oyster aquaculture may provide a much-needed economic boon for Virginia watermen, though it would be important to evaluate who and how many within the Virginia watermen community might benefit. A key question is the degree of access that traditional small-scale operators would have to the technology, capital, and markets required to grow triploid *C. ariakensis* in an open-water aquaculture setting. In Virginia many of the watermen may have difficulty switching to aquaculture of triploid *C. ariakensis*, owing to the higher production costs and related capital and informational needs of this form of aquaculture compared to those of the more traditional practice of rearing and harvesting native oysters on private leased bottoms. Although there could be employment in aquaculture for some watermen, who have primarily participated in the public-bottom fishery, it is unlikely that large numbers of public-bottom watermen would be so employed because this would conflict with summer crabbing and possibly with the limited oystering still available on public-bottom or leased beds.

If biologically and ecologically successful, the introduction of diploid nonnative oysters has the potential of restoring the commercial harvest of not only Maryland but also Virginia watermen. The Maryland Watermen's Association supports this option, arguing that the oyster fishery has declined to such a low level that something significant and different needs to be done. It is also relevant to consider how the introduction of a diploid nonnative oyster would affect the development of oyster aquaculture in Virginia. Key risk and benefit questions here include what would be the economic interactions between diploid nonnative aquaculture and public-bottom harvest of nonnative diploids. Which sector would be more profitable? Would continued small-scale public- or leased-bottom oystering compete with aquaculture of nonnative diploids? Would this result in an extensive and diverse industry that could support both aquaculture and public- or leased-bottom harvesting by watermen?

Baywide Social and Cultural Risk Factors

On a broader level, it is interesting to consider that interdiction of nonnative oysters could have a positive effect on the willingness of watermen, scientists, and oyster resource managers to form new partnerships to restore and profitably harvest from a smaller oyster fishery. The Oyster Recovery Partnership represents an initial framework and experiment along the lines of forming new relationships and exchanges among oyster scientists, watermen, and resource managers. While the challenges are great, these partners have begun the process of working together to manage reserves and sanctuaries with the goal of promoting the ecological and economic services of the oyster. There may also be employment opportunities for watermen to work alongside scientists and resource managers, undertaking such activities as moving spat, dredging bottoms, self-policing, etc. It might also provide a foundation for small-scale aquaculture by watermen who traditionally have worked public bottoms. However, as Chapter 6 suggests, current restoration efforts face serious challenges. While these challenges must be addressed, restoration of the native oyster fishery fits well within broader environmental ethics and values throughout the Chesapeake Bay area, values that emphasize restoration based on native species. The use of native species in restoration efforts can be an important environmental platform for innovative alliances among stakeholders that redefine traditional social roles and relationships toward resource management, including fisheries. (It may be true that the obverse is also the case: use of a nonnative species reduces the role of environmental values as a motivator for restoration and resource management.)

In the bay region there is a risk that triploid oysters could be perceived or publicly cast as "not natural," given the chromosomal changes required to induce sterility. Given the strong emphasis in the region on "native" and "pristine," any significant increase in public perception of triploid *ariakensis* as unnatural could reinforce existing concerns about the *C. ariakensis* being nonnative. It should be noted that the public perception of *C. ariakensis* will be largely influenced by such factors as how the product is marketed and labeled and the reaction of consumer and environmental organizations. Consumer preference for native versus nonnative oysters could reduce the market value of the *C. ariakensis* half-shell market, reducing the profitability of nonnative aquaculture (Grabowski et al., 2003).

Finally, introduction of a diploid nonnative oyster would likely run the risk of a public cultural-environmental backlash driven by ethics and values or preserving the bay's natural heritage. The important question would be who would benefit from such a backlash and how they might advance their concern in the public and policy arenas. It could be argued that the Chesapeake Bay is different than many coastal or estuarine areas in the strong cultural-environmental emphasis on restoration of native species and ecosystems.

Implementation Risk

Risk of Political Objection

Inception of triploid nonnative aquaculture or the sanctioned introduction of diploid nonnative oysters could be obstructed through objections raised in the regulatory approval process or through legal challenges brought by concerned parties. Several potential avenues for challenge were explored in Chapter 8. This section briefly explores two broadly constructed stakeholder classes—the fishing and environmental communities—that could challenge implementation of the options.

Fishing Community

As documented in Chapter 5 and above, the fishing community is heterogeneous across regions, modes of production, degree of concentration and integration, access to capital, level of financial stress, and ability to respond to the new opportunities represented by the options. Various segments of the fishing community could view themselves as advantaged or disadvantaged under one or more of the options. It is unlikely that members of the fishing community would mount legal or regulatory challenges to the ongoing native restoration plan unless the plan were revised to close public bottoms or otherwise further restrict harvest activities. Elements of the fishing community might object to aquaculture based on triploid nonnative oysters, particularly if they were concerned that triploid nonnative aquaculture might impinge on traditional public-bottom fishing zones or if they were concerned that production from triploid nonnative aquaculture operations might reduce access to product markets for their harvests of the native oyster. The first objection is more likely to affect decision making in Maryland than in Virginia, due to the limited extent of leased-bottom submerged lands in Maryland. The second objection is most likely to be raised if it is perceived that buyers will not differentiate between native and nonnative oysters. Elements of the fishing community are most likely to object to the intentional introduction of the nonnative oyster if there are regional differences in the introduction. Introduction of the nonnative offers the possibility of continuation of the traditional mix of leased- and public-bottom fisheries. If introduction is permitted in Virginia but not in Maryland, the Maryland fishing community may object to introductions in Virginia through opposition to permitting or through legal action. Alternatively, the Maryland fishing community might respond to introductions in Virginia with political and legal actions to allow introductions to occur in Maryland waters as well.

Environmental Community

The environmental community is also heterogeneous with multiple, potentially incompatible objectives. A large segment of the environmental community is primarily concerned with water quality issues associated with recreational activities (e.g,. swimming, boating, waterfowl viewing, recreational fishing) but who may not be politically well organized. These individuals may be more concerned with water quality improvements than with the means used to obtain those improvements. They may be unlikely to object to a continuation of the status quo, with restoration efforts utilizing broodstock selection. They may similarly be unlikely to object to the introduction of diploid nonnative oysters if such introduction is represented as accelerating water quality improvements. They may object to aquaculture of triploid nonnative oysters because they do not anticipate aquaculture being sufficiently extensive to result in large-scale water quality improvements and because they may perceive aquaculture as a competitive claimant on funds presently allocated to restoration efforts.

Other segments of the environmental community, in particular nongovernmental organizations with political clout, may be more concerned with the naturalness of the ecosystem and may object to actions that could result in changes in species composition and abundance. This segment of the environmental community could object to any of the options involving *C. ariakensis* and could mount regulatory, legal, or political challenges to continuation or implementation of the options. Objections to the introduction of diploid nonnative oysters could arise because of concerns about the ecological effects of *C. ariakensis* on native species, concerns that the introduction might serve as a vector for introducing exotic disease, and concerns that localized introductions could expand and displace otherwise healthy populations of the native oyster throughout the eastern seaboard and into the Gulf of Mexico. Objections to aquaculture of triploid nonnative oysters could be based on the risk that diploids will inevitably be introduced, owing to the imperfect fidelity, stability, and sterility of mated triploids (see Chapter 4). This organized environmental community could even object to the status quo option. Continued decline of native oyster stocks could stimulate interest in prohibiting commercial and recreational harvests. Objections could be mounted as well to certain restoration activities, such as those employing selectively bred, disease-resistant stocks to artificially supplement natural populations, based on the perceived risks to natural diversity.

Risk of Rogue Introductions

A rogue introduction would be a nonsanctioned direct release of diploid reproductive Suminoe oysters into the Chesapeake Bay, likely executed without benefit of adherence to the ICES protocols. The chief hazards of a rogue introduction are that the nonnative oyster would become established and pervasive; the incidental introduction of other nontarget plant, animal, or microbial species that could become invasive; or the incidental introduction of new pathogens and parasites that could attack native oysters or other bivalves both inside and outside the bay. Often the impact of associated species is as great as or greater than that of the species targeted for introduction. For example, Carlton (1999b) reports finding seven species of algae and invertebrates in a single container of hatchery-reared Pacific oyster seed imported for research purposes to the Woods Hole Oceanographic Institution and cites another study that found an additional 29 species of algae, diatoms, protozoans, and invertebrates in the water of other oyster shipments. Of the 30 nonindigenous molluscs found on the U.S. Pacific Coast, 10 gastropods and 10 bivalves were introduced along with the Eastern and Pacific oysters imported for commercial culture (Carlton, 1992b). Seaweeds and seagrasses, which were used as packing material for imported oysters, have transformed thousands of square kilometers of open mudflat habitat in Pacific Coast bays into stands of intertidal vegetation that harbor completely different invertebrate faunas than existed previously (Posey, 1988; Carlton, 1989).

Another hazard might be the introduction of a nontarget oyster species, caused by lack of care or ability to discriminate among morphologically similar species, subspecies, or physiologically distinct races. The taxonomy of Asian cupped oysters is poorly known and in a state of flux, depending on sporadic studies undertaken for various reasons (e.g., Buroker et al., 1979a, b; Banks et al., 1994; Ó Foighil et al., 1995, 1998; Boudry et al., 1998; Hedgecock et al., 1999; Day et al., 2000). In particular, two distinct geographic races of *C. ariakensis* have been uncovered through analysis of mitochondrial and nuclear DNA sequences since researchers at VIMS turned their attention to this species as a candidate for nonnative introduction (Francis et al., 2001). A nontarget oyster might not perform as well as the particular strains of *C. ariakensis* that have been tested to date in Chesapeake Bay field trials.

The likelihood of rogue behavior cannot be quantified but is judged to be substantial and depends on which management strategy is chosen (see below). Rogue introductions of the Pacific oyster have occurred previously in the Chesapeake Bay and elsewhere on the East Coast. This led Maryland to adopt legislation against the introduction of this species (see Table 3.2; Andrews, 1980). Hopes for the recovery of the oyster industry have been fueled by reports about the impressive survival and growth of *C. ariakensis* in experimental trials. The economic motive for carrying out an illegal rogue introduction is present and is likely to build over time if native oyster populations remain depressed. Although human behavior is unpredictable, shipment of live oysters from Asia to the United States would not present an obstacle to a rogue introduction. All life stages of cupped oysters (D-hinge or later larvae, spat, and adults) are readily shipped live via air courier. Adult Asian oysters could be readily obtained and imported live to the region, probably through normal channels of seafood supply. Hundreds of adults can be shipped live in a box no larger than a typical picnic cooler; if kept moist and refrigerated, adults can survive out of water for more than a week.

Obtaining larval and seed stages of *C. ariakensis* would require locating a cooperating hatchery in Asia, but these early life stages are attractive targets for rogue introduction because of the much larger numbers of oysters that could be imported. Commercial oyster hatcheries routinely supply farmers with late pediveliger or eyed larvae in very large numbers for remote setting; when screened from their culture water, for example, 2.5 million eyed larvae constitute a golf-ball-sized mass that is easily shipped in a small package. Eyed larvae remain competent for setting for about a week if kept moist and refrigerated. Remote setting requires little technology and minimum infrastructure.

The likelihood of a rogue introduction resulting in the establishment and spread of a population of nonnative oysters depends on the life stage introduced, the number and density of animals introduced, the spatial scale of introduction, the spawning and recruitment potential at the sites of introduction, and the frequency of introduction. The number of adults that could be shipped and introduced at any one time would likely be limited to a few hundred individuals. A small inoculum of adults could successfully found a population, in principle, owing to the high fecundity of the oyster. Nevertheless, the chances of establishment and spread would be governed by the likelihood that environmental conditions conducive to spawning, larval development, retention in a local area, and recruitment in sufficient density for successful spawning in the next generation were met. The chances of successful spawning and recruitment are classically difficult to predict for most marine animals, including the native oyster. It is noted that successful introduction of the Pacific oyster into France was made possible by massive importation of adults and spat (Chapter 3).

Much larger numbers of seed oysters could be introduced via a shipment of eyed larvae. The percentage of eyed larvae that can be successfully set is variable but probably in the range of 10 to 30%; of these, perhaps 10 to 20% might survive to a suitable size of about 8 mm. This means that from 2.5 million eyed larvae 100,000 seed could be reared and planted, of which thousands or tens of thousands might survive to reproductive maturity. With that size of inoculum, the chances of successful recruitment, establishment, and spread would be greatly increased though not guaranteed.

MANAGEMENT OPTIONS

The biological and social factors likely to be impacted by each of the three management options for introducing the Asian oyster, *C. ariakensis* into the Chesapeake Bay are listed in Table 9.2. The body of the table contains a qualitative assessment of potential outcomes for each factor under the three management options. Lack of information precludes definitive characterization of every hazard, particularly the ecological ones. Moreover, the very different ecological, economic, and social hazards cannot be weighted with respect to one another or summed to derive an overall relative risk for each management option. Table 9.2 primarily characterizes the various factors likely to be affected by the choice of management options. In the table the likelihood of a particular outcome (listed as positive, negative, or neutral) represents the committee's assessment for each management option under a short time frame (1 to 5 years). However, there are many uncertainties and potential scenarios that could af-

TABLE 9.2 Assessment of Potential Outcomes Under Each Management Option

Biological and Social Factors	No Introduction[a]	Triploid Introduction[b]	Diploid Introduction[c]
Ecological			
Disease introduction		-	-
Disease reservoir			
Susceptibility to endemic pathogens or parasites		-	-
Impacts on ecosystem		-	-
Competition with *C. virginica*—space, food, habitat		-	-
Competition with other species (relative to *C. virginica*)			
Invasion		-	-
Dispersal beyond the bay		-	-
Genetic interactions		-	-
Water quality	-	+	+
Reef structures and services	-	+	+
Economic/social/cultural			
Human health/pathogen			
Price			-
Viability of traditional fishery	-	-	+
Fishery employment	-	-	+
Viability of aquaculture	+	+	+
Aquaculture employment	+	+	+
Tourism, recreational, sports fishery		+	+
Public institutions			
Management effectiveness	-	-	-
Employment		+	+
Watermen communities	-	-	+
Watermen culture	-	-	+

continued

TABLE 9.2 (continued)

Biological and Social Factors	No Introduction[a]	Triploid Introduction[b]	Diploid Introduction[c]
Cultural perception of restoration, environment	+	-	-
Political impact			
Fishery	-	+	+
Environmental	+	-	-
Restoration efforts	+	+	-
Likelihood of rogue introduction	+	+	
Impact of rogue introduction	-	-	-

Biological and social factors are likely to be affected by selection of one of the three management options with regard to introducing the Suminoe oyster into the Chesapeake Bay. The assessments given here are developed for short-term (1 to 5 years) outcomes and are listed as +, positive; -, negative; and blank, no effect. The rationale for each of these values is explained in detail in the text. There are large uncertainties associated with each outcome; therefore, these values serve as an illustration, but not a prediction, of how the various management options might compare.

[a]Ecological and economic and social outcomes assume no rogue introduction.
[b]Outcomes assume eventual production of diploids and the establishment of small reproducing populations.
[c]Outcomes assume large managed introduction of diploids using ICES protocols.

fect outcomes. The qualitative characterization of risk presented in the table is meant to provide a starting point for reviewing the hazards associated with each of the three management options the committee was charged to consider.

Option 1. Status Quo, No Introduction of Nonnative Oysters

The first management option is simply to maintain the status quo by forbidding the introduction of all nonnative oysters into the Chesapeake Bay, whether diploid or triploid. The chief consequences likely to be associated with this action, were it successful in maintaining the status quo, would be:

- continued decline of the oyster fishery and erosion of the traditional economies and cultures of Chesapeake Bay watermen;
- possible increased pressure in the blue crab fishery;
- possible further declines in bay water quality, owing to loss of oyster filtering capacity, though scientific evidence that water quality is tightly coupled to oyster abundance is weak;
- possible continuing or accelerating losses of aquatic vegetation and oyster reef habitats, with cascading effects on the structure and stability of the bay's estuarine communities, though scientific evidence for these assumptions is lacking;
- possible reduction of bay acreage protected under the Clean Water Act's shellfish bed water quality preservation mandates; and
- erosion of confidence in governmental management of the living marine resources of the Chesapeake Bay.

The economic or ecological harm from these hazards can be reasonably extrapolated from recent trends in the fishing sector and in bay water quality and ecology.

The chief benefits of maintaining the status quo would be:

- avoidance of risks identified with either of the alternative options for introducing a nonnative oyster;
- increased emphasis on aquaculture of native oysters selectively bred for resistance to MSX and Dermo diseases;
- increased employment in the native oyster aquaculture sector, especially with new strains of disease-tolerant *C virginica*; and
- affirmation of cultural value on conserving native species and natural habitats.

Simply banning the introduction of nonnative oysters into the Chesapeake Bay, however, will not necessarily maintain the status quo. A no-introduction policy would increase the likelihood of a rogue introduction, that is, a nonsanctioned direct release of diploid reproductive Asian oysters, executed surreptitiously and without benefit of adherence to ICES protocols. The economic desperation created by the collapse of the traditional oyster fishery of the Chesapeake Bay, coupled with widespread awareness of the performance of triploid *C. ariakensis* in previous field trials and the ease with which live animals could be acquired through traditional fish markets, makes rogue introduction an easy response to the perception of management inaction. Industry representatives, who addressed the committee, made this hazard explicit. The risks associated with rogue introductions include the risks identified under sanctioned introductions that employ ICES protocols, as well as those incurred by circumventing ICES protocols, and would remain for as long as the population of a native oyster remained depressed. If a self-reproducing population of *C. ariakensis* were established as the result of a rogue introduction, the resulting harms and benefits would probably increase through time, with an increase in the abundance of the nonnative oyster. Unfortunately, the specific ecological, economic, or cultural harms or benefits of a rogue introduction cannot be specified nor can their magnitudes be predicted. Finally, under this option, management would presumably be burdened with monitoring for rogue introductions and with eradication of diploid nonnative oysters were they detected. Eradication of introduced marine species is extremely difficult or impossible, as recent experiences with the invasive seaweed *Caulerpa* in the Mediterranean Sea attest (Thibaut et al., 2001).

Any attempt to maintain the status quo should certainly be coupled with scrutiny of why the restoration of native oysters has failed so far. Such an examination was not part of the charge of this study. Clearly, however, successful restoration of native oysters and the traditional fishery would largely have precluded the present controversy over introduction of a nonnative oyster.

Option 2. Open-Water Aquaculture of Triploid Oysters

Because the fidelity, stability, and sterility of mated triploids are not likely to be 100%, expanding the introduction of mated triploid *C. ariakensis* in controlled aquaculture settings risks establishing a diploid self-reproducing population in the Chesapeake Bay. This hazard, however, can be broken down into components, most of which can be quantified and modeled, as attempted by Dew et al. (2003), for example, who simulated the population growth and local establishment of nonnative oysters introduced by triploid aquaculture under specified ecological conditions and management strategies. Furthermore, many of the hazards associated with open-water aquaculture of triploid nonnative oysters can be managed to reduce specific elements of risk. For example, increasing the containment of, and accountability for, planted stock could lessen the risk that triploids would remain in the bay long enough to revert to the diploid state. Likewise, the density of planted stocks could be managed to reduce the risk that gametes released by the small percentage of diploids that might be produced along with mated triploid seed would be able to find and fertilize each other. The number of triploids introduced could be constrained to reduce risk, but this would also reduce potential economic or ecological benefits.

Minimizing the duration and scale of the triploid culture effort would minimize the risks of this management option. Indeed, introduction of triploids could be used as a management strategy to buy time for restoration of native oysters, which could result either artificially (from the development of new and more successful approaches to restoration) or naturally (from a return to the more typical conditions of colder winters and wetter summers, which would inhibit parasite proliferation and provide the native oyster with more freshwater refuges from disease). Recovery of native oyster populations would reduce the incentive to introduce a nonnative species or would reduce the scale of any aquaculture sector based on the nonnative relative to the scale of a resuscitated traditional fishery.

Aside from the hazard of establishing a self-reproducing population of a nonnative oyster, some short- and long-term negative impacts of this management option are:

- continued declines in the traditional oyster fishery or possibly accelerated declines as hope for recovery is lost and extraction is maximized;
- economic hardships for watermen communities, unless they switch from fishing to aquaculture;
- no marked improvement in bay water quality in the near term, owing to only a marginal increase in oyster filtration capacity from triploid aquaculture;
- continued threat of rogue introductions of diploid nonnative oysters;
- potential introduction of pathogens that may not be excluded by adherence to ICES protocols;
- potential introduction of other pathogens owing to inadvertent breaches of ICES protocols;
- susceptibility of nonnative oysters to endemic pathogens or parasites;
- conflicts with cultural value placed on conservation of native species and habitats;

- erosion of confidence in resource management; and
- political resistance or legal challenges by environmentalists or states from outside the Chesapeake Bay.

As under the first option, management could face a considerable burden for monitoring bay waters for the establishment of diploid populations or for subsequent eradication of any diploids detected. Genetic markers could be profiled in all tetraploid and diploid stocks used to make triploids, so that the provenance of any diploids that might subsequently be detected or become established could be determined.

Some short- and long-term benefits of this option, aside from those attending the establishment of a diploid population, are:

- management control over most aspects of the authorized introduction;
- viability of aquaculture;
- aquaculture employment;
- possible retention of tourism, recreational, and sports fishery benefits associated with Chesapeake Bay oysters, even though nonnative; and
- increased incentive for restoring the native oyster, if it serves to rally the political constituents of restoration.

One important benefit to the controlled introduction of triploid *C. ariakensis* could be opportunities for research on the biology of *C. ariakensis* in the Chesapeake Bay. The likelihood of ecological harm or benefit could be more accurately assessed if basic information were available on the season of reproduction (triploids, though sterile, still go through an annual reproductive cycle), susceptibility to native pathogens and parasites, competition with *C. virginica* for space, and propensity to sustain old or restored reefs or to build new ones. The risks of expanded industrial trials could be partially offset by the inclusion of parallel ecological experiments designed to generate information critical to evaluating the risk that triploid aquaculture will eventually produce a diploid population.

Option 3. Introduction of Reproductive Diploid Oysters

This management option has strong local support because introduction of an oyster that can survive and grow in the Chesapeake Bay appears, to many, as the only hope of improving water quality and the bay's ecosystem, recovering the traditional fishing industry and sustaining watermen culture. Behind the hope is the assumption that purposeful introduction will quickly yield a large viable population of *C. ariakensis*, with little or no adverse effects on the remnant native oyster population or other species. However, introductions are not always successful. Initial trials with triploid Pacific oysters in the Chesapeake Bay showed, for example, that this nonnative oyster, though resistant to the diseases that kill native oysters, was susceptible to infestation with the shell-boring polychaete worm *Polydora*, which made them unacceptable in the market. Still, some introductions of oysters and other bivalves have been successful in establishing industries without untoward ecological harm. The introductions of the Pacific oyster to the west coasts of North America and Europe had positive impacts on fishing and farming industries. The Pacific oyster proved noninvasive on the West Coast of North America; hence, there were no pronounced ecological changes, with the important exception of problems stemming from cointroductions (e.g., *Spartina alterniflora* to the U.S. West Coast; Naylor et al., 2001). The risks of cointroductions, today, would be greatly reduced by the use of ICES protocols. Finally, opponents of diploid introduction can cite counterexamples of negative ecological impacts from introductions of oysters or other bivalves. The Pacific oyster, *C. gigas*, in New Zealand and Australia threatens endemic oyster species; the zebra mussel, *Dreissena polymorpha*, has caused widespread fouling problems in the Great Lakes and other regions in North America; and the Asian clam, *Potamocorbula*, has greatly modified the soft benthic fauna and primary productivity of the San Francisco Bay and delta. What mix of outcomes—no impact, positive impact, or negative impact—would follow a clean introduction of *C. ariakensis* into the Chesapeake Bay cannot be predicted.

Short- and long-term negative impacts of introducing diploid *C. ariakensis* into the Chesapeake Bay could include:

- disease introduction, though greatly reduced, would still present an unknown hazard from vertically transmitted pathogens even if ICES protocols are followed and perfectly effective;
- negative ecological impacts, such as competition with *C. virginica* or fouling of boats, marinas, and other marine structures;
- spread of nonnative oysters outside the bay, where competitive displacement of healthy native oyster populations might be possible;
- susceptibility to endemic pathogens or parasites;
- decreased management effectiveness;
- abandonment of attempts to restore the native oyster;
- conflicts with conservation ethic; and
- political resistance or legal challenges by environmentalists or states from outside the bay.

The chief benefits of a diploid introduction would ostensibly be the same as those deriving from recovery of the native oyster population, though hard scientific evidence supporting these presumed effects is limited or lacking:

- possible improvements in water quality;
- increases in aquatic vegetation;
- deposition of new reefs and increases in reef habitat for fish and other invertebrates;
- resuscitation of the traditional oyster fishery and fishery employment;
- continued viability of aquaculture and increased aquaculture employment;
- potential increases in tourism, sports fishery, and a recreational economy;
- maintenance of watermen communities and culture; and
- reduced likelihood of a rogue introduction.

All of these benefits assume that an introduction of diploid *C. ariakensis* would result in a large population of reef-building oysters, an outcome that is uncertain.

FINDINGS

- The three management options (no introduction of nonnative oysters, introduction of triploids for aquaculture, and introduction of diploids) entail differing arrays of ecological, socioeconomic, institutional, and implementation risks.
- The risk of a disease outbreak in either the native or nonnative oyster populations following an introduction is not zero, even if ICES protocols are followed. If ICES protocols are applied, the risk of disease outbreak has low probability but potentially high impact if it occurs.
- Assessing an array of ecological risks is severely constrained by lack of fundamental ecological information on the Suminoe oyster, *C. ariakensis*, and even by lack of sufficiently detailed ecological information for the native oyster and the Chesapeake Bay. Various ecological risks that can be postulated have unknown probabilities and unknown impacts.
- No human health risks are apparent. The risk to human health has a very low probability and a low impact.
- Assessment of the risks to institutions with responsibilities for managing the living resources of the Chesapeake Bay have unknown probabilities and unknown impacts.

- The risks of rogue introductions are likely high under the no-introduction management option; may remain high to moderate under the triploid aquaculture option, particularly among the "have not" stakeholders; and are likely low to moderate under the diploid introduction model. The potential impact of a rogue introduction is high, owing to the substantial ecological impacts that have been documented following the unintended cointroduction of other organisms besides the oyster.
- The breadth and quality of existing information on oysters and other introduced species are not sufficient to support a comprehensive risk assessment of the three management options.
- Comprehensive risk assessment is also not practicable, owing to the lack of well-defined and/or conflicting objectives and goals among Chesapeake Bay management agencies and users.

10

Decision Making and Recommendations for Future Research

TO INTRODUCE OR NOT TO INTRODUCE?

In deciding whether to approve or disapprove the introduction of a nonnative oyster, either open-water aquaculture of sterile triploid oysters or deliberate establishment of a reproductive population, policymakers must weigh the potential risks and benefits associated with introducing a nonnative species. For the Chesapeake Bay region, policymakers have been unable to reach agreement on whether nonnative oysters should be part of the response to the collapse of the native oyster population. Recent experiences with attempts to eradicate or even control invasions by nonnative nuisance species have left many resource managers averse to approving introductions of exotic species. On the other hand, there is growing impatience with the apparent failure of current efforts to rebuild stocks of the native oyster to support the oyster fishery. This study was commissioned to define and assess the risks and benefits of introducing the Suminoe oyster to help reconcile these two points of view.

In general, deliberate introductions of exotic species have had mixed success. Although many new species have been imported without major consequences, the exceptions include nuisance species that have precipitated dramatic and typically irreversible changes to ecosystems. To assess the potential threats associated specifically with oysters, the committee reviewed the results of previous oyster introductions (Chapter 3). The oysters themselves are a problem in some cases but not others. Introductions of the Pacific oyster, *Crassostrea gigas*, into several areas in Australia and New Zealand, for example, resulted in unanticipated losses of native oyster populations that had been commercially harvested. Introductions of this same oyster to the U.S. West Coast and France filled voids created by previous overharvesting or diseases of native species and have become the basis of extensive aquaculture industries. **It is impossible, given the present state of knowledge, to predict whether the Suminoe oyster will be a boon or an ecological disaster in this sense. The committee's review of case studies clearly indicates that greater ecological or economic harm typically arises from organisms that are inadvertently introduced with the foreign oyster.** With the exception of pathogens (including some viruses) that can be transmitted from adults to their progeny, unwelcome "hitchhiking" organisms can be avoided by strict adherence to the International Council for the Exploration of the Sea (ICES) protocols, which specify that only hatchery-reared nonnative oysters should be allowed in open waters. Current proposals to use the Suminoe oysters stipulate adherence to ICES protocols to minimize the risk of introducing a disease organism or other unwanted species. Although the nonnative oysters could become an unwelcome invasive species, this risk is probably lower than the risk of introducing a multitude of alien species incidental to an illegal rogue introduction. **Consequently, regulatory and enforcement measures should be taken to reduce the risk of a rogue introduction.**

Development of a quantitative risk assessment model for evaluating risks associated with the three management options would require a great deal of additional research that could take many years to complete. Moreover, while risk assessment is a tool for characterizing the likelihood of various outcomes, it does not provide a basis for determining whether an outcome or combination of outcomes is desirable or undesirable. That is, risk assessment may inform the decision-making process but cannot replace it. Nevertheless, a decision must be reached about whether or not to proceed with the use of the nonnative oyster despite uncertainty about the type and magnitude of the potential risks involved. Regulators and decision makers will need to consider all available information and weigh the often opposing interests of the various stakeholders to decide whether to allow introduction of the nonnative Suminoe oyster, *C. ariakensis*, into the Chesapeake Bay, either limited to aquaculture of the sterile triploid or as a deliberate inoculation of reproductive diploids. This is a particularly difficult circumstance because of the magnitude of uncertainty associated with each option and the perception, on all sides, that any decision will have lasting and serious consequences. Various decision analysis techniques have been used to clarify objectives and elucidate the effects of uncertainty on the possible outcomes of alternative management actions for similarly difficult decisions (e.g., Keeney and Raiffa, 1976; Hilborn and Walters, 1977; Walker et al., 1983; Bain, 1987; Saaty, 1990; Merritt and Criddle, 1993). The intention here is not to conduct a comprehensive decision analysis but to sketch a road map for assessing the likelihood of achieving the various goals under each of the three management options. In decision analysis the likely impact of an action is assessed relative to the uncertainty of the outcome, the reversibility of the action, and the likelihood of unintended consequences.

The first step in decision analysis requires clear definition of the management objectives and stakeholder concerns. At the study's October 2002 workshop, the committee heard presentations by various stakeholders involved in the question about how best to restore oysters (native or nonnative) to the Chesapeake Bay. Two major goals were expressed by this diverse assemblage of interested parties: (1) build a profitable oyster production and processing sector that is competitive in local, regional, and global markets and (2) improve ecological conditions in the Chesapeake Bay through restoration of oyster beds. These overarching goals share some common features but differ in other respects. For example, it may be possible to develop a profitable fishery without restoring oyster populations in the bay. Similarly, restoring oyster populations will not necessarily lead to the restoration of a profitable fishery or to meaningful changes in water quality. Different stakeholders support the major goals for different reasons. By decomposing the major goals into their constituent parts, the effects of a management action can be analyzed for more precisely defined subobjectives, thereby making the consequences of decisions more obvious and understandable. Moreover, elucidation of subobjectives often reveals common interests that can form the basis for decisions supported by multiple stakeholders. Each management option should be assessed for its efficacy in addressing these subobjectives in addition to the major goals. Again, the uncertainty of outcomes, reversibility of actions, and potential unintended consequences should be elucidated.

The committee was asked to assess whether or not there was sufficient information to support risk assessments of three management options. The committee concluded that quantitative risk assessments could not be conducted based on existing data but found that differences in the types of risks associated with each option could be described. On the basis of existing information about oysters in general and the Suminoe oyster in particular, Option 3—direct introduction of reproductive Suminoe oysters—would likely have a moderate to high chance of increasing the abundance of oysters in the Chesapeake Bay, particularly if it was done as a massive introduction, much as the Pacific oyster was introduced into France. To the extent that the introduction was successful, this action might reduce the likelihood of a rogue introduction. However, because of the lack of information on the biology and ecology of the Suminoe oyster, there is a high degree of uncertainty about the outcome of an introduction. The changed environment in many parts of the bay may no longer support the survival and growth of any oyster. Baywide, the Suminoe oyster may not perform as it has in the limited field tests conducted to date. It may prove susceptible to native pathogens, parasites, or predators. It may not form reefs but may instead spread as a thin layer on soft bottoms, smothering native fauna. It may become established and abundant but have low commercial value if, for example, there is a high incidence of mud blisters or if it has a short shelf life. The nonnative oyster may spread outside of the bay, where it could have negative impacts on still-healthy populations of the native Eastern oyster. Finally, it is unlikely that the Suminoe oyster could be eradicated after it was introduced; thus, any undesirable consequences that ensue would be permanent. **In sum, the irreversibility of introducing a reproductive nonnative oyster and the high level of uncertainty with regard to potential ecological hazards make Option 3 an imprudent course of action.**

Option 1—no use of nonnative oysters—is unlikely to result in significant changes in oyster abundance in the bay in the near term. This outcome has moderate uncertainty, owing to external events or actions that might favor the recovery of native oyster populations (e.g., favorable climate change or success in restoration efforts employing selectively bred, disease-resistant stocks). Although Option 1 is reversible in the sense that nonnative oysters could still be introduced at some time in the future, this option includes the risk of continued losses to the oyster industry and erosion of confidence in government action. One possible response to the latter is increased risk of a rogue introduction and the especially high hazards associated with unintentional cointroductions of disease organisms or nuisance species.

Option 2—sterile triploid aquaculture carried out with strict accountability and best management practices to minimize the risk of diploid escape—would probably have little impact on total oyster abundance because even expanded aquaculture operations are unlikely to produce a volume of oysters at the scale of the natural populations in the Bay. Uncertainty associated with this action would be low. The reversibility of this option depends on the effectiveness of measures taken to prevent reproduction; if aquaculture is practiced at a small scale, the probability of reproduction will be low and the reversibility of this management action will be high. The probability of introduction of reproductive oysters would increase if triploid aquaculture were to expand dramatically. Over the long term it is likely that some nonnative oyster larvae will be spawned; whether the larvae will survive, establish a nonnative population, and spread throughout the bay is unknown. The potential number of larvae released is directly related to the scale of triploid aquaculture. Under Option 2, dramatic expansion of aquaculture effectively grades into a small-scale diploid introduction as described for Option 3. The perception that Option 2 represents management action rather than inaction (Option 1) might reduce the likelihood of a rogue introduction. Over 6 or 7 years, controlled and strictly regulated triploid aquaculture could provide critical biological and ecological information now lacking for the Suminoe oyster. It would also allow time to assess whether climate variation or new approaches to restoration are effective at reversing the decline of the native oyster.

Option 2 has already received considerable scrutiny by the Chesapeake Bay Program and its member states and federal agencies. Limited field trials with triploid nonnative oysters have already been conducted in Virginia and outside the bay in North Carolina's coastal waters, and larger trials are in the advanced planning stages in both states. To a limited extent, the decision to introduce triploid Suminoe oysters has already been made. The risks of proceeding with triploid aquaculture in a responsible manner, using best management practices, are low relative to some of the risks posed under the other management options. **If regulators enforce strict protocols for accountability and require collection of the biological, economic, and social information necessary to evaluate the risks and benefits of culturing or introducing the nonnative oyster, this management option could provide useful information to support decision analyses and risk assessments regarding the future use of nonnative oysters in the Chesapeake Bay.**

UNREALISTIC EXPECTATIONS AND COMMON MISCONCEPTIONS

In evaluating the scientific evidence bearing on the potential risks and benefits of introducing a nonnative oyster into the Chesapeake Bay, the committee finds relatively little scientific support for many of the common assumptions that have shaped public discourse on this issue. These assumptions should be treated with healthy scientific skepticism if progress is to be made in resolving the problem. As a prelude to recommendations for action and future research, we portray five such misconceptions as "myths." These five myths may not be the only misconceptions and assumptions surrounding this complicated issue, but they do reveal major gaps in knowledge and uncertainties confronting a decision about whether or not to introduce a nonnative oyster into the Chesapeake Bay.

Myth I: Declines in the Oyster Fishery and Water Quality of the Chesapeake Bay Can Be Quickly Reversed

This is an overarching myth with respect to hopes for the success of restoration efforts and to the issue of nonnative oyster introduction, rooted perhaps in fundamental cultural beliefs in the power of technology and control over nature (McPhee, 1989). The oyster fishery in the Chesapeake Bay has been in decline for about a century; the native oyster is on the verge of local economic extinction. Also, water quality has declined as a consequence of numerous activities over a long period of time, including urbanization, deforestation, agriculture, overfishing, shell mining, and waste disposal. The belief that these trends will be readily reversed, restoring the bay to some semblance of its pristine past, is unrealistic. The failure of a succession of corrective actions to reverse the decline in the fishery—relaying of spat on cultch, restoration of shell-reef habitat, establishment of sanctuaries, hatchery enhancement—each of which was once thought to be the solution to the bay's problems, is testimony to the absence of easy answers and quick fixes. Progress on reversing the long-term declines in oyster populations and water quality will be achieved only when unrealistic expectations for a quick fix are replaced with a long-term commitment to systematic approaches for addressing the bay's complex multidimensional problems. Progress may also depend on climate variation. The native oyster has been rendered particularly vulnerable to disease by a series of drought years with milder than usual winters, conditions that increase infection rates and mortality. The native oyster would benefit enormously from a reversal of this climate pattern and a return to wetter years with colder winters, which would inhibit proliferation of parasites in the oyster and provide oyster populations with more low-salinity refuges from diseases. Variation in climate patterns, on interannual and perhaps interdecadal timescales, must be incorporated into expectations about the recovery of the native oyster and the ecology of the bay.

Myth II: Oyster Restoration, Whether Native or Nonnative, Will Dramatically Improve Water Quality in the Chesapeake Bay

The role of filter-feeding bivalves in estuarine ecosystems has been well documented (e.g., Dame, 1993), but the conventional wisdom that either a disease-free native oyster or a nonnative oyster could repopulate the Chesapeake Bay to an extent that would restore water quality and dramatically improve the condition of the bay's living resources is unreasonably optimistic given current environmental conditions. The connection between oysters and water quality was popularized by the oft-cited calculation suggesting that the native oyster was abundant enough, even in the late 19th century, to filter a volume of water equal to that of the Chesapeake Bay every 3.3 days, while today's depleted population takes nearly a year to filter the equivalent volume (Newell, 1988). Although this type of calculation dramatically illustrates how oysters may have contributed to the ecology of the bay in the 19th century, it is a fault of logic to assume that all of the bay's current water quality problems are the consequence of the loss of the oyster population and therefore can be corrected by efforts to restore this population. Other ecosystem stressors have intensified since the late 19th century, including increased nutrient runoff, higher sediment loads, climate shifts, and more toxic chemical pollution. These stressors could delay or prevent the expected improvements in water quality predicted with restoration of the oyster population.

Gerritsen et al. (1994) conclude that in regions of the bay with exceptionally high numbers of suspension feeders (e.g., upper shallow reaches, oligohaline regions of the bay), most of the primary production (phytoplankton biomass) is consumed by zooplankton and bivalves. Although the oyster population has declined in these areas, several species of invasive bivalves now provide a similar functional role as the oysters. In the middle and deeper regions of the Chesapeake, less than 50% of the primary production is consumed, and physical attributes of the bay's deeper waters will always limit the ability of suspension feeders to reduce the volume of phytoplankton biomass. These authors conclude that restoring oyster populations could improve water clarity in shallow zones, but for the large volume of water in the main stem of the bay, much more comprehensive efforts will be necessary to achieve significant improvements in water quality.

Increasing the biomass of suspension-feeding bivalves (in this case oysters) has the potential to enhance biofiltration. Under controlled triploid aquaculture, however, the effects will be spatially limited to the immediate grow-out areas. Intensive triploid aquaculture, similar to some of the coastal bays and lagoons in France where the Pacific oyster is raised, could increase filtration enough to improve water clarity in some of the shallow, retention-type tributaries of the bay. If diploid nonnative oysters were introduced and populated the bay, water clarity in shallow reaches could improve, an outcome similar to the zebra mussel invasion of the Great Lakes. However, high fishing pressure and mortality from predators of the native species (e.g., fish, crabs) could limit the population size and spread of an introduced oyster. In contrast, zebra mussels have no commercial value in the Great Lakes and are often found in areas where there are no natural predators.

Lastly, increased sedimentation of prime oyster settlement areas could prevent any oyster from achieving as large an abundance as the historical population of native oysters. This myth, though it has served to make political bedfellows of diverse stakeholders who share the goal of increasing the oyster population in the bay, should be replaced by the more realistic assumption that declining water quality results from multiple stressors that cannot be reversed by simply stocking more oysters in the bay. Oysters are only one part of the solution to the complex problems that affect water quality in this region.

Myth III: Restoration of Native Oyster Populations Has Been Tried and Will Not Work

Disenchantment with the failure of restoration efforts is evident in press accounts and the testimony received by the committee at its public meetings. While current advocates of restoration may claim that past expectations were too high and that signs of limited success exist, it is clear that despite the considerable resources that have been expended, large-scale reversals in the decline of native oyster populations have not been achieved. Nevertheless, it would be erroneous to infer that future restoration efforts will be entirely ineffectual. Restoration efforts to date have not corrected or ameliorated the multiple stressors that contributed to the loss of native oysters. For example, restoration efforts that address reef habitat but do not address disease are not likely to succeed. Similarly, restoration efforts that serve only to create a put-and-take fishery are unlikely to yield meaningful increases in baywide oyster populations. The idea of using disease-resistant oyster stocks for restoration rather than aquaculture has only recently been explored (Allen and Hilbish, 2000). The incidence and severity of oyster diseases have been exacerbated by drought conditions for the past 4 years. A series of more typical colder and wetter winters could give restoration efforts a substantial boost relative to progress in the recent past.

Myth IV: *C. ariakensis* Will Rapidly Populate the Bay, Increasing Oyster Landings and Improving Water Quality

The assumption that the Suminoe oyster will rapidly invade the Chesapeake Bay is heralded by those who see this as the salvation of a fishery on the verge of extinction and abhorred by those who fear ecological disaster. Champions of introduction base their assumption of rapid proliferation on the results of small field experiments with triploid *C. ariakensis*, in which the nonnative oyster grew rapidly and was resistant to or tolerant of the two diseases affecting the native oyster (Calvo et al., 2001; Thompson, 2001). The specter of an ecological disaster, on the other hand, is made by analogy to the negative consequences of invasions by the zebra mussel and other nuisance species. However, there are no data to suggest that reproductive populations of *C. ariakensis* will necessarily enjoy a rapid rate of population growth in the Chesapeake Bay. The Suminoe oyster has not invaded the West Coast of the United States, where it was inadvertently introduced with *C. gigas* in the 1970s (Langdon and Robinson, 1996), but this may not be a reliable indicator of its likelihood of becoming invasive on the East Coast. For example, the Pacific oyster has not been invasive on the U.S. West Coast, but in New Zealand and Australia this same species has invaded and displaced the native rock oyster. Indeed, review of the consequences of oyster introductions around the world suggests the difficulty if not the impossibility of predicting whether or not the Suminoe oyster will be invasive in the Chesapeake Bay and beyond. This myth is also based on the assumption that there has been no loss of suitable oyster habitat in the bay and that the current environment is capable of supporting oyster populations as large as those in the past. Increased sedimentation in the bay from upland sources already limits settlement of native oysters; the settlement of nonnative oysters would be similarly affected by sedimentation.

Myth V: Aquaculture of Triploid *C. ariakensis* Will Solve the Economic Problems of a Devastated Fishery and Restore the Ecological Services Once Provided by the Native Oyster

The primary motivation for pursuing triploid aquaculture of *C. ariakensis* is to reduce the risk of establishing a reproductive population of nonnative oysters in the bay. The use of triploids, however, does not completely eliminate the risk of nonnative oyster introduction. Indeed, the release of fertile diploids from aquaculture of triploid oysters is considered inevitable by a consensus of scientists, watermen and producers, regulators, nongovernmental organizations, and concerned citizens from Virginia, Maryland, North and South Carolina, Delaware, and New Jersey (Hallerman et al., 2001). However, this risk can be managed and reduced through application of stringent aquaculture protocols as described in the section below. A primary risk management strategy is to contain triploid oysters so as to ensure their removal from the water before they are likely to reproduce. Triploids may reproduce either through spawning of the small percentage of diploid oysters that occur in each batch of triploids, through spawning of triploid oysters that can produce diploid offspring, or through age-dependent reversion of triploid reproductive tissue to the diploid condition. Contained aquaculture, however, is likely to be limited by the relatively high costs of materials and labor. On the U.S. West Coast, single oysters are produced in contained aquaculture for the half-shell market, in which premium prices offset higher production costs. Contained aquaculture may not be competitive if the product is destined for the traditional shucked-meat market. Thus, contained aquaculture of triploid oysters would likely involve only a fraction of the bay's oyster industry, and the scale would be too small to make a noticeable difference in the water quality of the Chesapeake Bay. Success with triploid aquaculture could lead, however, to a push by industry for more economical on-bottom culture, which might be more competitive with wild harvest fisheries but would increase the risk of diploid escape, reproduction, and establishment.

RECOMMENDATIONS

Biosecurity Against Rogue Introductions

Regulations against rogue introductions and enforcement of existing regulations should be reviewed by responsible state agencies and the Chesapeake Bay Program. The likely routes of introduction and critical points where interdiction might be achieved should be identified, along with the resources required to achieve various levels of biosecurity. Genetic profiling of natural populations of *C. ariakensis* has begun (Francis et al., 2001) but should be expanded to make available a suitable set of diagnostic markers for forensic applications, such as determining the provenance of any unsanctioned introductions that are discovered. Most importantly, the public should be educated about the risks of unregulated introduction of nonnative oysters to increase awareness and vigilance.

Development of Standards for Regulating Nonnative Oyster Aquaculture

Before the commencement of open-water aquaculture (or pilot-scale field trials), the committee recommends the development of a stringent protocol to minimize and monitor the unintentional release of sexually competent *C. ariakensis*. The Hazard Analysis Critical Control Point system used in the field of food safety provides a framework for the development of the type of protocol that will be required. The protocol should include:

- assessment of the security of confinement and certification of adult brood stock, larvae, and seed in the hatchery producing the triploids;
- sampling to detect the frequency of diploid oysters in triploid seed lots produced by tetraploid by diploid crosses (chemical induction should not be used to induce triploidy because of the relatively high percentage of diploid larvae produced);
- statistical determination of the density of culture that reduces the risk of cross-fertilization of gametes from these diploids to an acceptable level;
- sampling to monitor gonadal maturation and to detect reversion;

- standards and protocols for removing nonnative oysters in the event of catastrophe, such as a storm, a disease outbreak, or unusual rates of gonadal maturation or reversion;
- accurate accounting of all animals, from hatchery through grow-out and disposal, including the use of genetic markers for identification in case nonnative oysters are found in open waters outside the grow-out containers;
- sampling protocol to detect morbidity and to ascertain the causes of unusual mortalities;
- documentation and prior approval of the transfer or relaying of nonnative oysters;
- estimation of the probability of control system failure, resulting in the escape or disappearance of subject oysters; and
- identification of parties responsible for monitoring and enforcement.

Bonding should be considered as an incentive to enhance compliance with the control system protocols.

Biological Research

Large gaps in biological knowledge exist for both native and nonnative oysters, and the biology of both species needs to be understood in the broader context of long-term environmental change in the Chesapeake Bay. The committee thus recommends a spectrum of biological research to provide information for risk assessments of the three management options, from specific near-term objectives to broader longer-term research goals. Many of these recommendations have been the subjects of previous or current research, but because the problems addressed are extremely complex, answers to some of these research needs will require significant effort over a period of several years.

The largest and most immediate gaps in knowledge concern the nonnative oyster. Some of this research could be conducted in quarantined experimental facilities; some of it could be conducted in conjunction with experimental field trials using triploid nonnative oysters. If introduction of triploid or diploid oysters does proceed, it should be done with risk-averse management practices based on a foundation of biological information. Data from such research should be integrated immediately into models of the biological risks associated with introduction of nonnative oysters as either triploids or diploids. Several short-term research objectives are needed to address critical gaps in knowledge of the biology of the nonnative oyster:

- continue the development and refinement of assay methods and hatchery protocols for the detection and elimination of pathogens that might be introduced, even using ICES protocols, via vertical transmission in oyster gametes;
- determine the susceptibility of *C. ariakensis* to *Bonamia ostreae* through challenge experiments, by obtaining DNA from the *Bonamia*-like organism associated with a *C. ariakensis* mortality in France (archived material available from IFREMER, La Tremblade) for comparison with known *B. ostreae* sequences, or both;
- develop a better understanding of *C. ariakensis* biology in the Chesapeake Bay, particularly its growth rate, gametogenic cycle, larval behavior, and settlement patterns in different hydrodynamic regimes; size-specific postsettlement mortality rates and susceptibility to native parasites, pathogens, and predators incorporating salinity and temperature dependencies of all features;
- develop sufficient data on the fidelity, stability, and fertility of mated triploid *C. ariakensis* to permit estimates not only of mean parameters, such as the proportions of triploids, diploids, or mosaics in lots of mated triploid seed, the proportion of triploid individuals undergoing gametogenesis, the fecundity of triploids, the types and proportions of gametes produced by triploids, the fertility of these gametes, etc., but also estimates of uncertainty in these parameters;
- determine the distance and range of concentrations over which fertilization can be achieved under various conditions of water flow;
- determine the ecological interactions of *C. ariakensis* and *C. virginica* at the juvenile, adult, and gametic life stages; and
- determine the genetic and phenotypic diversity of different geographic populations of *C. ariakensis* and other closely related Asian species of the genus *Crassostrea* and the extent to which they might respond differently to the Chesapeake Bay environment.

Longer-term research goals, though not immediately applicable to a decision about introducing the nonnative oyster within the next few months or years, are needed, nevertheless, to address larger questions about the ecological role and future abundance or success of native and nonnative oysters in the Chesapeake Bay:

- develop an integrated approach to understanding the responses of native and nonnative oysters to environmental change and multiple stressors, including naturally occurring or introduced dis-

eases, climate change, land use, nutrient loading, sedimentation, pollutants, and the interactions of these factors;
- develop an integrated, science-based approach to restoration of the native oyster. This would include development of an effective breeding program for building genetically diverse native oyster stocks that are tolerant of or resistant to Dermo and MSX. Integrate the resulting disease-tolerant or disease-resistant stocks into the restoration of protected sanctuary areas that favor local recruitment, as recommended by Allen and Hilbish (2000). Perform a cost-benefit analysis of an integrated restoration program;
- develop a model of oyster larval dispersion based on a detailed circulation model for the Chesapeake Bay; incorporate information about differences in the larval behavior or physiological performance between the native and nonnative oysters to predict their dispersal throughout and outside the Chesapeake Bay;
- determine the causes of recruitment success or failure for *C. virginica*;
- determine the success of sanctuaries for the native oyster and their relationship to recruitment;
- determine the genetic and physiological bases for disease tolerance or resistance of native oysters;
- determine why native oysters are not developing natural resistance in the Chesapeake Bay but do seem to be developing resistance in Delaware Bay; and
- determine the ecological role of oyster reefs, whether they are simply attractants or whether they provide essential services to species.

The number and difficulty of these big questions highlight the complete uncertainty about the outcome of such a dramatic ecological experiment as introducing a diploid nonnative oyster into the Chesapeake Bay, either directly or indirectly as the consequence of escape from triploid aquaculture.

Decision Making and Regulatory Framework

Characterize Public Versus Private Use of Submerged Lands

As discussed in Chapter 8, states differ in the extent of submerged lands offered for leasing, the security and enforcement of leaseholders' exclusive use rights, the conditions and constraints imposed on leaseholders, and the duration and conditions for renewal of leases. These institutional differences have shaped the development of oyster fisheries in Virginia and Maryland and will affect the economic and sociocultural impacts of the three management options. A better understanding of the institutional differences, their consequences, and their possible evolution will be critical to predicting the outcomes of the proposed options, the ability of managers to oversee and control production practices, and the potential for Maryland and Virginia oystermen to compete with producers in other regions. Institutional differences, their consequences, and their possible evolution need to be assessed. A better understanding of these factors will be critical to predicting the outcomes of the proposed options, the ability of managers to oversee and control production practices, and the potential for Maryland and Virginia oystermen to compete with producers in other regions.

Evaluation of Alternative Institutional Structures

The rules that govern access to fishery resources and the utilization of those resources have been shown to exert an important influence on the level of excess capital investment, the dissipation of economic profits, and the incentive to weigh the consequences of excessive harvests. Unlike other shellfish and finfish fisheries, the oyster fishery has not been carefully examined with regard to the attributes of alternative institutional designs. During this period of evolution in the oyster fishery, the current institutional structure and alternative structures should be reassessed. This analysis could consider the applicability of individual transferable quotas or harvester cooperatives, individual and community territorial use rights in fisheries (TURFs),[1] limited entry, long-term leases, and other institutional designs.

Review the Chesapeake Bay Program's Decision Making Authority

In Chapter 8, the existing regulatory and institutional framework was found to be inadequate for monitoring or overseeing the interjurisdictional aspects of open-water aquaculture or direct introduction of *C. ariakensis*. While the Chesapeake Bay Program (CBP) appears well positioned to assume monitoring and oversight authority, its decisions are nonbinding. Moreover, the CBP does not appear to have sufficient budget or personnel resources to support analyses of the ecological, economic, or sociocultural consequences of alternative management actions.

[1]TURFs are spatially based individual or collective harvest privileges that have often been established in less industrialized and smaller-scale coastal fisheries where management has been based on restricting participation to a localized population in a limited geographical area (Christy, 1982; Seijo, 1993).

The committee recommends a review of the CBP, by parties to the Chesapeake Bay Agreement, to determine whether the CBP could serve as a venue for interjurisdictional decision making and to identify what institutional changes would be required for the CBP or another multistate entity to have binding authority over management decisions that could affect the bay.

Economic and Sociocultural Analyses

Development of Baseline Data

The current economic and sociocultural characteristics of the oyster fishery and fishery-dependent communities need to be carefully documented to serve as a baseline for postimplementation analyses of the impacts of whatever actions are taken. Recent trends in the economic and sociocultural data series and likely ongoing trends also need to be documented to help isolate postimplementation changes that are a direct result of management action as opposed to the continuation of prior trends.

Economic Feasibility of Alternative Production Systems

Restoring oyster production to 1985 production levels using the same old production technologies operating under the same old institutional arrangements is unlikely to yield profits comparable to those realized in 1985; much of the rest of the world has moved past the hunter-gatherer stage of oyster production. The economic viability of different production systems should be examined for each management option and by production region. For example, under the status quo, estimates should be developed for the profitability of public and leased-bottom fisheries in Virginia and Maryland for tong, dredge, diver harvest, and power-dredge production modes. These traditional modes should be compared with the profitability of relaying to manage disease exposure, seeding of selected stock, and so forth. Similarly, examination of triploid nonnative aquaculture should consider the likely profitability of various contained and on-bottom culture systems, including the use of antipredator strategies.

Impacts of Alternative Production Systems on Communities and Households

Structured interviews should be conducted with watermen and coastal community members, during which they are provided with descriptions of the management options and alternative production systems and in which their opinions about the likely effects of the various combinations of management structure and production system are solicited. Information gained in the interviews will be useful in modeling the impacts of different management options and production systems on households and communities. These outcomes will depend in part on the economic consequences but also depend on whether the production systems associated with the options are compatible with current sociocultural norms and household production patterns. In addition to structured and systematic interviews, existing literature on changes in fishery production systems should be reviewed for information on the effects on fishers, processors (including workers), and socioeconomic and cultural characteristics of local communities.

Cost Minimization Model

While development of a comprehensive risk assessment model is daunting, development of a model to minimize the conditional costs of meeting biological risk assessment objectives is relatively straightforward and should be pursued. Although this approach would not integrate the full suite of risks and would not address risk tradeoffs, it would provide a criterion for selecting among alternative management decisions with similar levels of risk. For example, if the risk assessment model suggested that two alternatives had equal but acceptably low likelihoods of the unintended establishment of reproductive populations of *C. ariakensis*, managers could select the alternative that resulted in the least adverse impacts on watermen and fishery-dependent communities.

Restoration Activities

The proposed restoration activities should be assessed for probable level of success and the near- and intermediate-term economic consequences posed for watermen. The analysis should include expected net benefits and the variance of expected net benefits for harvesting and processing sectors in Maryland and Virginia for probable levels of recovery at different time points on the recovery schedule.

Establish an Ongoing Economic and Sociocultural Data Collection Program

The ability of social scientists to predict or evaluate the outcomes of contemplated management actions is impaired by a lack of baseline data and a lack of postimplementation data. The contemplated actions are likely to engender substantial changes in the economic organization of the fishery and fishing communities. Therefore, the states of Virginia and Maryland should establish programs to collect baseline economic and sociocultural data. Such data should include economic information on production costs, including capital and labor expenditures, market trends and marketing practices, and changes in economic strategies and decision making in response to changes in the fishery. Sociocultural information should be collected on household- and community-level responses to changes in the oyster fishery and how such changes modify traditional sociocultural norms of such communities. The collection of economic and sociocultural data should be coordinated to maximize integration and complementarity. The data should be collected at different levels of scale, ranging from baywide to subregions and communities where existing industry structures (e.g., public versus leased), ecological conditions (e.g., salinity), and harvesting practices (e.g., power dredging versus patent tonging) could result in different sociocultural and economic consequences.

Without good baseline data and consistent data collection over the next 5 to 10 years, it is unlikely that the effects of management action can be separated from the effects of unrelated changes in the fishery. While there is a tradition of this type of observation in some social sciences, there have been few longitudinal studies of fisheries. This research program could be organized through local Sea Grant offices or through a multistate entity but should be designed and budgeted for the full 5- to 10- year period with research conducted by a team of social scientists.

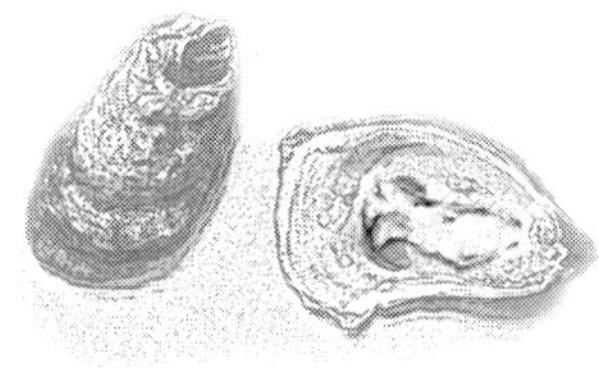

References

Agnello, R.J., and L.P. Donnelly. 1975. Property rights and efficiency in the oyster industry. *Journal of Law and Economics* 18:521-533.

Agnello, R.J., and L.P. Donnelly. 1976. Externalities and property rights in the fisheries. *Land Economics* 52:518-529.

Agnello, R.J., and L.P. Donnelly. 1984. Regulation and the structure of property rights: The case of the U.S. oyster industry. *Research in Law and Economics* 6:267-281.

Ahmed, M., S. Barkati, and M. Sanaullah. 1987. Spatfall of oysters in the Gharo-phitti salt water creek system near Karachi Pakistan. *Pakistan Journal of Zoology* 19(3):245-252.

Ajuzie, E.I., and M.A. Altobello. 1997. Property rights and pollution: Their implications for Long Island Sound and the oyster industry. *Review of Agricultural Economics* 19(2):242-251.

Allen, S.K., Jr. 1993. Triploids for field tests? The good, the bad, and the ugly. *Journal of Shellfish Research* 12(1):125.

Allen, S.K., Jr. 1999. Population control using triploid bivalves and the special problem of chromosome set instability. *Bulletin of National Research Institute of Aquaculture* (Suppl. 1):149.

Allen, S.K., S.L. Downing, and K.K. Chew. 1989. *Hatchery Manual for Producing Triploid Oysters*. University of Washington Press, Seattle, Washington.

Allen, S.K., Jr., and S.L. Downing. 1990. Performance of triploid Pacific oysters *Crassostrea Gigas* Gametogenesis. *Canadian Journal of Fisheries & Aquatic Sciences* 47(6):1213-1222.

Allen, S.K., Jr., P.M. Gaffney, J. Scarpa, and D. Bushek. 1993. Inviable hybrids of *Crassostrea virginica* (Gmelin) with *C. rivularis* (Gould) and *C. gigas* (Thunberg). *Aquaculture* 113:269-289.

Allen, S.K., Jr., and T.J. Hilbish. 2000. *Genetic Considerations for Hatchery-Based Restoration of Oyster Reefs*. Workshop Summary, September 21-22, 2000. Virginia Institute of Marine Science, Gloucester Point.

Allen, S.K., Jr., K. Sellner, and M. Zhou. 2002. Recommendations to the Chesapeake Scientific Community: Contacts and Opportunities for *C. ariakensis* Research in China. Unpublished report. For copy, contact S.K. Allen, Virginia Institute of Marine Science, Gloucester Point.

Alpine, A.E., and J.E. Cloern. 1992. Trophic interactions and direct physical effects control phytoplankton biomass and production in an estuary. *Limnology & Oceanography* 37(5):946-955.

Amemiya, I. 1928. Ecological studies of Japanese oyster, with special reference to the salinity of their habitats. Imperial University, Tokyo. *Journal of the College of Agriculture* 9:333-382.

Andrews, J.D. 1968. Oyster mortality studies in Virginia. VII. Review of epizootiology and origin of *Minchinia nelsoni*. *Proceedings of the National Shellfisheries Association* 58: 23-36.

Andrews, J.D. 1980. A review of introductions of exotic oysters and biological planning for new importations. *Marine Fisheries Review* 42(12):1-11.

Andrews, J.D. 1983. *Minchinia nelsoni* MSX infections in the James River Virginia USA seed oyster area and their expulsion in spring. *Estuarine Coastal & Shelf Science* 16(3):255-270.

Andrews, J.D., and W.G. Hewatt. 1957. Oyster mortality studies in Virginia II. The fungus disease caused by *Dermocystidium marinum* in oysters of Chesapeake Bay. *Ecological Monographs* 27(1):1-25.

Arve, J. 1960. Preliminary report on attracting fish by oyster-shell plantings in Chincoteague Bay, Maryland. *Chesapeake Science* 1:58-65.

Arzul, I., J.L. Nicolas, A.J. Davison, and T. Renault. 2001. French scallops: A new host for Ostreid herpesvirus-1. *Virology* 290(2):342-349.

Arzul, I., T. Renault, and C. Lipart. 2001. Experimental herpes-like viral infections in marine bivalves: Demonstration of interspecies transmission. *Diseases of Aquatic Organisms* 46(1):1-6.

Arzul, I., T. Renault, A. Thebault, and A. Gerard. 2002. Detection of oyster herpesvirus DNA and proteins in asymptomatic *Crassostrea gigas* adults. *Virus Research* 84(1-2):151-160.

Asif, M. 1979. Hermaphroditism and sex reversal in the 4 common oviparous species of oysters from the coast of Karachi Pakistan. *Hydrobiologia* 66(1):49-56.

Audemard, C., F. Le Roux, A. Barnaud, C. Collins, B. Sautour, P.G. Sauriau, X. De Montaudouin, C. Coustau, C. Combes, and F. Berthe. 2002. Needle in a haystack: Involvement of the copepod *Paracartia grani* in the life-cycle of the oyster pathogen *Marteilia refringens*. *Parasitology* 124(3):315-323.

Awati, P.R., and H.S. Rai. 1931. *Ostrea cucullata: The Bombay oyster. The Indian Zoological Memoirs on Indian Animal Types*. Methodist Publishing House, Lucknow, India. 107p.

Ayres, P. 1991. Introduced Pacific oysters in Australia. Pp. 3-8 in *The ecology of* Crassostrea gigas *in Australia, New Zealand, France and Washington State,* M.C. Greer, J.C. Leffler, and Maryland Sea Grant College Program, College Park.

Bacher, C., P. Duarte, J.G. Ferreira, M. Héral, and O. Raillard. 1997a. Assessment and comparison of the *Marennes Oleron* Bay (France) and *Carlingford Lough* (Ireland) carrying capacity with ecosystem models. *Aquatic Ecology* 31(4):379-394.

Bacher, C., B. Millet, and A. Vaquer. *1997b*. Modeling the impact of cultivated filter-feeders on phytoplanktonic biomass of the Thau lagoon (France). [French]. *Comptes Rendus De L'Academie Des Sciences - Serie Iii, Sciences De La Vie* 320(1):73-81.

Bahr, L.M. 1976. Energetic aspects of the intertidal oyster reef community at Sapelo Island, Georgia (USA). *Ecology* 57(1):121-131.

Bahr, L.M., and W.P. Lanier. 1981. *The Ecology of Intertidal Oyster Reefs of the South Atlantic Coast: A Community Profile*. U.S. Fish and Wildlife Service, Office of Biological Service, Washington, D.C., 105p.

Bain, M.B. 1987. Structured decision making in fisheries management: Trout fishing regulations on the Au Sable River, Michigan. *North American Journal of Fisheries Management* 7(4):475-481.

Baird, D., and R.E. Ulanowicz. 1989. The seasonal dynamics of the Chesapeake Bay ecosystem. *Ecological Monographs* 59(4):329-364.

Balouet, G. 1979. *Marteilia refringens* – Considerations of the life cycle and development of Abers disease in *Ostrea edulis*. *Marine Fisheries Review* 41(1-2):64-66.

Balouet, G., C. Chastel, A. Cahour, A. Quillard, and M. Poder. 1979. Étude épidémiologique et pathologique de la maladie de l'huître plate en Bretagne. *Bulletin Peches Maritimes* 289:13-25.

Balouet, G., M. Poder, and A. Cahour. 1983. *Haemocytic parasitosis*: Morphology and pathology of lesions in the French flat oyster, *Ostrea edulis*. *Aquaculture* 34(1-2):1-14.

Banks, M.A., D. Hedgecock, and C. Waters. 1993. Discrimination between closely related Pacific oyster species (*Crassostrea*) via mitochondrial DNA sequences coding for large subunit rRNA. *Molecular Marine Biology & Biotechnology* 2(3):129-136.

Banks, M.A., D.J. McGoldrick, W. Borgeson, and D. Hedgecock. 1994. Gametic incompatibility and genetic divergence of Pacific and Kumamoto oysters, *Crassostrea gigas* and *C. sikamea*. *Marine Biology* 121(1):127-135.

Barber, B.J. 1997. Impacts of bivalve introductions on marine ecosystems: A review. *Bulletin of National Research Institute of Aquaculture* (Supp. 3):141-153.

Barber, B.J., S.E. Ford, and R.N. Wargo. 1991. Genetic variation in the timing of gonadal maturation and spawning of the Eastern oyster, *Crassostrea virginica* (Gmelin). *Biological Bulletin* 181(2):216-221.

Barber, B.J., and R. Mann. 1994. Comparative physiology of Eastern oyster, *Crassostrea virginica* (Gmelin, 1791), and Pacific oyster, *Crassostrea gigas* (Thunberg, 1793): Growth, mortality and effects of the parasite, *Perkinsus marinus*. *Journal of Shellfish Research* 13(1):109-114.

Barrett, E.M. 1963. The California oyster industry. *California Department of Fish and Game Fishery Bulletin* 123:1-103.

Bartol, I.K., and R. Mann. 1997. Small-scale settlement patterns of the oyster *Crassostrea virginica* on a constructed intertidal reef. *Bulletin of Marine Science* 61(3):881-897.

Bartol, I.K., and R. Mann. 1999. Small-scale patterns of recruitment on a constructed intertidal reef: The role of spatial refugia. Pp. 159-170 in *Oyster Reef Habitat Restoration: A Synopsis and Synthesis of Approaches*, M.W. Luckenbach, R. Mann, and J.A. Wesson. Virginia Institute of Marine Science, Gloucester Point.

Bayne, B.L. 2002. A physiological comparison between Pacific oysters *Crassostrea gigas* and Sydney Rock oysters *Saccostrea glomerata*: Food, feeding and growth in a shared estuarine habitat. *Marine Ecology Progress Series* 232:163-178.

Bayne, B.L., and R.M. Warwick. 1998. Feeding and growth of bivalves: Observations and models. *Journal of Experimental Marine Biology and Ecology, Special Issue* 219. Elsevier, Amsterdam. 270p.

Bean, M.J. 1996. Legal authorities for controlling alien species: A survey of tools and their effectiveness. Pp. 271-281 in *Invasive Species and Biodiversity Management*, O.T. Sandlund, P.J. Schei, and A. Viken. Kluwer Academic, Dordrecht, Netherlands.

Beck, M.W., K.L. Heck Jr., K.W. Able, D.L. Childers, D.B. Eggleston, B.M. Gillanders, B. Halpern, C.G. Hays, K. Hoshino, and T.J. Minello. 2001. The identification, conservation, and management of estuarine and marine nurseries for fish and invertebrates. *Bioscience* 51(8):633-641.

Belval, D.L., and Sprague, L.A. 1999. Monitoring nutrients in the major rivers draining to Chesapeake Bay. U.S. Department of Interior, U.S. Geological Survey Water-Resources Investigations Report 99-4238.

Bergelson, J. 1994. Changes in fecundity do not predict invasiveness: A model study of transgenic plants. *Ecology (Tempe)* 75(1):249-252.

Berkman, P.A., D.W. Garton, M.A. Haltuch, G.W. Kennedy, and L.R. Febo. 2000. Habitat shift in invading species: Zebra and Quagga mussel population characteristics on shallow soft substrates. *Biological Invasions* 2(1):1-6.

Berman, M., S. Killeen, R. Mann, and J. Wesson. 2002. *Virginia Oyster Reef Restoration Map Atlas*. Virginia Institute of Marine Science, Center for Coastal Resources Management, Gloucester Point.

Berthe, F. 2002. Taxinomie et épidémiologie moléculaires du parasite *Marteilia refringens*: Intérêt pour la gestion du risque zoosanitaire en conchyliculture. PHD dissertation, Université Blaise Pascal, Clermont-Ferrand, 127p.

Berthe, F.C., F. Le Roux, E. Peyretaillade, P. Peyret, D. Rodriguez, M. Gouy, and C.P. Vivares. 2000. Phylogenetic analysis of the small subunit ribosomal RNA of *Marteilia refringens* validates the existence of phylum Paramyxea (Desportes and Perkins, 1990). *Journal of Eukaryotic Microbiology* 47(3):288-293.

Biber, E. 1999. Exploring regulatory options for controlling the introduction of non-indigenous species to the United States. *Virginia Environmental Law Journal* 18:375-465.

Bigley, R.E., and P.G. Harrison. 1986. Shoot demography and morphology of *Zostera japonica* and *Ruppia maritime* from British Columbia, Canada. *Aquatic Botany* 24:69-82.

Bigliardi, E., and A. Carapelli. 2002. Microsporidia in the springtail *Isotomurus fucicolus* (Collembola, Isotomidae) and possible pathways of parasite transmission. *Italian Journal of Zoology* 69(2):109-113.

Biosecurity Australia. 2002. *Import Risk Analysis (IRA) of Non-Viable Bivalve Molluscs*. Technical Issues Paper. Australian Government, Department of Agriculture, Fisheries and Forestry, Canberra, Australia, 67p.

Bockstael, N.E., K.E. McConnell, and I.E. Strand. 1987a. *Benefits from Improvements in Chesapeake Bay Water Quality. Benefit Analysis Using Indirect or Imputed Market Methods, Volume II*. EPA 811043-01-0. U.S. Environmental Protection Agency, Washington, D.C.

Bockstael, N.E., W.M. Hanemann, and C.L. Kling. 1987b. Estimating the value of water quality improvements in a recreational demand framework. *Water Resources Research* 23(5):951-960.

Bockstael, N.E., K.E. McConnell, and I.E. Strand. 1988. Benefits from improvement in Chesapeake Bay water quality. *Benefit Analysis Using Indirect or Imputed Market Methods, Volume III*. U.S. EPA-811043-01-0. U.S. Environmental Protection Agency, Washington, D.C.

Bockstael, N.E., K.E. McConnell, and I.E. Strand. 1989. Measuring the benefits of improvements in water quality: The Chesapeake Bay. *Marine Resource Economics* 6(1):1-18.

Boicourt, W.C., M. Kuzmić, and T.S. Hopkins. 1999. The inland sea: Circulation of Chesapeake Bay and the Northern Adriatic. Pp. 81-130 in *Coastal and Estuarine Studies: Ecosystems at the Land-Sea Margin: Drainage Basin to Coastal Sea*, T.C. Malone, A. Malej, L.W. Harding, Jr., N. Smodlaka, and R.E. Turner, eds. American Geophysical Union, Washington, D.C.

Bosch, D.J., and L.A. Shabman. 1989. The decline of private sector oyster culture in Virginia: Causes and remedial policies. *Marine Resource Economics* 6(3):227-243.

Bosch, D.J., and L.A. Shabman. 1990. Simulation modeling to set priorities for research on oyster production. *American Journal of Agricultural Economics* 72(2):371-381.

Boudry, P., S. Heurtebise, B. Collet, F. Cornette, and A. Gerard. 1998. Differentiation between populations of the Portuguese oyster, *Crassostrea angulata* (Lamark) and the Pacific oyster, *Crassostrea gigas* (Thunberg), revealed by mtDNA RFLP analysis. *Journal of Experimental Marine Biology and Ecology* 226(2):279-291.

Bower, S.M. 1987. *Labyrinthuloides-haliotidis* new-species protozoa labyrinthomorpha a pathogenic parasite of small juvenile Abalone in a British Columbia Canada mariculture facility. *Canadian Journal of Zoology* 65(8):1996-2007.

Bower, S.M., S.E. McGladdery, and I.M. Price. 1994. Synopsis of infectious diseases and parasites of commercially exploited shellfish. *Annual Review of Fish Diseases* 4(0):1-199.

Boynton, W.R., J.H. Garber, R. Summers, and W.M. Kemp. 1995. Inputs, transformations, and transport of nitrogen and phosphorus in Chesapeake Bay and selected tributaries. *Estuaries* 18(1B):285-314.

Boynton, W.R., and W.M. Kemp. 2000. Influence of river flow and nutrient loads on selected ecosystem processes: A synthesis of Chesapeake Bay data. Pp. 269-298 in *Estuarine Science: A Synthetic Approach to Research and Practice,* J.E. Hobbie. Island Press, Washington, D.C.

Branch, G.M. 1984. Competition between marine organisms: Ecological and evolutionary implications. *Oceanography and Marine Biology: An Annual Review* 64:429-593.

Breese, W.P., and R.E. Malouf. 1977. Hatchery rearing techniques for the oyster *Crassostrea rivularis. Aquaculture* 12(2):123-126.

Breitburg, D.L. 1992. Episodic hypoxia in Chesapeake Bay: Interacting effects of recruitment, behavior, and physical disturbance. *Ecological Monographs* 62(4):525-546.

Breitburg, D.L. 1999. Are three-dimensional structure and healthy oyster populations key to an ecologically interesting and important fish community? Pp. 239-250 in *Oyster Reef Habitat Restoration: A Synopsis and Synthesis of Approaches,* M.W. Luckenbach, R. Mann, and J.A. Wesson. Virginia Institute of Marine Science, Gloucester Point.

Breitburg, D.L., M.A. Palmer, and T. Loher. 1995. Larval distributions and the spatial patterns of settlement of an oyster reef fish: Responses to flow and structure. *Marine Ecology Progress Series* 125:45-60.

Breitburg, D., and T.J. Miller. 1999. *Importance of Vertical Relief to Use of Oyster Reefs by Commercially and Recreationally Important Fishes and Blue Crab: Final Report.* UMCES TS-200-99. University of Maryland Center for Environmental Science, Chesapeake Biological Laboratory, Chesapeake Bay Program, Solomons, Maryland, 32p.

Bricelj, V.M., S.E. Ford, F.J. Borrero, F.O. Perkins, G. Rivara, R.E. Hillman, R.A. Elston, and J. Chang. 1992. Unexplained mortalities of hatchery-reared, juvenile oysters, *Crassostrea virginica* (Gmelin). *Journal of Shellfish Research* 11:(2)331-347.

Brush, G.S., and W.B. Hilgartner. 2000. Paleoecology of submerged macrophytes in the upper Chesapeake Bay. *Ecological Monographs* 70(4):645-667.

Buroker, N.E., W.K. Hershberger and K.K. Chew. 1979a. Population genetics of the family Ostreidae. I. Intraspecific studies of *Crassostrea gigas* and *Saccostrea commercialis. Marine Biology* 54:157-169.

Buroker, N.E., W.K. Hershberger and K.K. Chew. 1979b. Population genetics of the family Ostreidae. II. Interspecific studies of the genera *Crassostrea* and *Saccostrea. Marine Biology* 54:171-184.

Buroker, N.E. 1983. Population genetics of the American oyster *Crassostrea virginica* along the Atlantic coast and the Gulf of Mexico. *Marine Biology* 75(1):99-112.

Burrell, V.G., Jr, M.Y. Bobo, and J.J. Manzi. 1984. A comparison of seasonal incidence and intensity of Perkinsus marinus between subtidal and intertidal oyster populations in South Carolina. *Journal of the World Mariculture Society* 14:301-309.

Burreson, E.M. 1991. Effects of *Perkinsus marinus* infection in the Eastern oyster I: Susceptibility of native and MSX-resistant stocks. *Journal of Shellfish Research* 10:(2)417-424.

Burreson, E.M. 1994. Further evidence of regular sporulation by *Haplosporidium nelsoni* in small oysters, *Crassostrea virginica*. *Journal of Parasitology* 80(6):1036-1038.

Burreson, E.M. 2000. Status of the major oyster diseases in the Chesapeake Bay in 1999 and considerations for future restoration efforts. Pp. 8-9 in *Chesapeake Bay Program Oyster Restoration, Workshop Proceedings and Agreement Statement* held in Waldorf, Maryland, January 13-14, 2000. Chesapeake Bay Program, Annapolis.

Burreson, E.M., and J.D. Andrews. 1988. Unusual intensification of Chesapeake Bay oyster diseases during recent drought conditions. Pp. 799-802 in *Oceans' 88 Proceedings* held in Baltimore, Maryland, October 31-November 2, 1988. The IEEE Service Center, Piscataway, New Jersey.

Burreson, E.M., and L.M. Ragone-Calvo. 1996. Epizootiology of *Perkinsus marinus* disease of oysters in Chesapeake Bay, with emphasis on data since 1985. *Journal of Shellfish Research* 15(1):17-34.

Burreson, E.M., N.A. Stokes, and C.S. Friedman. 2000. Increased virulence in an introduced pathogen: *Haplosporidium nelsoni* (MSX) in the Eastern oyster *Crassostrea virginica*. *Journal of Aquatic Animal Health* 12:1-8.

Bushek, D., and S.K. Allen, Jr. 1996. Host-parasite interactions among broadly distributed populations of the Eastern oyster *Crassostrea virginica*, and the protozoan *Perkinsus marinus*. *Marine Ecology Progress Series* 139:127-141.

Cahn, A.R. 1950. *Oyster culture in Japan*. Report. General Headquarters, Supreme Commander for the Allied Powers, Natural Resources Section, Tokyo, 80p.

Cahour, A. 1979. *Marteilia refringens* and *Crassostrea gigas*. *U.S. National Marine Fisheries Service Marine Fisheries Review* 41(1-2):19-20.

Cai, Y., C. Deng, and Z. Lui. 1992. Studies on the ecology of *Crassostrea rivularis* (Gould) in Zhanjiang Bay. [Chinese] *Tropic Oceanology/Redai Haiyang*. Guangzhou 11(3):37-44.

Cai, Y., and X. Li. 1990. Oyster culture in the People's Republic of China. *World Aquaculture* 21(4): 67-73.

Cai, Y., Y. Zhang, and R. Wei. 1979. *Bei lei xue gai lun. Di 1 ban ed. Shanghai: Shanghai ke xue ji shu chu ban she*. Xin hua shu dian Shanghai fa xing suo fa xing, 6, 303p.

Calvo, G.W., M.W. Luckenbach, S.K. Allen, Jr., and E.M. Burreson. 1999. Comparative field study of *Crassostrea gigas* (Thunberg, 1893) and *Crassostrea virginica* (Gmelin, 1791) in relation to salinity in Virginia. *Journal of Shellfish Research* 18(2):465-473.

Calvo, G.W., M.W. Luckenbach, S.K. Allen, Jr., and E.M. Burreson. 2001. A Comparative field study of *Crassostrea ariakensis* (Fujita, 1913) and *Crassostrea virginica* (Gmelin 1791) in relation to salinity in Virginia. *Journal of Shellfish Research* 20(1):221-229.

Carlton, J.T. 1985. Transoceanic and interoceanic disperal of coastal marine organisms: The biology of ballast water. *Oceanography Marine Biology Annual Reviews* 23:313-374.

Carlton, J.T. 1987. Patterns of transoceanic marine biological invasions in the Pacific Ocean. *Bulletin of Marine Science* 41:452-465.

Carlton, J.T. 1989. Man's role in changing the face of the ocean biological invasions and implications for conservation of near-shore environments. *Conservation Biology* 3(3):265-273.

Carlton, J.T. 1992a. Dispersal of living organisms into aquatic ecosystems as mediated by aquaculture and fisheries activities. Pp. 13-46 in *Dispersal of Living Organisms Into Aquatic Ecosystems*, A. Rosenfield and R. Mann, eds. Maryland Sea Grant College Program, College Park.

Carlton, J.T. 1992b. Introduced marine and estuarine mollusks of North America: An end-of-the 20th century perspective. *Journal of Shellfish Research* 11(2):489-505.

Carlton, J.T. 1999a. Molluscan invasions in marine and estuarine communities. *Malacologia* 41(2):439-454.

Carlton, J.T. 1999b. The scale and ecological consequences of biological invasions in the world's oceans. Pp. 195-212 in *Invasive Species and Biodiversity Management*, O.T. Sandlund, P.J. Schei, and Å.e. Viken. Kluwer Academic Publishers, Dordrecht, Netherlands.

Carlton, J.T. 2001. *Introduced Species in U.S. Coastal Waters: Environmental Impacts and Management Priorities*. Pew Oceans Commission, Arlington, Virginia, 28p.

Carlton, J.T., and J.B. Geller. 1993. Ecological roulette: the global transport of nonindigenous marine organisms. *Science* 261(5117):78-82.

Carpenter, S.R., and J.F. Kitchell. 1988. Consumer control of lake productivity. *Bioscience* 38(11):764-769.

Carriker, M.R., and P.M. Gaffney. 1996. A Catalogue of Selected Species of Living Oysters *(Ostreacea)* of the World. Pp. 1-18 in *The Eastern Oyster*: Crassostrea virginica, V.S. Kennedy, R.I.E. Newell, and A.F. Eble. Maryland Sea Grant College Program, College Park.

Case, T.J. 1990. Invasion resistance arises in strongly interacting species-rich model competition communities. *Proceedings of the National Academy of Sciences of the United States of America* 87(24):9610-9614.

Chambers, J.R. 1992. Coastal degradation and fish population losses. Pp. 45-51 in *Stemming the Tide of Coastal Fish Habitat Loss*, R.H. Stroud. National Coalition for Marine Conservation, Savannah, Georgia.

Cheney, D.P., B.F. MacDonald, and R.A. Elston. 2000. Summer mortality of Pacific oysters, *Crassostrea gigas* (Thunberg): Initial findings on multiple environmental stressors in Puget Sound, Washington, 1998. *Journal of Shellfish Research* 19(1):353-359.

Chesapeake Bay Program. 1993. Chesapeake Bay Policy for the Introduction of Non-indigenous Aquatic Species. U.S. Environmental Protection Agency, Washington, D.C.

Chesapeake Bay Program. 2000. Chesapeake 2000 and the Bay: Where Are We and Where Are We Going? [Online] Available: http://www.chesapeakebay.net/pubs/Snapc2k.pdf [May 1, 2003].

Chesapeake Bay Program. 2002. Comprehensive Oyster Management Plan Draft. [Online] Available: http://www.chesapeakebay.net/info/oyster_plan.cfm [October 13, 2003].

Chesapeake Research Consortium. 1999. Consensus of a Meeting of Scientific Experts. Virginia Institute of Marine Science, Glousester Point.

Chew, K.K. 1984. Recent advances in the cultivation of molluscs in the Pacific United States and Canada. *Aquaculture* 39 (1-4):69-81.

Chew, K.K. 1989. Manila clam biology and fishery development in western North America. Pp. 234-261 in *Clam Mariculture in North America*, J.J. Manzi, and M. Castagna, eds. (Developments in Aquaculture and Fisheries Science Vol. 19) Elsevier, New York, 461p.

Christy, F.T. 1964. The Exploitation of a Common Property Natural Resource: The Maryland Oyster Industry. Ph.D. Dissertation, University of Michigan, Ann Arbor.

Christy, F.T. 1982. *Territorial Use Rights in Marine Fisheries: Definitions and Conditions*. FAO Fisheries Technical Paper No. 227. Food and Agriculture Organization of the United Nations, Rome.

Chu, F.-L.E., and F. La Peyre. 1993. *Perkinsus marinus* susceptibility and defense-related activities in eastern oysters *Crassostrea virginica*: Temperature effects. *Diseases of Aquatic Organisms* 16(3):223-234.

Chu, F.-L.E., J.F. La Peyre, and C.S. Burreson. 1993. *Perkinsus marinus* susceptibility and defense-related activities in eastern oysters, *Crassostrea virginica*: Salinity effects. *Journal of Invertebrate Pathology* 62:226-232.

Cigarria J., and R. Elston. 1997. Independant introduction of *Bonamia ostreae*, a parasite of *Ostrea edulis* to Spain. *Diseases of Aquatic Organisms* 29:157-158.

Clark, E. 1964. *The oysters of Locmariaquer*. Pantheon Books, New York, NY, 203p.

Cloern, J.E. 1982. Does the benthos control phyto plankton biomass in south San Francisco Bay, California, USA? *Marine Ecology-Progress Series* 9(2):191-202.

Cloern, J.E. 1996. Phytoplankton bloom dynamics in coastal ecosystems: A review with some general lessons from sustained investigation of San Francisco Bay, California. *Reviews of Geophysics* 34(2):127-168.

Cloern, J.E., A.E. Alpine, B.E. Cole, and T. Heller. 1992. Seasonal Changes in the Spatial Distribution of Phytoplankton in Small Temperate-Zone Lakes. *Journal of Plankton Research* 14(7):1017-1024.

Clout, M.N. 1999. Biodiversity conservation and the management of invasive animals in New Zealand. Pp. 349-361 in *Invasive species and biodiversity management,* O.T. Sandlund, P.J. Schei, and Å.e. Viken. Kluwer Academic, Dordrecht, Netherlands.

Clout, M.N., and S.J. Lowe. 2000. Invasive species and environmental changes in New Zealand. Pp. 369-383 in *Invasive Species in a Changing World,* H.A. Mooney and R.J. Hobbs. Island Press, Washington, D.C.

Coan, E.V., P.V. Scott, and F.R. Bernard. 2000. *Bivalve seashells of western North America : Marine bivalve mollusks from Arctic Alaska to Baja California.* 1st ed. Santa Barabara Museum of Natural History Monographs. Studies in Biodiversity, No. 2. Santa Barbara Museum of Natural History, Santa Barbara, CA, 764p.

Cochennec, N., T. Renault, P. Boudry, B. Chollet, and A. Gerard. 1998. *Bonamia*-like parasite found in the Suminoe oyster *Crassostrea rivularis* reared in France. *Diseases of Aquatic Organisms* 34(3):193-197.

Coen, L.D., M.W. Luckenbach, and D.L. Breitburg. 1999. The role of oyster reefs as essential fish habitat: A review of current knowledge and some new perspectives. *American Fisheries Society Symposium* 22:438-454.

Coen, L.D., and M.W. Luckenbach. 2000. Developing success criteria and goals for evaluating oyster reef restoration: Ecological function or resource exploitation? *Ecological Engineering* 15(3-4):323-343.

Cohen, A.N. 2001. *Washington State Exotics Expedition: A rapid survey of exotic species in the shallow waters of Elliott Bay, Totten and Eld Inlets, and Willapa Bay.* Washington State Department of Natural Resources, Nearshore Habitat Program, Olympia, 47p.

Cohen, A.N., and J.T. Carlton. 1995. *Nonindigenous Aquatic Species in a United States Estuary: A Case Study of the Biological Invasions of the San Francisco Bay and Delta: A report for the United States Fish and Wildlife Service, Washington, D.C. and the National Sea Grant College Program.* United States Fish and Wildlife Service, Washington, D.C., 25p.

Cohen, A.N., and J.T. Carlton. 1998. Accelerating invasion rate in a highly invaded estuary. *Science* 279(5350):555-558.

Cohen, A.N., H.D. Berry, C.E. Mills, D. Milne, K. Britton-Simmons, M.J. Wonham, D.J. Secord, J.A. Barkas, B. Bingham, B.E. Bookheim, J.E. Byers, J.W. Chapman, J.R. Cordell, B. Dumbauld, A. Fukuyama, L.H. Harris, A.J. Kohn, K. Li, T.F. Mumford Jr., V. Radashevsky, A.T. Sewell, and K. Welch. 2001. *Washington State Exotics Expedition 2000: A Rapid Survey of Exotic Species in the Shallow Waters of Elliot Bay, Totten and Eld Inlets, and Willapa Bay.* Washington State Department of Natural Resources, Olympia, Washington, Nearshore Habitat Program. 52p. [Online]. Available: http://faculty.washington.edu/cemills/WSX2000.pdf [October 13, 2003].

Comps, M. 1988. Epizootic diseases of oysters associated with viral infections. *American Fisheries Society* 18(Spec. Pub):23-37.

Comps, M., M.S. Park, and I. Desportes. 1986. Etude ultrastructurale de *Marteilioides chungmuensis* n.g. n.sp., parasite des ovocytes de l'huître *Crassostrea gigas* Th. *Protistologica* 22:279-285.

Conte, F. S., S.C. Harbell, R. L. RaLonde. 1996. *Oyster culture: fundamentals and technology of the West Coast industry*. WRAC Publication No. 94-101. Western Regional Aquaculture Center, Seattle, Washington.

Cook, T., M. Folli, J. Klinck, S. Ford, and J. Miller. 1998. The relationship between increasing sea-surface temperature and the northward spread of *Perkinsus marinus* (Dermo) disease epizootics in oysters. *Estuarine, Coastal and Shelf Science* 46(4):587-597.

Cooper, S.R., and G.S. Brush. 1991. Long-term history of Chesapeake Bay anoxia. *Science* 254(5034):992-996.

Cooper, S.R., and G.S. Brush. 1993. A 2,500-year history of anoxia and eutrophication in Chesapeake Bay. *Estuaries* 16:617-626.

Craig, R.K. 2002. The other side of sustainable aquaculture: Mariculture and nonpoint source pollution. *Washington Journal of Law and Policy* 9:163-203.

Crawford, C.M., C.K.A. Macleod, and I.M. Mitchell. 2003. Effects of shellfish farming on the benthic environment. *Aquaculture* 224(1-4):117-140.

Crawley, M.J. 1989. Chance and timing in biological invasions. Pp. 407-423 in *Biological Invasions: A Global Perspective,* J.A. Drake, H.A. Mooney, F. di Castri, R.H. Groves, F.J. Kruger, M. Rejmanek, and M. Williamson. John Wiley and Sons, Chichester, UK.

Criddle, K.R. 1996. Predicting the consequences of alternative harvest regulations in a sequential fishery. *North American Journal of Fisheries Management* 16(1):30-40.

Criddle, K.R., and A.Y. Streletski. 2000. Multiple criterion management of a sequential fishery. *Annals of Operations Research* 94:259-273.

Cunningham, C.W., and T.M. Collins. 1994. Developing model systems for molecular biogeography: Vicariance and interchange in marine invertebrates. Pp. 406-433 in *Molecular Ecology and Evolution: Approaches and applications*, B. Schierwater, B. Streit, G.P. Wagner, and R. DeSalle, eds. Birkhauser Verlag, Basel, Switzerland, 405-433.

Dame, R.F. 1976. Energy flow in an intertidal oyster population. *Estuarine Coastal and Marine Sciences* 4:243-253.

Dame, R.F. 1979. The abundance, diversity and biomass of macrobenthos on North Inlet, South Carolina, intertidal oyster reefs. *Proceedings of the National Shellfish Association* 68:6-10.

Dame, R.F. 1993. *Bivalve filter feeders in estuarine and coastal ecosystem processes*. NATO ASI Series, No. 33, Series G, Ecological Sciences. Springer-Verlag, Berlin, 579p.

Dame, R.F. 1996. *Ecology of Marine Bivalves: An Ecosystem Approach*. CRC Press, Boca Raton, Florida.

Dame, R.F., and B.C. Patten. 1981. Analysis of energy flows in an intertidal oyster reef. *Marine Ecology Progress Series* 5(2):115-124.

Dame, R.F., J.D. Spurrier, and T.G. Wolaver. 1989. Carbon nitrogen and phosphorus processing by an oyster reef. *Marine Ecology Progress Series* 54(3):249-256.

Daniel, P., T. Cole, and D. Rittschof. 1983. Chemoreception and Life History of *Stylochus ellipticus*. *Journal of Shellfish Research* 3(1):84-88.

Darnell, R.M., and T.M. Soniat. 1981. Nutrient enrichment and estuarine health. Pp. 225-245 in *Estuaries & Nutrients*. CRC Publication No. 90 (1981). Chesapeake Research Consortium, Clifton, New Jersey.

Davis, H.C. 1950. On Interspecific Hybridization in *Ostrea*. *Science* 111(2889):522.

Davis, H.C. 1958. Survival and growth of clam and oyster larvae at different salinities. *Biological Bulletin* 114:296-307.

Davis, H.C., and A. Calabrese. 1964. Combined effects of temperature and salinity on development of eggs and growth of larvae of *M. mercenaria* and *C. virginica*. *Fishery Bulletin* 63:643-655.

Davis, J.P. 1988. Growth rate of sibling diploid and triploid Pacific oysters, *Crassostrea gigas*. *Journal of Shellfish Research* 7(1):202.

Day, A.J., A.J.S. Hawkins, and P. Visootiviseth. 2000. The use of allozymes and shell morphology to distinguish among sympatric species of the Rock oyster *Saccostrea* in Thailand. *Aquaculture* 187(1-2):51-72.

DeAlteris, J.T. 1988. The geomorphic development of Wreck Shoal, a subtidal oyster reef of the James River, Virginia. *Estuaries* 11(4):240-249.

Dean, D. 1979. Introduced species and the Maine situation. Pp. 149-164 in *Exotic Species in Mariculture: Proceedings of a Symposium on Exotic Species in Mariculture: Case histories of the Japanese oyster, Crassostrea gigas (Thunberg), with Implications for Other Fisheries*, R. Mann, ed. Woods Hole Oceanographic Institution, Massachusetts, September 18 -20, 1978. MIT Press, Cambridge, 363p.

Dekshenieks, M.M., E.E. Hofmann, J.M. Klinck, and E.N. Powell. 1996. Modeling the vertical distribution of oyster larvae in response to environmental conditions. *Marine Ecology Progress Series* 136:97-110.

Dentler, J.L. 1993. Noah's farce: The regulation and control of exotic fish and wildlife. *University of Puget Sound Law Review* 17:191-242.

Deslous-Paoli J.-M. 1981. *Mytilicola orientalis* Mori, *Crassostrea gigas* Thunberg's parasite, in the basin of Marennes-Oleron: Impact on the condition and the biochemical composition of oysters during rearing. *ICES Council Meeting 1981 (Collected Papers).* International Council for the Exploration of the Sea, Copenhagen, Denmark, 16p.

Dew, J.R., J. Berkson, E.M. Hallerman, and S.K. Allen, Jr. 2003. A model for assessing the likelihood of self-sustaining populations resulting from commercial production of triploid Suminoe oysters (*Crassostrea ariakensis*) in Chesapeake Bay. *Fishery Bulletin* 101:758-768.

Dinamani P. 1991a. The northern rock oyster *Saccostrea glomerata* (Gould, 1850) in New Zealand. In *Estuarine and marine bivalve mollusk culture*, E.W. Menzel. CRC Press Boca Raton, Florida.

Dinamani, P. 1991b. Introduced Pacific oysters in New Zealand. Pp. 9-12 in *The ecology of* Crassostrea gigas *in Australia, New Zealand, France and Washington State*, M.C. Greer and J.C. Leffler, eds. Maryland Sea Grant College Program, College Park, Maryland.

Dittman, D.E. 1997. Latitudinal compensation in oyster ciliary activity. *Functional Ecology* 11(5):573-578.

Dittman, D.E., S.E. Ford, and H.H. Haskin. 1998. Growth patterns in oysters, *Crassostrea virginica*, from different estuaries. *Marine Biology (Berlin)* 132(3):461-469.

Dow, R.L., and D.E. Wallace. 1971. Japanese oysters *(crassostrea gigas)* in Maine. *Fisheries Circular*. Maine Department of Marine Resources, Augusta, Maine, 5p.

Downing, S.L., and S.K. Allen, Jr. 1987. Induced triploidy in the Pacific oyster *Crassostrea-Gigas* optimal treatments with Cytochalasin B depend on temperature. *Aquaculture* 61(1):1-16.

Dunn, A.M., R.S. Terry, and J.E. Smith. 2001. Transovarial transmission in the microsporidia. *Advances in Parasitology* 48:57-100.

Dunning, R.D., and C.M. Adams. 1995. Economic analysis of the potential for Eastern oyster *Crassostrea virginica* (Gmelin, 1791) depuration in coastal northwest Florida. *Journal of Shellfish Research* 14:113-119.

Easley Jr., J.R. 1982. Property rights in shellfish relaying: managing fisheries for higher economic returns. *North American Journal of Fisheries Management* 4:343-350.

Edwards, C. 1976. A study in erratic distribution: The occurrence of the medusa *Gonionemus* in relation to the distribution of oysters. *Advances in Marine Biology* 14:251-284.

Ehrlich, P.R. 1986. Which animal will invade? *Ecological Studies Analysis & Synthesis* 58:79-95.

Elston, R.A. 1984. Prevention and management of infectious diseases in intensive mollusc husbandry. *Journal of the World Mariculture Society* 15:284-300.

Elston, R.A. 1990. Mollusc Diseases: Guide for the Shellfish Farmer. University of Washington, Washington Sea Grant, Seattle, Washington, 73p.

Elston, R.A. 1999. *Health Management, Development and Histology of Seed Oysters.* World Aquaculture Society, Baton Rouge, Louisiana.

Elston, R.A., and M.T. Wilkinson. 1985. Pathology management and diagnosis of oyster velar virus disease. *Aquaculture* 48(3-4):189-210.

Elston, R.A., C.A. Farley, and M.L. Kent. 1986. Occurrence and significance of Bonamiasis in European flat oysters *Ostrea-edulis* in North America. *Diseases of Aquatic Organisms* 2(1):49-54.

Elston, R.A., J.H. Beattie, C. Friedman, R. Hedrick and M.L. Kent. 1987. Pathology and significance of fatal inflammatory bacteraemia in the Pacific oyster, *Crassostrea gigas* Thunberg. *Journal of Fish Diseases* 10:121-132.

Everett, R.A., G.M. Ruiz, and J.T. Carleton. 1995. Effect of oyster mariculture on submerged aquatic vegetation: An experimental test in a Pacific Northwest estuary. *Marine Ecology Progress Series* 125:205–221.

Ewel, J.J., D.J. O'Dowd, J. Bergelson, C.C. Daehler, C.M. D'Antonio, L.D. Gomez, D.R. Gordon, R.J. Hobbs, A. Holt, and K.R. Hopper. 1999. Deliberate introductions of species: Research needs. *Bioscience* 49(8):619-630.

Farley, C.A. 1975. Epizootic and enzootic aspects of *Minchinia nelsoni* (Haplosporida) disease in Maryland oysters. *Journal of Protozoology* 22:418-427.

Farley, C.A. 1978. Viruses and virus-like lesions in marine mollusks. *Marine Fisheries Review* 40(10):18-20.

Farley, C.A. 1992. Dispersal of pathogens, parasites, pests, predators and competitors. Pp. 139-154 in *Dispersal of living organisms into aquatic ecosystems,* A. Rosenfield and R. Mann, eds. Maryland Sea Grant College Program, College Park, Maryland.

Farley, C.A., W.G. Banfield, G. Kasnic, Jr., and W.S. Foster. 1972. Oyster herpes-type virus. *Science* 178(62):759-760.

Farley, C.A., P.H. Wolf, and R.A. Elston. 1988. A long term study of "Microcell" disease in oysters with a description of a new genus — *Mikrocytos* (g.n.) and two new species — *Mikrocytos mackini* (sp. n.) and *Mikrocytos roughleyi* (sp. n.). *U.S. National Marine Fish Service Bulletin* 86:581-593.

Fegley, S.R., S.E. Ford, J.N. Kraeuter, and D.R. Jones. 1994. *Relative Effects of Harvest Pressure and Disease Mortality on Eastern Oyster Population Dynamics in Delaware Bay.* Final Report No. NA26FL0588. National Oceanic and Atmospheric Administration, National Marine Fisheries Service Oyster Disease Research Program, Haskin Shellfish Research Laboratory, Port Norris, New Jersey.

Figueras, A.J. 1991. *Bonamia* status and its effects in cultured flat oysters in the Ria de Vigo Galicia northwestern Spain. *Aquaculture* 93(3):225-234.

Fisher, W.S. 1988. *Disease processes in marine bivalve molluscs.* American Fisheries Society Special Publication No. 18. American Fisheries Society, Bethesda, Maryland, 315p.

Fitt, W.K., S.L. Coon, M. Walch, R.M. Weiner, R.R. Colwell, and D.B. Bonar. 1990. Settlement behavior and metamorphosis of oyster larvae (*Crassostrea gigas*) in response to bacterial supernatants. *Marine Biology* 106(3):389-394.

Fitt, W.K., M.P. Labare, W.C. Fuqua, M.Walch, S.L. Coon, D.B. Bonar, R.R. Colwell, and R.M. Weiner. 1989. Factors influencing bacterial production of inducers of settlement behavior of larvae of the oyster *Crassostrea Gigas. Microbial Ecology* 17(3):287-298.

Flemer, D.A., G.B. Mackiernan, W. Nehlsen, V.K. Tippie, R.B. Biggs, D. Blaylock, N.H. Burger, L.C. Davidson, D. Haberman, K.S. Price, and J.L. Taft. 1983. Chesapeake Bay: A Profile of Environmental Change. Chesapeake Bay Program, Annapolis, MD, 200p.

Foighil, D.O., P.M. Gaffney, and T.J. Hilbish. 1995. Differences in mitochondrial 16S ribosomal gene sequences allow discrimination among American [*Crassostrea virginica* (Gmelin)] and Asian [*C. gigas* (Thunberg) *C. ariakensis* Wakiya] oyster species. *Journal of Experimental Marine Biology and Ecology* 192(2):211-220.

Foighil, D.O., P.M. Gaffney, A.E. Wilbur, and T.J. Hilbish. 1998. Mitochondrial cytochrome oxidase I gene sequences support as Asian origin for the Portuguese oyster *Crassostrea angulata*. *Marine Biology* 131(3):497-503.

Food and Agriculture Organization. 1996. *Technical Consultation on the Precautionary Approach to Capture Fisheries (Including Species Introductions)*: FAO Technical Guidelines for Responsible Fisheries. Food and Agriculture Organization of the United Nations, Rome, 53p.

Food and Agriculture Organization. 1997. *Review of the State of World Aquaculture.* FAO Fisheries Circular 0429-9329. Food and Agriculture Organization of the United Nations, Rome, 163p.

Food and Agriculture Organization. 1998. Aquaculture Production Statistics 1987-1996. FAO Fisheries Circular 0429-9329. Food and Agriculture Organization of the United Nations, Rome, 197p.

Food and Agriculture Organization. 2001. Yearbooks of Fishery Statistics Summary tables – 2001. [Online]. Available: http://www.fao.org/fi/statist/summtab/default.asp [October 13, 2003].

Forcella, F., J.T. Wood, and S.P. Dillon. 1986. Characteristics distinguishing invasive weeds within *Echium* (Bugloss). *Weed Research* 26(5):351-364.

Ford, S.E. 1985. Effects of salinity on survival of the MSX parasite *Haplosporidium nelsoni* (Haskin, Stauber, and Mackin) in oysters. *Journal of Shellfish Research* 2:85-90.

Ford, S.E. 1996. Range extension by the oyster parasite *Perkinsus marinus* into the northeastern US: Response to climate change? *Journal of Shellfish Research* 15(1):45-56.

Ford, S.E. 2001. Pests, parasites, diseases, and defense mechanisms. Pp. 591-628 in *The Biology of the Hard Clam,* Mercenaria mercenaria (Linné), J.N. Kraeuter and M. Castagna, eds. Elsevier Scientific Publishers, Amsterdam.

Ford, S.E., and H.H. Haskin. 1982. History and epizootiology of *Haplosporidium nelsoni* (MSX), an oyster pathogen in Delaware Bay, 1957-1980. *Journal of Invertebrate Pathology* 40(1):118-141.

Ford, S.E., and H.H. Haskin. 1987. Infection and mortality patterns in strains of oysters *Crassostrea Virginica* selected for resistance to the parasite *Haplosporidium Nelsoni* MSX. *Journal of Parasitology* 73(2):368-376.

Ford, S.E., and H.H. Haskin. 1988. Management strategies for MSX (*Haplosporidium nelsoni*) disease in eastern oysters. Pp. 249-256 in *Disease Processes in Marine Bivalve Molluscs,* W. S. Fisher, ed. American Fisheries Society, Bethesda, Maryland.

Ford, S.E., and M.R. Tripp. 1996. Diseases and Defense Mechanisms. Pp. 581-660 in *The Eastern Oyster*: Crassostrea virginica, V.S. Kennedy, R.I.E. Newell, and A.F. Eble, eds. Maryland Sea Grant College Program, College Park, Maryland.

Ford, S.E., Z. Xu, and G. Debrosse. 2001. Use of particle filtration and UV irradiation to prevent infection by *Haplosporidium nelsoni* (MSX) and *Perkinsus marinus* (Dermo) in hatchery-reared larval and juvenile oysters. *Aquaculture* 194(1-2):37-49.

Ford, S.E., J.N. Kraeuter, R.D. Barber, and G. Mathis. 2002. Aquaculture-associated factors in QPX disease of hard clams: Density and seed source. *Aquaculture* 208(1-2):23-38.

Francis E.A., P.M. Gaffney, S.K. Allen, and K.S. Reece. 2001. Species designation among sympatric oysters *Crassostrea ariakensis, C. gigas* and *C. sikamea*. *Journal of Shellfish Research* 20(1):544-545.

Friedman, C.S. 1996. Haplosporidian infections of the Pacific oyster, *Crassostrea gigas* (Thunberg), in California and Japan. *Journal of Shellfish Research* 15(3):597-600.

Friedman, C.S., and R.P. Hedrick. 1991. Pacific oyster nocardiosis: Isolation of the bacterium and induction of laboratory infections. *Journal of Invertebrate Pathology* 57(1):109-120.

Friedman, C.S., and F.O. Perkins. 1994. Range extension of *Bonamia ostreae* to Maine, USA. *Journal of Invertebrate Pathology* 64(3):179-181.

Galtsoff, P.S. 1964. The American oyster, *Crassostrea virginica* Gmelin. *Fishery Bulletin of the Fish and Wildlife Service*, V. 64. U.S. Government Printing Office, Washington, D.C., 480p.

Gangnery, A., C. Bacher, and D. Buestel. 2001. Assessing the production and the impact of cultivated oysters in the Thau lagoon (Mediterranee, France) with a population dynamics model. *Canadian Journal of Fisheries and Aquatic Sciences* 58:1012-1020.

Gardner, G.R., J.C. Harshbarger, J.L. Lake, T.K. Sawyer, K.L. Price, M.D. Stephenson, P.L. Haarker, and H.A. Togstad. 1995. Association of prokaryotes with symptomatic appearance of withering syndrome in black abalone *Haliotis cracherodii*. *Journal of Invertebrate Pathology* 66:111-120.

Garnier-Gere, P.H., Y. Naciri-Graven, S. Bougrier, A. Magoulas, M. Héral, G. Kotoulas, A. Hawkins, and A. Gerard. 2002. Influences of triploidy, parentage and genetic diversity on growth of the Pacific oyster *Crassostrea gigas* reared in contrasting natural environments. *Molecular Ecology* 11(8):1499-1514.

Gätje, C., and K. Reise. 1998. *Ökosystem Wattenmeer: Austausch-, Transport- und Stoffumwandlungsprozesse.* [Translation]: *The Wadden Sea Ecosystem: Exchange, Transport and Transformation Processes.* Springer Verlag, Berlin, 570p.

Gerritsen, J., A.F. Holland, and D.E. Irvine. 1994. Suspension-feeding bivalves and the fate of primary production: An estuarine model applied to Chesapeake Bay. *Estuaries* 17:403-416.

Gibbons, M.C., and F.-L.E. Chu. 1989. Does tidal zonation affect the intensity and incidence of *Perkinsus Marinus* in juvenile American oysters in Virginia USA. *Journal Of Shellfish Research* 7(3):572.

Gobster, P.H., and R.B. Hull. 2000. *Restoring Nature: Perspectives from the Social Sciences and Humanities.* Island Press, Washington, D.C., 321p.

Goldburg, R., M.S. Elliott, and R. Naylor. 2001. *Marine Aquaculture in the United States: Environmental Impacts and Policy Options.* Pew Oceans Commission, Arlington, Virginia, 33p.

Gordon, D.G., N.E. Blanton, and T.Y. Nosho. 2001. *Heaven on the Half Shell: The Story of the Northwest's Love Affair with the Oyster.* University of Washington, Washington Sea Grant Program, Seattle, Washington, 160p.

Gordon, D.R., and K.P. Thomas. 1997. Florida's invasion by nonindigenous plants: history, screening and regulation. Pp. 21-37 in *Strangers in Paradise: Impacts and Management of Non-indigenous Species in Florida*, D. Simberloff, D.C. Schmitz, and T.C. Brown. Island Press, Washington, D.C.

Goulletquer P., M. Héral, C. Béchemin, and P. Richard. 1989. Anomalies de calcification chez la palourde japonaise Ruditapes philippinarum: Caractérisation et comparaison des compositions en acides aminés de différentes parties de la coquille analysées par HPLC. *Aquaculture* 81:169-183.

Goulletquer, P., and M. Héral. 1991. Introduced Pacific oysters in New Zealand. Pp. 13-20 in *The ecology of* Crassostrea gigas *in Australia, New Zealand, France and Washington State,* M.C. Greer, J. C. Leffler, and Maryland Sea Grant Program, 13-20.

Goulletquer, P., and M. Héral. 1997. Marine molluscan production trends in France: From fisheries to aquaculture. *NOAA (National Oceanic & Atmospheric Administration) Technical Report NMFS (National Marine Fisheries Service)* 0(129):137-164.

Goulletquer P., M. Héral, and B.J. Rothschild. 1994a. Causes of decline of oyster production *Crassostrea virginica* in the Maryland portion of the Chesapeake Bay: A literature study. *Haliotis* 23:87-112.

Goulletquer P., M. Héral, and J. Prou, 1994b. Combined effects of temperature-salinty on larval survival of the Eastern oyster *Crassostrea virginica* in the Maryland portion of Chesapeake Bay. *Haliotis* 23:71-86.

Goulletquer P., G. Bachelet, P.G. Sauriau, and P. Noel. 2002. Pp. 276-290 in *Open Atlantic Coast of Europe: A Century of Introduced Species into French Waters*, E. Leppäkoski, S. Olenin, S. Gollasch. Kulwer Academic Publishers, Dordrecht, Boston.

Grabowski, J.H., S.P. Powers, C.H. Peterson, M.J. Powers, and D.P Green. 2003. Consumer ratings of non-native (*Crassostrea gigas* and *Crassostrea ariakensis*) vs. native (*Crassostrea virginica*) oysters. *Journal of Shellfish Research* 22:21-30.

Grant, J., M. Dowd, K. Thompson, C. Emerson, and A. Hatcher. 1993. Perspectives on field studies and related biological models of bivalve growth and carrying capacity. Pp. 371-420 in *Bivalve Filter Feeders*, Richard F. Dame ed. Springer-Verlag, New York.

Greenley, D.A., R.G. Walsh, and R.A. Young. 1982. *Economic benefits of improved water quality: Public perceptions of option and preservation values*. Westview Press, Boulder, Colorado.

Grizel, H. 1997. Risks and prevention of diseases of bivalve molluscs. *Revue Scientifique Et Technique Office International Des Epizooties* 16(1):161-171.

Grizel, H., and M. Héral. 1991. Introduction into France of the Japanese oyster (*Crassostrea gigas*). *Journal Conseil International pour l'Exploration de la Mer* 47:399-403.

Grosholz, E. 2002. Ecological and evolutionary consequences of coastal invasions. *Trends in Ecology & Evolution* 17(1):22-27. [Online]. http://journals.bmn.com/journals/list/latest?jcode=tree.

Grosholz, E.D., G.M. Ruiz, C.A. Dean, K.A. Shirley, J.L. Maron, and P.G. Connors. 2000. The impacts of a non-indigenous marine predator in a California bay. *Ecology* 81:1206-1224.

Groves, R.H., and J.J. Burdon. 1986. *Ecology of Biological Invasions: An Australian Perspective.* Cambridge University Press, Cambridge, Massachusetts. 166p.

Gulka, G., and P.W. Chang. 1984. Pathogenicity and infectivity of a rickettsia-like organism in the sea scallop, *Placopecten magellanicus*. *Journal of Fish Diseases* 8:309-318.

Guo, X., W.K. Hershberger, K. Cooper, and K. Chew. 1994. Tetraploid induction with mitosis I inhibition and cell fusion in the Pacific oyster *Crassostrea gigas* (Thunberg, 1793). *Journal of Shellfish Research* 13(1):193-198.

Guo X., and S.K. Allen, Jr. 1994a. Viable tetraploids in the Pacific oyster (*Crassostrea gigas* (Thunberg) produced by inhibiting polar body 1 in eggs from triploids. *Molecular Marine Biology and Biotechnology* 3(1):42-50.

Guo, X., and S.K. Allen, Jr. 1994b. Reproductive potential and genetics of triploid Pacific oysters, *Crassostrea gigas* (Thunberg). *Biological Bulletin* 187(3):309-318.

Guo, X., G.A. DeBrosse, and S.K. Allen, Jr. 1996. All-triploid Pacific oysters (*Crassostrea gigas* Thunberg) produced by mating tetraploids and diploids. *Aquaculture* 142:149-161.

Guo, X., S.E. Ford, and F. Zhang. 1999. Molluscan aquaculture in China. *Journal of Shellfish Research* 18:(1)19-31.

Hallerman, E., M. Leffler, S. Mills, and S. Allen, Jr. 2001. *Aquaculture of Triploid Crassostrea ariakensis in Chesapeake Bay: A Symposium Report.* Virginia Sea Grant Publication VSG-02-03, College of William and Mary, Williamsburg, Virginia, 20p.

Hallock, Z.R., and G.O. Marmorino. 2002. Observations of the response of a buoyant estuarine plume to upwelling favorable winds. *Journal of Geophysical Research* 107(C7):3-13.

Ham, L.C., and M. Irvine. 1975. Techniques for determining seasonality of shell middens from marine mollusk remains. *Syesis* 8:363-373.

Harding, L.W., Jr., D. Degobbis, and R. Precali. 1999. Production and fate of phytoplankton: annual cycles and interannual variability. Pp. 131-172 in *Coastal and Estuarine Studies: Ecosystems at the Land-Sea Margin Drainage Basin to Coastal Sea,* T.C. Malone, A. Malej, L.W. Harding, Jr., N. Smodlaka, and R.E. Turner, eds. American Geophysical Union, Washington, D.C.

Hargis, W.J., Jr. 1999. The evolution of the Chesapeake Bay oyster reef system during the Holocene epoch. Pp. 5-23 in *Oyster Reef Habitat Restoration: A Synopsis and Synthesis of Approaches,* M.W. Luckenbach, R. Mann, and J.A. Wesson, eds. Virginia Institute of Marine Science, Gloucester Point.

Hargis, W.J., Jr., and D.S. Haven. 1988. Rehabilitation of the troubled oyster industry of the lower Chesapeake Bay. *Journal of Shellfish Research* 7:271-279.

Hargis, W.J., Jr. and D.S. Haven. 1999. Chesapeake oyster reefs, their importance, destruction and guidelines for restoring them. Pp. 329-358 in *Oyster Reef Habitat Restoration: A Synopsis and Syntheses of Approaches.* M.W. Luckenbach, R. Mann, and J.A. Wesson. Virginia Institute of Marine Science, Gloucester Point.

Harry, H.W. 1985. Synopsis of the supraspecific classification of living oysters *Bivalvia Gryphaeidae* and *Ostreidae. Veliger* 28(2):121-158.

Hasan, S.A. 1960. Oyster fishing sources in Pakistan. Pp. 167-171 in *Proceedings of fourth Pan-Indian ocean science association congress, PIOSA,* Section B, Karachi, Pakistan. Special Report for the Food and Agriculture Organization of the United Nations, Rome, Italy.

Haskin, H.H., and S. Tweed. 1976. *Oyster Setting and Early Spat Survival at Critical Salinity Levels on Natural Seed Oyster Beds of Delaware Bay: Water Problems in an Urbanizing State.* U.S. Department of the Interior Office of Water Research and Technology, Water Resources Research Institute, Rutgers University, New Brunswick, New Jersey, 66p.

Haskin, H.H., and S.E. Ford. 1979. Development of resistance to *Minchinia nelsoni* (MSX) mortality in laboratory-reared and native oyster stocks in Delaware Bay. *Marine Fisheries Review* 41(1-2):54-63.

Haskin, H.H., and S.E. Ford. 1986. Report to the Bureau of Shellfisheries, New Jersey Department of Environmental Protection on the Delaware Bay Oyster Research Program 1984-1986. Rutgers University. Port Norris, New Jersey.

Haskin, H.H., and J.D. Andrews. 1988. Uncertainties and speculations about the life cycle of the Eastern oyster pathogen *Haplosporidium nelsoni* (MSX). Pp. 5-22 in *Disease Processes in Marine Bivalve Molluscs,* W.S. Fisher, ed. American Fisheries Society Special Publication 18, Bethesda, Maryland.

Haven D., 1972. The oyster industry. Pp. 27-63 in *A study of Commercial and Recreational Fisheries of the Eastern Shore of Virginia: Accomac and Northampton counties,* V.G. Burrell, M. Castagna, R.K. Dias. Special Report in Applied Marine Science and Ocean Engineering, No. 20. Virginia Institute of Marine Science, Gloucester Point, Virginia.

Haven, D.S., W.J. Hargis, and P.C. Kendall. 1978. *The Oyster Industry of Virginia: Its Status, Problems, and Promise: A Comprehensive Study of the Oyster Industry in Virginia.* VIMS Special Papers in Marine Science, No. 4. Virginia Institute of Marine Science, Gloucester Point, Virginia. 1024p.

Haven, D.S., J.P. Whitcomb, and P.C. Kendall. 1981. *The Present and Potential Productivity of the Baylor Grounds in Virginia, Volume I: Rappahannock, Corrotoman, Great Wicomico, Piankatank, York and Poquoson Rivers, and Mobjack Bay and Its Tributaries.* Special Report in Applied Marine Science and Ocean Engineering No. 243. Virginia Institute of Marine Science, Gloucester Point, Virginia, 167p.

Hay, B. 1999. Outline *Pacific Oyster Farming in New Zealand: A Self-paces Learning Package for Unit No. 16672.* Prepared by B. Hay, Aquabio Consultants and Framework Solutions, Ltd., Wellington, New Zealand, 92p.

Healey, M.C. 1984. Multiattribute analysis and the concept of Optimum Yield. *Canadian Journal of Fisheries & Aquatic Sciences* 41(9):1393-1406.

Heck, K.L., Jr., and R.J. Orth. 1980. Seagrass habitats: The roles of habitat complexity, competition and predation in structuring associated fish and motile macroinvertebrate assemblages. Pp. 449-464 in *Estuarine Perspectives, Proceedings of the 1979 biennial meeting of the Estuarine Research Federation,* V.S. Kennedy, ed. Academic Press, New York.

Hedgecock, D., M.A. Banks, and D.J. McGoldrick. 1993. The status of the Kumamoto oyster *Crassostrea sikamea* (Amemiya 1928) in U.S. commercial brood stocks. *Journal of Shellfish Research* 12:215-221.

Hedgecock, D., G. Li, M.A. Banks, and Z. Kain. 1999. Occurrence of the Kumamoto oyster *Crassostrea sikamea* in the Ariake Sea, Japan. *Marine Biology* 133:65-68.

Héral, M. 1993. Why carrying capacity models are useful tools for management of bivalve mollusc culture. Pp. 455-477 in *Bivalve Filter Feeders in Estuarine and Coastal Ecosystem Processes.* R. Dame, ed. Springer-Verlag, Berlin.

Héral, M., and J.M. Deslous-Paoli. 1991. Oyster culture in European countries. Pp. 153-190 in *Estuarine and Marine Bivalve Mollusk Culture,* W. Menzel, ed. CRC Press, Boca Raton, Florida.

Hickey, J.M. 1979. Culture of Pacific oyster, *Crassostrea gigas,* in Massachusetts waters. Pp.1 29-148 in *Exotic Species in Mariculture: Proceedings of a Symposium on Exotic Species in Mariculture: Case Histories of the Japanese oyster, Crassostrea gigas (Thunberg), with Implications for Other Fisheries,* R. Mann, ed. Symposium held at Woods Hole Oceanographic Institution, Woods Hole, Massachusetts, September 18-20, 1978. MIT Press, Cambridge, Massachusetts.

Higgins, P.J. 1980. Effects of food availability on the valve movements and feeding behavior of juvenile *Crassostrea virginica* (Gmelin). II. Feeding rates and behavior. *Journal of Experimental Marine Biology and Ecology* 46(1):17-27.

Hilborn, R., and C. Walters. 1977. Differing goals of salmon management on the Skeena River. *Journal of the Fisheries Research Board of Canada* 34:64-72.

Hildreth, R.G. 1991. Regional oceans resources management. Pp. 75-87 in *Coastal Zones '91: Proceedings of the 7th Symposium on Coastal & Ocean Management ASCEI* held in Long Beach, California, July 8-12, 1991. Center for the Study of Marine Policy, University of Delaware, Newark, Delaware.

Hine, P.M., B. Wesney, and P. Besant. 1998. Replication of a herpes-like virus in larvae of the flat oyster *Tiostrea chilensis* at ambient temperatures. *Diseases of Aquatic Organisms* 32(3):161-171.

Hirase, S. 1930. On the classification of Japanese oysters. *Japanese Journal of Zoology* 3:1-65.

His, E. 1977. Observations relatives à l'infestation de *Crassostrea gigas* (Thunberg) par le copépode parasite Mytilicola orientalis Mori dans le bassind 'Arcachon. ICES Communication. C.M. K:33, 16p.

Hixon, M.A., and J.P. Beets. 1993. Predation, prey refuges, and the structure of coral-reef fish assemblages. *Ecological Monographs* 63(1):77-101.

Hobbs, R.J. 1989. The nature and effect of disturbance relative to invasions. Pp. 389-405 in *Biological Invasions: A Global Perspective,* J.A. Drake, H.A. Monney, F. DiCastri, R.H. Groves, M. Kruger, M. Rejmanek, and W. Williamson. John Wiley and Sons, New York.

Hofmann, E., S. Ford, E. Powell, and J. Klinck. 2001. Modeling studies of the effect climate variability on MSX disease in eastern oyster (*Crassostrea virginica*) populations. *Hydrobiologia* 460:195-212.

Horton, T., and W.M. Eichbaum. 1991. *Turning the Tide: Saving the Chesapeake Bay.* Island Press, Washington, D.C., 327p.

Hu, Y.P., R.A. Lutz, and R.C. Vrijenhoek. 1993. Overdominance in early life stages of an American oyster strain. *Journal of Heredity* 84(4):254-258.

Huanxin, W., Z. Lejun, and B.J. Presley. 2000. Bioaccumulation of heavy metals in oyster (*Crassostrea virginica*) tissue and shell. *Environmental Geology* 39(11):1216-1226.

Hurley, L.M. 1991. Submerged aquatic vegetation. Pp.2-4 – 2-19 in *Habitat Requirements for Chesapeake Bay Living Resources*, 2nd ed., S.L. Funderburk, S.J. Joran, J.A. Mihursky, and D. Riley. Solomons, Chesapeake Research Consortium.

Ingersoll, E. 1881. *The History and Present Conditions of the Fisheries Industries: The Oyster Industry.* United States Commission of Fish and Fisheries, Washington, D.C., 251p.

International Council for the Exploration of the Seas (ICES). 1988. *Report of the Working Group on Environmental Impacts of Mariculture.* Hamburg, Germany.

International Council for the Exploration of the Seas (ICES). 1994. ICES Code of Practice on the Introductions and Transfers of Marine Organisms. [Online]. Available: http://www.ices.dk/pubs/itmo.pdf [August 1, 2003].

Jackson, J.B.C., M.X. Kirby, W.H. Berger, K.A. Bjorndal, L.W. Botsford, B.J. Bourque, R.H. Bradbury, R. Cooke, J. Erlandson, J.A. Estes, T.P. Hughes, S. Kidwell, C.B. Lange, H.S. Lenihan, J.M. Pandolfi, C.H. Peterson, R.S. Steneck, M.J. Tegner, and R.R. Warner. 2001. Historical overfishing and the recent collapse of coastal ecosystems. *Science* 293(5530):629-638.

Jakus P., M. McGuinness, and A. Krupnick. 2002. *The Benefits and Costs of Fish Consumption Advisories for Mercury.* Discussion Paper 02-55. Resources for the Future, Washington, D.C.

Johnson, P.J. 1988. *Working the Water: The Commercial Fisheries of Maryland's Patuxent River.* Calvert Marine Museum, Solomons, Maryland and University Press of Virginia, Charlottesville, Virginia, 218p.

Johnson, R.W. 1989. Water pollution and the public trust doctrine. *Environmental Law* 19(3):485-513.

Jonas, R.B., and J.H. Tuttle. 1990. Bacterioplankton and organic carbon dynamics in the lower mesohaline Chesapeake Bay. *Applied & Environmental Microbiology* 56(3):747-757.

Jones, C.G., J.H. Lawton, and M. Shachak. 1994. Organisms as ecosystem engineers. *Oikos* 69(3):373-386.

Jones, G., and B. Jones. 1982. *Methods for Setting Hatchery Produced Oyster Larvae.* Information Report No. 4. Marine Resources Branch, Victoria, B.C., 61p.

Kahn, J.R., and W.M. Kemp. 1985. Economic losses associated with the degradation of an ecosystem the case of submerged aquatic vegetation in Chesapeake Bay. *Journal of Environmental Economics & Management* 12(3):246-263.

Kang, P.A. 1980. On the Minchinia sp. infection in the oysters (*Crassostrea virginica*) from Chungmu area. *Acta Chimica* 25:25-28.

Katkansky, S.C., W.A. Dahlstrom, and W.W. Ronald. 1969. Observations on survival and growth of the european Flat oyster *Ostrea Edulis* in California USA. *California Fish & Game* 55(1):69-74.

Katkansky, S.C., and R.W. Warner. 1970. The occurrence of a haplosporidian in Tomales Bay California. *Journal of Invertebrate Pathology* 16(1):144.

Katkansky, S.C., and R.W. Warner. 1974. *Pacific Oyster Disease and Mortality Studies in California.* Marine Resources Technical Report No. 25. State of California, Resources Agency, Department of Fish and Game, Sacramento, 51p.

Kavanagh, L.D. 1940. Mud blisters in Japanese oysters imported to Louisiana. *Louisiana Conservation Review* 31-34.

Keck, R., D. Maurer, J.C. Kauer, and W.A. Sheppard. 1971. Chemical stimulants affecting larval settlement in the American oyster. *Proceedings National Shellfisheries Association* 61:24-28.

Keeney, R.L., and H. Raiffa. 1976. *Decisions with Multiple Objectives: Preferences And Value Tradeoffs*. Wiley Series in Probability and Mathematical Statistics. Wiley, New York, 569p.

Keithly, W.R., and H. Diop. 2001a. The impact pf risk information on the demand for Gulf of Mexico and Chesapeake oysters. *Journal of Shellfish Research* 20:1285-1291.

Keithly, W.R., and H. Diop. 2001b. The demand for eastern oysters, *Crassostrea virginica*, from the Gulf of Mexico in the presence of *Vibrio vulnicus*. *Marine Fisheries Review* 63: 47-53.

Kemp, W.M., and W.R. Boynton. 1992. Benthic-pelagic interactions: Nutrients and oxygen dynamics. Pp. 149-221 in *Oxygen Dynamics in the Chesapeake Bay: A Synthesis of Recent Research*, D.E. Smith, M. Leffler, and G. Mackiernan, eds. Maryland Sea Grant College Program, College Park, Maryland.

Kemp, W.M., J. Faganeli, S. Ouskaric, E.M. Smith, and W.R. Boynton. 1999. Pelgic-benthic coupling and nutrient cycling. Pp. 295-340 in *Coastal and Estuarine Studies: Ecosystems at the Land-Sea Margin Drainage Basin to Coastal Sea*, T.C. Malone, A. Malej, L.W. Harding, Jr., N. Smodlaka, and R.E. Turner, eds. American Geophysical Union, Washington, D.C.

Kemp, W.M., E.M. Smith, M. Marvin-DiPasquale, and W.R. Boynton. 1997. Organic carbon balance and net ecosystem metabolism in Chesapeake Bay. *Marine Ecology Progress Series* 150:229-248.

Kennedy, V.S. 1996. Biology of larvae and spat. Pp. 383-450 in *The Eastern Oyster* Crassostrea virginica, R.I.E. Newell, V.S. Kennedy, and A.F. Eble, eds. Maryland Sea Grant College Program, College Park.

Kennedy, V.S. 2002. *History of the Oyster Fishery in Chesapeake Bay*. Presented to the National Research Council Committee on Nonnative Oysters in the Chesapeake Bay, Fredericksburg, Virginia, October 8, 2002. National Research Council, Washington, D.C.

Kennedy, V.S., and L.L. Breisch. 1981. *Maryland's Oysters: Research and Management*. UM-SG-TS-81-04. University of Maryland, College Park, 286p.

Kennedy, V.S., and L.L. Breisch. 1983. Sixteen decades of political management of the oyster fishery in Maryland's Chesapeake Bay. *Journal of Environmental Management* 16 (2):153-171.

Kennedy, V.S., R.I.E. Newell, and A.F. Eble. 1996. *The Eastern Oyster*: Crassostrea virginica. Maryland Sea Grant College Program, College Park, 734p.

Kent, B.W. 1988. *Making dead oysters talk: Techniques for analyzing oysters from archeological sites*. Maryland Historical Trust, Jefferson Patterson Park and Museum Report. Crownsville, Maryland.

Kern, F.G. 1976. Sporulation of *Minchinia sp. Haplosporida Haplosporidiidae* in the Pacific oyster *Crassostrea Gigas* from Korea. *Journal of Protozoology* 23(4):498-500.

Kimmerer, W.J., E. Gartside, and J.J. Orsi. 1994. Predation by an introduced clam as the likely cause of substantial declines in zooplankton of San Francisco Bay. *Marine Ecology-Progress Series* 113(1-2):81-93.

Kirkley, J.E. 1997. *Virginia's Commercial Fishing Industry: Its Economic Performance and Contributions*. Virginia Sea Grant Publication No. 337. Virginia Institute of Marine Science, Gloucester Point, Virginia, 77p.

Kirtley, D.W., and W.F. Tanner. 1968. Sabellariid worms: builders of a major reef type. *Journal of Sedimentary Petrology* 38:73-78.

Kneib, R.T. 1984. Patterns of invertebrate distribution and abundance in the intertidal salt marsh: Causes and questions. *Estuaries* 7(4A):392-412.

Koganezawa, A. 1974. *Present status of studies on the mass mortality of cultured oysters in Japan and its prevention*. Proceedings of the Third U.S.-Japan Meeting on Aquaculture held in Tokyo, Japan, October 15-16, 1974. Special Publication of Fishery Agency, Japanese Government and Japan Sea Regional Fisheries Research Laboratory, under the Aquaculture Panel, U.S.-Japan Cooperative Program in Natural Resources (UJNR).

Kolar, C.S., and D.M. Lodge. 2001. Progress in invasion biology: Predicting invaders. *Trends in Ecology & Evolution* 16(4):199-204.

Kolar, C.S., and D.M. Lodge. 2002. Ecological predictions and risk assessment for alien fishes in North America. *Science* 298(5596):1233-1236.

Krupnick, A., V. McConnell, D. Austin, M. Cannon, T. Soessell, and B. Morton. 1998. *The Chesapeake Bay and the Control of NOx Emissions: A Policy Analysis*. Discussion Paper 98-46. Resource for the Future, Washington, D.C.

Lacoste, A., F. Jalabert, S. Malham, A. Cueff, F. Gelebart, C. Cordevant, M. Lange, and S.A. Poulet. 2001. A Vibrio splendidus strain is associated with summer mortality of juvenile oysters *Crassostrea gigas* in the Bay of Morlaix (North Brittany, France). *Diseases of Aquatic Organisms* 46(2):139-145.

Lang, J. 1973. Interspecific aggression by Scleractinian corals: part 2, why the race is not only to the swift. *Bulletin of Marine Science* 23(2):260-279.

Lang, J.C., and E.A. Chornesky. 1990. Competition between scleractinian reef corals: A review of mechanisms and effects. Pp. 209-252 in *Coral Reefs. Ecosystems of the World*, Z. Dubinsky. Elsevier, Amsterdam.

Langdon, C.J., and A.M. Robinson. 1996. Aquaculture potential of the Suminoe oyster (*Crassostrea ariakensis* Fugita 1913). *Aquaculture* 144(4):321-338.

Langland, M.J., R.E. Edwards, S.A. Sprague, and S. Yochum. 2001. *Summary of Trends and Status Analysis for Flow, Nutrients, and Sediments at Selected Nontidal Sites, Chesapeake Bay Basin, 1985-99*. USGS Report 01-73. U.S. Department of the Interior, Washington, D.C.

Le Deuff, R.M., T. Renault, and A. Gerard. 1996. Effects of temperature on herpes-like virus detection among hatchery-reared larval Pacific oyster *Crassostrea gigas*. *Diseases of Aquatic Organisms* 24:(2)149-157.

Le Gall, G.G., D. Chagot, E. Mialhe, and H. Grizel. 1988. Branchial rickettsiales-like infection associated with a mass mortality of sea Scallop *Pecten-Maximus*. *Diseases of Aquatic Organisms* 4(3):229-232.

Le Roux, F., G. Lorenzo, P. Peyret, C. Audemard, A. Figueras, C. Vivares, M. Gouy, and F. Berthe. 2001. Molecular evidence for the existence of two species of *Marteilia* in Europe. *Journal of Eukaryotic Microbiology* 48(4):449-454.

Lenihan, H.S., and C.H. Peterson. 1998. How habitat degradation through fishery disturbance enhances impacts of hypoxia on oyster reefs. *Ecological Applications* 8(1):128-140.

Lenihan, H.S., C.H. Peterson, J.E. Byers, J.H. Grabowski, G.W. Thayer, and D.R. Colby. 2001. Cascading of habitat degradation: Oyster reefs invaded by refugee fishes escaping stress. *Ecological Applications* 11(3):764-782.

Leppäkoski, E., S. Gollasch, and S. Olenin, eds. 2002. Invasive aquatic species of Europe: distribution, impacts, and management. Kluwer Academic Publishers, Boston, 583p.

Levitan, D.R., M.A. Sewell, and F.S. Chia. 1992. How distribution and abundance influence fertilization success in the Sea Urchin *Strongylocentrotus-Franciscanus*. *Ecology* 73(1):248-254.

Lindsay C.E., and D. Simons. 1997. The fisheries for Olympia oysters, *Ostreola conchaphila*; Pacific oysters *Crassostrea gigas*, and Pacific razor clams *Siliqua patula*, in the State of Washington. Pp 89-114 in *The history, present condition and future of molluscan fisheries of North and Central America and Europe, Volume 2, Pacific coast and supplemental topics*, C.L. MacKenzie Jr., V.G. Burrell, A. Rosenfield, and W.L. Hobart (eds.). National Oceanic and Atmospheric Administration Technical Report. National Marine Fisheries Service 128, Department of Commerce, Washington, D.C. 217p.

Linville, R.G., S.N. Luoma, L. Cutter, and G.A. Cutter. 2002. Increased selenium threat as a result of invasion of the exotic bivalve *Potamocorbula amurensis* into the San Francisco Bay-Delta. *Aquatic Toxicology* 57(1-2):51-64.

Lipton, D.W. 2002. *Chesapeake Bay Oyster Economics*. Presented to the National Research Council Committee on Nonnative Oysters in the Chesapeake Bay, Fredericksburg, Virginia, October 2002. National Research Council, Washington, D.C.

Lipton, D.W., E.F. Lavan, and I.E. Strand. 1992. Economics of molluscan introductions and transfers: The Chesapeake Bay dilemma. *Journal of Shellfish Research* 11(2):511-519.

Lipton, D.W., J.E. Kirkley. 1994. *A Profile of the Oyster Industry: Northeastern United States*. Maryland Sea Grant Marine Advisory Program; Virginia Sea Grant Marine Advisory Program, College Park, Maryland, and Gloucester Point, Virginia, 68p.

Littlewood, D.T.J., R.N. Wargo, J.N. Kraeuter, and R.H. Watson. 1992. The influence of intertidal height on growth mortality and *Haplosporidium Nelsoni* infection in MSX mortality-resistant Eastern oysters *Crassostrea-Virginica* (Gmelin 1791). *Journal Of Shellfish Research* 11(1):59-64.

Livingston, R.J., R.L.I. Howell, X. Niu, F.G.I. Lewis, and G.C. Woodsum. 1999. Recovery of oyster reefs (*Crassostrea virginica*) in a Gulf estuary following disturbance by two hurricanes. *Bulletin of Marine Science* 64(3):465-483.

Lodge, D.M. 1993. Biological invasions: Lessons for ecology. *Trends in Ecology & Evolution* 8(4):133-137.

Lohrer, A.M., R.B. Whitlatch, K. Wada, and Y. Fukui. 2000. Home and away: Comparisons of resource utilization by a marine species in native and invaded habitats. *Biological Invasions* 2(1):41-57.

Longshaw, M., S.W. Feist, R.A. Matthews, and A. Figueras. 2001. Ultrastructural characterisation of *Marteilia* species (Paramyxea) from *Ostrea edulis*, *Mytilus edulis* and *Mytilus galloprovincialis* in Europe. *Diseases of Aquatic Organisms* 44(2):137-142.

Loosanoff, V.L. 1955. The European oyster in American waters. *Science* 121(3135):119-121.

Loosanoff, V.L., and C.A. Nomejko. 1951. Existence of physiologically different races of oysters, *Crassostrea virginica*. *Biological Bulletin* 101:151-156.

Lough, R.G. 1975. A reevaluation of the combined effects of temperature and salinity on survival and growth of bivalve larvae using response surface techniques. *Fishery Bulletin* 73(1):86-94.

Luckenbach, M.W. 2001. *Status report on oyster restoration in Chesapeake Bay*. Testimony to House of Representatives Subcommittee on Fisheries Conservation, Wildlife and Oceans. October 22, 2001. [Online]. Available: http://www.house.gov/resources/107cong/fisheries/2001oct22/luckenbach.htm. [October 10, 2003].

Luckenbach, M.W., R. Mann, and J.A. Wesson. 1999. *Oyster Reef Habitat Restoration: A Synopsis and Synthesis of Approaches; Proceedings from the Symposium, Williamsburg, Virginia, April 1995*. Virginia Institute of Marine Science, Gloucester Point, 366p.

Luckenbach, M.W., F.X. O'Beirn, and J. Taylor, 1999. *An Introduction to Culturing Oysters in Virginia*. Virginia Institute of Marine Science, Gloucester Point, 20p.

Lunz, G.R. 1938. Comparison between Pre-Colonial and Present-Day Oysters. *Science* 87(2260):367.

Macinko, S. 1993. Public or private?: United States commercial fisheries management and the public trust doctrine, reciprocal challenges. *Natural Resources Journal* 33(4):919-995.

MacIsaac, H.J. 1996. Potential abiotic and biotic impacts of zebra mussels on the inland waters of North America. *American Zoologist* 36(3):287-299.

MacKenzie, C.L., Jr. 1969. Causes of oyster spat mortality conditions of oyster setting beds and recommendations for oyster bed management. *Proceedings National Shellfisheries Association* 60:59-67.

MacKenzie, C.L., Jr. 1970. Causes of oyster spat mortality, conditions of oyster setting beds, and recommendations for oyster bed management. *Proceedings of the National Shellfisheries Association* 60:59-67.

MacKenzie, C.L., Jr. 1983. To increase oyster production in the northeastern United States. *Marine Fisheries Review* 45(3):1-22.

MacKenzie, C.L., Jr. 1997. *The History, Present Condition, and Future of the Molluscan Fisheries of North and Central America and Europe.* NOAA Technical Report 127-129. U.S. Government Printing Office, Washington, D.C.

MacKenzie, C.L., and V.G. Burrell, Jr. 1997. Trends and status of molluscan fisheries in North and Central America and Europe—a synopsis. Pp. 1-14 in *The History, Present Condition, and Future of Molluscan Fisheries of North and Central America and Europe,* Volume 1, Atlantic and Gulf Coasts; U.S. Department of Commerce, NOAA Technical Report 127, Washington, D.C.

Mackett, D.J. 1985. Strategic planning for research and management of the Albacore Tuna fishery. *Systems Research* 2(3):201-210.

Mackin, J.G., and S.H. Hopkins. 1962. *Studies on Oyster Mortality in Relation to Natural Environments and to Oil fields in Louisiana.* The University of Texas, Institute of Marine Science, Port Aransas, Texas, 131p.

Maes, P., and C. Paillard. 1992. Effect du Vibrio P1, pathogene de Ruditapes philippinarum, sur d'autres espèces de bivalves : Les Molluscques Marins, Biologies et Aquaculture. *Actes de Colloques* 14:141-148.

Mahadevan, S. 1987. Oyster resources of India. Pp. 14-16 in *Oyster Culture-Status and Prospects,* K.N. Nayar and S. Mahadevan. *Central Marine Fisheries Research Institute Bulletin* 38.

Malej, A., L.W. Harding, Jr., N. Smodlaka, and T.C. Malone. 1999. Summary and Conclusions. Pp. 367-378 in *Coastal and Estuarine Studies: Ecosystems at the Land-Sea Margin Drainage Basin to Coastal Sea,* T.C. Malone, A. Malej, L.W. Harding, Jr., N. Smodlaka, and R.E. Turner. American Geophysical Union, Washington, D.C.

Malone, T.C., A. Malej, L.W. Harding, Jr., N. Smodlaka, and R.E. Turner, eds. 1999. *Ecosystems at the land-sea margin: Drainage basin to coastal sea. Coastal and Estuarine Studies* 55. American Geophysical Union, Washington, D.C., 381p.

Malone, T.C., W.M. Kemp, H.W. Ducklow, W.R. Boynton, J.H. Tuttle, and R.B. Jonas. 1986. Lateral variation in the production and fate of phytoplankton in a partially stratified estuary. *Marine Ecology Progress Series* 32(2-3):149-160.

Mann, R. 1979. *Proceedings of a Symposium on Exotic Species in Mariculture: Case Histories of the Japanese Oyster,* Crassostrea gigas *(Thunberg), with Implications for Other Fisheries,* Woods Hole, Massachusetts, September 18 to 20, 1978. MIT Press, Cambridge Press, 363p.

Mann, R. 2000. Restoring oyster reef communities in the Chesapeake Bay: A commentary. *Journal of Shellfish Research* 19(1):335-339.

Mann, R., E.M. Burreson, and P.K. Baker. 1991. The decline of the Virginia oyster fishery in Chesapeake Bay: Considerations for introduction of a non-endemic species, *Crassostrea gigas* (Thunberg, 1793). *Journal of Shellfish Research* 10:(2)379-388.

Mann, R., and J.M. Harding. 1998. *Continuing Trophic Studies on Constructed "Restored" Oyster Reefs.* Virginia Institute of Marine Science, Gloucester Point, Virginia, 58p.

Manzi, J.J., and M. Castagna. 1989. Clam mariculture in North America. *Developments in Aquaculture and Fisheries Science* 19. Elsevier Science Publishing Company, New York, 461p.

Marmorino, G.O., and C.L Trump. 2000. Gravity current structure of the Chesapeake Bay outflow plume. *Journal of Geographic Research* 105:28, 847-28, 861.

Marshall, N. 1954. Changes in the physiography of oyster bars in the James River, Virginia. *Virginia Journal of Science* 5(Series):173-181.

Maurin C., and J. LeDantec. 1979. The culture of Crassostrea gigas in France. Pp. 106-120 in *Exotic Species in Mariculture: Proceedings of a Symposium on Exotic Species in Mariculture: Case Histories of the Japanese Oyster,* Crassostrea gigas *(Thunberg), with Implications for Other Fisheries,* R. Mann, ed. Woods Hole, Massachusetts, September 18-20, 1978. MIT Press, Cambridge, Massachusetts, 363p.

McCay, B.J. 1998. *Oyster Wars and the Public Trust: Property, Law, and Ecology in New Jersey History.* University of Arizona Press, Tucson, Arizona, 246p.

McCoy, H.D. 2000. *American and International Aquaculture Law: A Comprehensive Legal Treatise and Handbook Covering Aquaculture Law, Business and Finance of Fishes, Shellfish, and Aquatic Plants.* Supranational Publishing Company, Peterstown, West Virginia, 476p.

McPhee, J.A. 1989. *The control of nature.* 1st ed. Farrar, Straus, Giroux, New York, New York, 72p.

Medcof, J.G. 1961. Trial introduction of European oysters (*Ostrea edulis*) to Canadian east coast. *Proceedings of the National Shellfisheries Association* 50:113-124.

Menzel, W. 1991. *Estuarine and marine bivalve mollusk culture.* CRC Press, Boca Raton, Florida, 362p.

Merritt, M.A., and K.R. Criddle. 1993. Multiple criterion decision theory for judging management strategies and resolving conflict: a case study of the Kenai River recreational fisheries. Pp. 683-704 in *Proceedings of the International Symposium on Management Strategies for Exploited Fish Populations,* G. Kruse, D.M. Eggers, R.J. Marasco, C. Pautzke, and T.J. Quinn II, eds. University of Alaska, Alaska Sea Grant Program, Fairbanks, Alaska.

Meyer, D.L., E.C. Townsend, and G.W. Thayer. 1997. Stabilization and erosion control values of oyster cultch for intertidal marsh. *Restoration Ecology* 5(1):93-99.

Meyers, T. 1979. Preliminary studies on a chlamydial agent in the digestive diverticular epithelium of hard clams *'Mercenaria mercenaria'* (L.) from Great South Bay, New York. *Journal of Fish Diseases* 2:179-189.

Miller, D.C., R.J. Geider, and H.L. MacIntyre. 1996. Microphytobenthos: The ecological role of the "secret garden" of unvegetated, shallow-water marine habitats. II. Role in sediment stability and shallow-water food webs. *Estuaries* 19(2A):202-212.

Moazzam, M., and S.H.N. Rizvi. 1983. Settlement of oyster larvae in Pakistani waters and its possible implication for setting up oyster culture. *Proceedings of the Symposium on Coastal Aquaculture, Part 2: Molluscan Culture.* Marine Biological Association of India, Cochin, India.

Mooney, H.A., and R.J. Hobbs. 2000. *Invasive Species in a Changing World.* Island Press, Washington, D.C., 457p.

Moore, J.D., T.T. Robbins, and C.S. Friedman. 2000. Withering syndrome in farmed Red Abalone *Haliotis rufescens*: Thermal induction and association with a gastrointestinal Rickettsiales-like prokaryote. *Journal of Aquatic Animal Health* 12:(1)26-34.

Morton, B. 1987. Recent marine introductions into Hong Kong. *Bulletin of Marine Science* 41:503-513.

Morton, B. 1997. The aquatic nuisance species problem: A global perspective and review. Pp. 1-53 in *Zebra Mussel and Aquatic Nuisance Species,* F. D'Itri. Ann Arbor Press, Chelsea, Michigan.

Moser, F.D. 2002. Invasive species in the Chesapeake Bay watershed. UM-SG-TS-2002-03. University of Maryland, Maryland Sea Grant, 121p.

Muth, M., D.W. Anderson, S.A. Karns, B.C. Murray, and J.L. Domanico. 2000. *Economic Impacts of Requiring Post-Harvest Treatment of Oysters. Report of the Interstate Shellfish Sanitation Conference* held in Scotsdale, Arizona, July 15 to 20, 2000. For copies contact K. Moore, Executive Director, ISSC.

Nadol, V. 1999. Aquatic invasive species in the coastal west: An analysis of state regulation within a federal framework. *Environmental Law* 29(2):339-376.

National Marine Fisheries Service (NMFS). 1976. *Fisheries of the United States.* Current Fishery Statistics No. 6900. U.S. Department of Commerce, National Oceanic and Atmospheric Administration, National Marine Fisheries Service, Government Printing Office, Washington, D.C.

National Marine Fisheries Service (NMFS). 2002. *A Code of Conduct for Responsible Aquaculture Development in the U.S. Exclusive Economic Zone.* [Online]. Available: http://www.nmfs.noaa.gov/trade [October 31, 2003].

National Marine Fisheries Service (NMFS). 2003. *Commercial Fisheries Landings.* [Online]. Available: http://www.nmfs.noaa.gov/Commercial.htm [October 13, 2003].

National Research Council. 1983. *Risk Assessment in the Federal Government: Managing The Process.* National Academy Press, Washington, D.C., 191p.

National Research Council. 1993. *Issues in Risk Assessment.* National Academy Press, Washington, D.C., 356p.

National Research Council. 1995. *Understanding Marine Biodiversity.* National Academy Press, Washington, D.C.

National Research Council. 1996a. *Understanding Risk: Informing Decisions in a Democratic Society.* National Academy Press, Washington, D.C., 249p.

National Research Council. 1996b. *Stemming the Tide: Controlling Introductions of Nonindigenous Species by Ships' Ballast Water.* National Academy Press, Washington, D.C., 141p.

National Research Council. 2002a. *Environmental Effects of Transgenic Plants: The Scope and Adequacy of Regulation.* National Academy Press, Washington, D.C., 320p.

National Research Council. 2002b. *Animal Biotechnology: Science-Based Concerns.* National Academy Press, Washington, D.C., 181p.

Naylor, R.L., R.J. Goldburg, J.H. Primavera, N. Kautsky, M.C.M. Beveridge, J. Clay, C. Folke, J. Lubchenco, H. Mooney, and M. Troell. 2000. Effect of aquaculture on world fish supplies. *Nature* 405(6790):1017-1024.

Naylor, R.L., S.L. Williams, and D.R. Strong. 2001. Aquaculture: A gateway for exotic species. *Science* 294(5547):1655-1656.

Needler, A.W.H., and R.R. Logie. 1947. Serious mortalities in Prince Edward Island oysters caused by a contagious disease. *Transactions of the Royal Society of Canada,* 4(V - Series III):73-89.

Nell, J.A. 2002. Farming triploid oysters. *Aquaculture* 210(1-4):69-88.

Nelson, T.C. 1946. On the need for developing new strains of oysters through selective breeding of domestics stocks, cross breeding with other species, and the introduction of species from other areas. *Presentation to the 1946 Joint Annual Convention of the Oyster Growers and Dealers Association of North America, National Shellfisheries Association,* held in New York, New York, June 5 to 7, 1946. For copies, contact K. Sellner, Chesapeake Research Consortium, Edgewater, Maryland.

Newell, R.I.E. 1988. Ecological changes in Chesapeake Bay: Are they the result of overharvesting the Eastern oyster (*Crassostrea virginica*). Pp. 536-546 in *Understanding the Estuary: Advances in Chesapeake Bay Research,* M.P. Lynch and E.C. Krome, eds. Chesapeake Research Consortium Publication 129. CBP/TRS 24/88, Solomons, Maryland.

Newell, R.I.E., and C.J. Langdon. 1996. Mechanisms and physiology of larval and adult feeding. Pp. 185-230 in *The Eastern Oyster* Crassostrea virginica, R.I.E. Newell, V.S. Kennedy, and A.F. Eble, eds. Maryland Sea Grant College Program, College Park, Maryland.

Newell, R.I.E., and J.A. Ott. 1999. Macrobenthic communities and eutrophication. Pp. 265-293 in *Ecosystems at the Land-Sea Margin: Drainage Basin to Coastal Sea,* T.C. Malone, A. Malej, L.W. Harding, Jr., N. Smodlaka, and R.E. Turner, eds. *Coastal and Estuarine Studies* 55. American Geophysical Union, Washington, D.C.

Newell, R.I.E., G.S.J. Alspach, V.S. Kennedy, and D. Jacobs. 2000. Mortality of newly metamorphosed eastern oysters (*Crassostrea virginica*) in mesohaline Chesapeake Bay. *Marine Biology* 136(4):665-676.

Newell, R.I.E., J.C. Cornwell, and M.S. Owens. 2002. Influence of simulated bivalve biodeposition and microphytobenthos on sediment nitrogen dynamics: A laboratory study. *Limnology & Oceanography* 47(5):1367-1379.

Nichols, F.H., J.K. Thompson, and L.E. Schemel. 1990. Remarkable invasion of San Francisco Bay (California, USA) by the Asian clam *Potamocorbula amurensis.* II. Displacement of a former community. *Marine Ecology Progress Series* 66: 95-101.

Nie, Z.Q. 1991. The culture of marine bivalve mollusks in China. Pp. 261-276 in *Estuarine and Marine Bivalve Mollusk Culture,* W. Menzel. Chesapeake Research Consortium Press, Boston, Massachusetts.

Nixon, S.W. 1995. Coastal marine eutrophication: A definition, social causes, and future concerns. *Ophelia* 41(0):199-219.

Norton, J.H., M.A. Shepherd, M.R. Abdon-Naguit, and S. Lindsay. 1993. Mortalities in the giant clam Hippopus *hippopus* associated with rickettsiales-like organisms. *Journal of Invertebrate Pathology* 62(2):207-209.

Nose, T. 1985. Recent advances in aquaculture in Japan. *Geojournal* 10(3):261-276.

Ó Foighil, D., P.M. Gaffney, and T.J. Hilbish. 1995. Differences in mitochondrial 16S ribosomal gene sequences allow discrimination among American [*Crassostrea virginica* (Gmelin)] and Asian [*C. gigas* (Thunberg), *C. ariakensis* (Wakiya)] oyster species. *Journal of Experimental Marine Biology and Ecology* 192:211-220.

Ó Foighil, D., P.M. Gaffney, A.E. Wilbur, and T.J. Hilbish. 1998. Mitochondrial cytochrome oxidase I gene sequences support an Asian origin for the Portuguese oyster *Crassostrea angulata. Marine Biology* 131:497-503.

O'Beirn, F.X., M.W. Luckenbach, R. Mann, J. Harding, and J. Nesterode. 1999. Ecological Functions of Constructed Oyster Reefs Along an Environmental Gradient in Chesapeake Bay. Final Report. The Virginia Institute of Marine Science, Gloucester Point, Virginia, 69p.

O'Clair, R.M., and S.C. Lindstrom. 2000. *North Pacific Seaweeds.* Plant Press, Auke Bay, Alaska, 162p.

O'Connor, R.J. 1986. Biological characteristics of invaders among bird species in Britain. *Philosophical Transactions of the Royal Society of London B, Biological Sciences* 314(1167):583-598.

Office Internationale des Epizooties. 2002. MSX disease (*Haplosporidium nelsoni*) in Canada. [Online]. Available: http://www.oie.int/eng/info/hebdo/AIS_48.HTM#Sec1 [October 25, 2002].

Officer, C.B., T.J. Smayda, and R. Mann. 1982. Benthic Filter Feeding a Natural Eutrophication Control. *Marine Ecology-Progress Series* 9(2):203-210.

Officer, C.B., R.B. Biggs, J.L. Taft, L.E. Cronin, and M.A. Tyler. 1984. Chesapeake Bay anoxia: Origin, development, and significance. *Science* 223(4631):22-27.

Orth, R.J., and K.A. Moore. 1983. Chesapeake Bay USA: An unprecedented decline in submerged aquatic vegetation. *Science* 222(4619):51-53.

Orth, R.J., K.L. Heck, Jr. and J. van Montfrans. 1984. Faunal communities in seagrass beds: A review of the influence of plant structure and prey characteristics on predator-prey relationships. *Estuaries* 7:339-350.

Orton, J.H. 1924. An account of investigations into the cause or causes of the unusual mortality among oysters in English oyster beds during 1920 and 1921. Fisheries Investigations Series II. Marine Fisheries of Great Britain, Ministry of Agriculture. *Fisheries and Food* 6(3):1-199.

Osborn, K. 1999. Effects of *Sargassum muticum* on native marine communities. Presented at the Pacific Forum on Wednesday, August 25, 1999. Monterey Bay Aquarium Research Institute, Moss Landing, California. [Online]. Available: www.mbari.org/seminars/1999/aug25_osborn.html [October 21, 2003].

Osman, R.W. 1977. The establishment and development of a marine epifaunal community. *Ecological Monographs* 47(1):37-63.

Osman, R.W., R.B. Whitlatch, and R.N. Zajac. 1989. Effects of resident species on recruitment into a community larval settlement versus post-settlement mortality in the oyster *Crassostrea Virginica*. *Marine Ecology-Progress Series* 54(1-2):61-73.

Paillard, C., L. Percelay, M. Le Pennec, and D. Le Picard. 1989. Pathogenic origin of the "brown ring" in *Tapes philippinarum* (Mollusca, Bivalvia). [French]. *Comptes Rendus De L'Academie Des Sciences, Series III, Sciences De La Vie* 309(7):235-241.

Paillard, C., P. Maes, and R. Oubella. 1994. Brown ring disease in clams. *Annual Review of Fish Diseases* 4(0):219-240.

Parmesan, C., and G. Yohe. 2003. A globally coherent fingerprint of climate change impacts across natural systems. *Nature* 421(6918):37-42.

Patel, S.K., and K.L. Jetani. 1991. Survey of edible oysters from the Saurashtra coast. *Journal of Current Biosciences* 8(3):79-82.

Paynter, K.T. 1999. Managing around oyster diseases in Maryland and Maryland oyster roundtable strategies. Pp. 317-328 in *Oyster Reef Habitat Restoration: A Synopsis and Synthesis of Approaches*, M.W. Luckenbach, R. Mann, and J.A. Wesson. Virginia Institute of Marine Science, Gloucester Point, Virginia.

Perdue, J.A., and G. Erickson. 1984. A comparison of the gametogenic cycle between the Pacific oyster *Crassostrea gigas* and the Suminoe oyster *Crassostrea rivularis* in Washington State. *Aquaculture* 37 (3):231-237.

Peterson, B.J., and K.L. Heck, Jr. 2001. An experimental test of the mechanism by which suspension feeding bivalves elevate seagrass productivity. *Marine Ecology-Progress Series* 218:115-125.

Peterson, C.H., H.C. Summerson, M. Dozier, D. Gaskill. 1999. *Application for a Permit to Conduct Aquaculture Trials with Non-native Oyster Species in North Carolina.* Presented to the North Carolina Division of Marine Fisheries, June 15, 1999. North Carolina Division of Marine Fisheries, Morehead City, North Carolina.

Phelps, H.L. 1994. The Asiatic clam (*Corbicula fluminea*) invasion and system-level ecological change in the Potomac River estuary near Washington, D.C. *Estuaries* 17(3):614-621.

Pinit, P.T., R.J. Bellmer, and G.W. Thayer. 1999. NOAA Fisheries Technical Guidance Manual for Success Criteria in Restoration Projects. [Draft Manuscript] NOAA, Office of Habitat Conservation, Silver Spring, Maryland. [Online]. Available: www.nmfs.noaa.gov/habitat/restoration/projects_programs/research/SuccessCriteria Guidance.pdf [October 14, 2003].

Posey, M.H. 1988. Community changes associated with the spread of an introduced seagrass *Zostera-Japonica*. *Ecology* 69(4):974-983.

Posey, M.H., T.D. Alphin, C.M. Powell, and E. Towsend. 1999. Use of oyster reefs as habitat for epibenthic fish and decapods. Pp. 229-237 in *Oyster Reef Habitat Restoration: A Synopsis and Synthesis of Approaches*, M.W. Luckenbach, R. Mann, and J.A. Wesson. Virginia Institute of Marine Science, Gloucester Point, Virginia.

Powell, E.N., E.E. Hofmann, J.M. Klinck, and S.M. Ray. 1992. Modeling oyster populations I. A commentary on filtration rate. Is faster always better? *Journal of Shellfish Research* 11:387-398.

Power, G. 1970. More about oysters than you wanted to know. *Maryland Law Review* 30:199-225.

Pratt, J.W., H. Raiffa, and R. Schlaifer. 1995. *Introduction to Statistical Decision Theory*. MIT Press, Cambridge, 875p.

Prins, T.C., V. Escaravage, A.C. Smaal, and J.C.H. Peeters. 1995. Nutrient cycling and phytoplankton dynamics in relation to mussel grazing in a mesocosm experiment. *Ophelia* 41(0):289-315.

Prins, T.C., A.C. Smaal, and R.F. Dame. 1997. A review of the feedbacks between bivalve grazing and ecosystem processes. *Aquatic Ecology* 31:349-359.

Pritchard, D.W. 1967. Observations of circulation in coastal plain estuaries. Pp. 37-44 in *Estuaries*, G.H. Lauff. AAAS Publication No. SS-83., Washington, D.C.

Qualman, A. 1990. *Blood on the Half Shell*. Binford and Mort Publishing, Hillsboro, Oregon.

Quayle, D.B. 1938. *Paphia bifurcata*, a new molluscan species from Ladysmith Harbor, B.C. *Journal of the Fisheries Research Board of Canada* 4:53-54.

Quayle, D.B. 1988. Pacific oyster culture in British Columbia, Canada. *Canadian Bulletin of Fisheries & Aquatic Sciences* (218):I-VI; 1-241.

Quick, J.A.J., and J.G. Mackin. 1971. Oyster parasitism by Labyrinthomyxa marina in Florida. Florida Department of Natural Resources Marine Research Laboratory, *Professional Papers Series* (13):1-55.

Ragone-Calvo, L.M., and E.M. Burreson. 1994. Characterization of overwintering infections of *Perkinsus marinus* (Apicomplexa) in Chesapeake Bay oysters. *Journal of Shellfish Research* 13(1):123-130.

Ragone-Calvo, L.M., and E.M. Burreson. 2002. QPX susceptibility in hard clams varies with geographic origin of broodstock, Report No. VSG-02-18. Virginia Institute of Marine Science, Gloucester Point, Virginia.

Ragone-Calvo, L.M., C.F. Dungan, B.S. Roberson, and E.M. Burreson. 2003a. Systematic evaluation of factors controlling *Perkinsus marinus* transmission dynamics in the lower Chesapeake Bay. *Diseases of Aquatic Organisms* 56(1):75-86.

Ragone Calvo, L.M., Gustavo W. Calvo, and Eugene M. Burreson. 2003b. Dual disease resistance in a selectively bred eastern oyster, *Crassostrea virginica*, strain tested in Chesapeake Bay. *Aquaculture* 220(1-4):69-87.

Raillard, O., and A. Menesguen. 1994. An ecosystem box model for estimating the carrying capacity of a macrotidal shellfish system. *Marine Ecology-Progress Series* 115(1-2):117-130.

Raina, S.K., S. Das, M.M. Rai, and A.M. Khurad. 1995. Transovarial transmission of *Nosema locustae* in the migratory locust *Locusta migratoria migratorioides*. *Parasitology Research* 81(1):38-44.

Rao, K.S. 1987. Taxonomy of Indian oysters. *Central Marine Fisheries Research Institute Bulletin* 38:1-6.

Ray, S.M. 1996. Historical perspective on *Perkinsus marinus* disease of oysters in the Gulf of Mexico. *Journal of Shellfish Research* 15(1):9-11.

Reeb, C.A., and J.C. Avise. 1990. A genetic discontinuity in a continuously distributed species: Mitochondrial DNA in the American oyster, *Crassostrea virginica*. *Genetics* 124(2):397-406.

Reichard, S.H., and C.W. Hamilton. 1997. Predicting invasions of woody plants introduced into North America. *Conservation Biology* 11(1):193-203.

Rejmanek, M., and D.M. Richardson. 1996. What attributes make some plant species more invasive? *Ecology* 77(6):1655-1661.

Renault, T., C. Lipart, and I. Arzul. 2001. A herpes-like virus infecting *Crassostrea gigas* and *Ruditapes philippinarum* larvae in France. *Journal of Fish Diseases* 24(6):369-376.

Renault, T., R.M. Le Deuff, B. Chollet, N. Cochennec, and A. Gerard. 2000. Concomitant herpes-like virus infections in hatchery-reared larvae and nursery-cultured spat *Crassostrea gigas* and *Ostrea edulis*. *Diseases of Aquatic Organisms* 42:173-183.

Rennie, S.E., J.L. Largier, and S.J. Lentz. 1999. Observations of a pulsed buoyancy current downstream of Chesapeake Bay. *Journal of Geophysical Research* 104(C8):18227-18240.

Ricciardi, A., and J.B. Rasmussen. 1998. Predicting the identity and impact of future biological invaders: A priority for aquatic resource management. *Canadian Journal of Fisheries & Aquatic Sciences* 55(7):1759-1765.

Riemann, B., T.G. Nielsen, S.J. Horsted, P. Koefoed Bjoernsen, and J. Pock-Steen. 1988. Regulation of phytoplankton biomass in estuarine enclosures. *Marine Ecology-Progress Series* 48(3):205-215.

Riisgard, H.U. 1988. Efficiency of particle retention and filtration rate in six species of northeast American bivalves. *Marine Ecology-Progress Series* 45:217-223.

Roegner, G.C., and R. Mann. 1995. Early recruitment and growth of the American oyster *Crassostrea virginica* with respect to tidal zonation and season. *Marine Ecology-Progress Series* 117(1-3):91-101.

Root, T.L., J.T. Price, K.R. Hall, S.H. Schneider, C. Rosenzweig, and J.A. Pounds. 2003. Fingerprints of global warming on wild animals and plants. *Nature* 421(6918):57-60.

Rosenfield, A., and F.G. Kern. 1979. Molluscan imports and the potential for introduction of disease organisms. Pp. 165-191 in *Exotic Species in Mariculture: Proceedings of a Symposium on Exotic Species in Mariculture: Case Histories of the Japanese Oyster, Crassostrea gigas (Thunberg), with Implications for Other Fisheries*, R. Mann, ed. held in Woods Hole Oceanographic Institution, Woods Hole, Massachusetts, September 18-20, 1978. Cambridge, MIT Press, 363p.

Rosenfield, A., and R. Mann. 1992. *Dispersal of Living Organisms into Aquatic Ecosystems*. Maryland Sea Grant College Program, College Park, Maryland, 471p.

Rosenthal, H., D. Weston, R. Gowen, and E. Black. 1988. Report of the ad hoc Study Group on "Environmental Impact of Mariculture." Cooperative Research Report No. 154, 83p.

Rothschild, B.J., J.S. Ault, P. Goulletquer, and M. Héral. 1994. Decline of the Chesapeake Bay oyster population: A century of habitat destruction and overfishing. *Marine Ecology-Progress Series* 111(1-2):29-39.

Roughley, T.C. 1926. An investigation of the cause of an oyster mortality on the Georges River, New South Wales, 1924-25. *Proceedings of the Linnean Society of New South Wales* 51:446-491.

Ruiz, G.M., J.T. Carlton, E.D. Grosholz, and A.H. Hines. 1997. Global invasions of marine and estuarine habitats by non-indigenous species: Mechanisms, extent, and consequences. *American Zoologist* 37(6):621-632.

Ruiz, G.M., P. Fofonoff, A.H. Hines, and E.D. Grosholz. 1999. Non-indigenous species as stressors in estuarine and marine communities: Assessing invasion impacts and interactions. *Limnology & Oceanography* 44(3 PART 2):950-972.

Ruiz, G.M., P.W. Fofonoff, J.T. Carlton, M.J. Wonham, and A.H. Hines. 2000. Invasion of coastal marine communities in North America: Apparent patterns, processes, and biases. *Annual Review of Ecology and Systematics* 31:481-531

Russell, K.M., J.N. Galloway, S.A. Macko, J.L. Moody, and J.R. Scudlark. 1998. Sources of nitrogen in wet deposition to the Chesapeake Bay region. *Atmospheric Environment* 32(14-15):2453-2465.

Ryman, N., and L. Laikre. 1991. Effects of supportive breeding on the genetically effective population size. *Conservation Biology* 5(3):325-329.

Saaty, T.L. 1990. *Multicriteria Decision Making: The Analytic Hierarchy Process*. RWS Publications, Pittsburgh, Pennyslvania.

Sagasti, A., L.C. Schaffner, and J.E. Duffy. 2001. Effects of periodic hypoxia on mortality, feeding and predation in an estuarine epifaunal community. *Journal of Experimental Marine Biology & Ecology* 258(2):257-283.

Samonte-Tan, G.P.B., and G.C. Davis. 1998. Economic analysis of stake and rack-hanging methods of farming oysters *(Crassostrea iredalei)* in the Philippines. *Aquaculture* 160(3-4):239-249.

Sandlund, O.T., P.J. Schei, and Å.e. Viken. 1999. *Invasive species and biodiversity management.* Population and Community Biology Series: 24. Kluwer Academic Publishers, Boston, Massachusetts, 431p.

Santopietro, G.D., and L.A. Shabman. 1992a. Can privatization be inefficient?: The case of the Chesapeake Bay oyster fishery. *Journal of Economic Issues* 26(2):407-419.

Santopietro, G.D., and L.A. Shabman. 1992b. Property rights to the Chesapeake Bay oyster fishery: History and implications. *Society & Natural Resources* 5(2):165-178.

Sauriau, P.G. 1991. Spread of Cyclope-neritea Mollusca Gastropoda along the northeastern Atlantic coasts in relation to oyster culture and to climatic fluctuations. *Marine Biology* 109(2):299-310.

Scholz, A.J., C. Jones, R.E. Westley, and D.F. Tufts. 1984. *Improved Techniques for Culture Pacific Oysters,* Crassostrea gigas*: A Summary of Studies Conducted by the Washington Department of Fisheries Since 1955.* Washington Department of Fisheries Completion Report. Washington Department of Fish and Wildlife, Olympia.

Seagle, S.W., R. Pagnotta, and F.A. Cross. 1999. The Chesapeake Bay and Northern Adriatic Sea drainage basins: land-cover and nutrient export. Pp. 7-27 in *Coastal and Estuarine Studies: Ecosystems at the Land-Sea Margin Drainage Basin to Coastal Sea,* T.C. Malone, A. Malej, L.W. Harding, Jr., N. Smodlaka, and R.E. Turner, eds. American Geophysical Union, Washington, D.C.

Seijo, J.C. 1993. Individual transferable grounds in a community managed artisanal fishery. *Marine Policy* 18:78-81.

Seliger, H.H., and J.A. Boggs. 1988. Evidence for loss of suitable habitats for oysters in tributaries of the Chesapeake Bay. Pp. 111-127 in *Understanding the Estuary, Advances in Chesapeake Bay Research,* M.P. Lynch and E.C. Krome, eds. Chesapeake Research Consortium Publication 129 CBP/TRS 24/88. Chesapeake Research Consortium, Edgewater, Maryland.

Seliger, H.H., J.A. Boggs, and W.H. Biggley. 1985. Catastrophic anoxia in the Chesapeake Bay in 1984. *Science* 228:70-73.

Shaw, W.N. 1997. The shellfish industry of California—past, present and future. In *The History, Present Condition, and Future of Molluscan Fisheries of North and Central America and Europe,* Volume 2, Pacific Coast and supplemental topics; U.S. Department of Commerce, NOAA Technical Report 127, Washington, D.C.

Shumway, S.E., T.L. Cucci, R.C. Newell, and C.M. Yentsch. 1985. Particle selection, ingestion, and absorption in filter-feeding bivalves. *Journal of Experimental Marine Biology and Ecology* 91(1-2):77-92.

Shumway, S.E. 1996. Natural environmental factors. Pp. 467-513 in *The Eastern Oyster:* Crassostrea Virginica, V.S. Kennedy, R.I E. Newell, and A.F. Eble. Maryland Sea Grant College Program, College Park, Maryland.

Simberloff, D. 1989. Which insect introductions succeed and which fail? Pp. 61-75 in *Biological Invasions: A Global Perspective,* J.A. Drake, H.A. Monney, F. DiCastri, R.H. Groves, M. Kruger, M. Rejmanek, and W. Williamson. John Wiley and Sons, New York.

Simenstad, C.A., S.B. Brandt, A. Chalmers, R. Dame, L.A. Deegan, R. Hodson, and E.D. Houde. 2000. Habitat-biotic interactions. Pp. 427-459 in *Estuarine Science: A Synthetic Approach to Research and Practice,* J.E. Hobbie, ed. Island Press, Washington, D.C.

Sindermann, C.J. 1990. *Principal Diseases of Marine Fish and Shellfish.* 2nd ed. Academic Press, San Diego.

Sindermann, C.J., and A. Rosenfield. 1968. Principal diseases of commercially important marine bivalve mollusca and *Crustacea* oysters mussels copepod trematodes bacteria protozoa parasites. *Fishery Bulletin* 66(2):335-385.

Smaal, A.C., and T.C. Prins. 1993. The uptake of organic matter and the release of inorganic nutrients by bivalve suspension feeder bed. Pp. 273-298 in *Bivalve Filter Feeders in Estuarine and Coastal Ecosystem Processes,* R.F. Dame. Springer Verlag, Berlin.

Smith, V.K., and W.H. Desvousges. 1986. *Measuring water quality benefits.* International Series in Economic Modeling. Kluwer Academic Publishers, Boston, Massachusetts.

Smith, V.K., and W.H. Desvousges. 1987. An empirical analysis of the economic value of risk changes. *Journal of Political Economy* 95(1):89-114.

Smith, D.E., M. Leffler, and G. Mackiernan, eds. 1992. *Oxygen Dynamics in the Chesapeake Bay—A Synthesis of Recent Results.* Maryland Sea Grant College Program, College Park, Maryland, 234p.

Smith P.J., H. Ozaki, Y. Fujio. 1986. No evidence for reduced genetic variation in the accidentally introduced oyster *C. gigas* in New Zealand. *New Zealand Journal of Marine and Freshwater Research* 20:569-574.

Smodlaka, N., T.C. Malone, A. Malej, and L.W. Harding, Jr. 1999. Introduction. Pp. 1-6 in *Ecosystems at the Land-Sea Margin: Drainage Basin to Coastal Sea,* T.C. Malone, ed. Coastal and Estuarine Studies 55. American Geophysical Union, Washington, D.C.

Sprague, L.A., M.J Langland, S.E. Yochum, R.E. Edwards, J.D. Blomquist, S.W. Phillips, G.W. Shenk, and S.D. Preston. 2000. *Factors Affecting Nutrient Trends in Major Rivers of the Chesapeake Bay Watershed.* U.S. Geological Survey Water-Resources Investigations Report 00-4218. U.S. Geological Survey, Reston, Virginia.

Southworth, M., and R. Mann. 1998. Oyster reef broodstock enhancement in the Great Wicomico River, Virginia. *Journal of Shellfish Research* 17(4):1101-1114.

Stachowicz, J.J., R.B. Whitlatch, and R.W. Osman. 1999. Species diversity and invasion resistance in a marine ecosystem. *Science* 286(5444):1577-1579.

Stachowicz, J.J., H. Fried, R.W. Osman, and R.B. Whitlatch. 2002a. Biodiversity, invasion resistance, and marine ecosystem function: Reconciling pattern and process. *Ecology* 83 (9):2575-2590.

Stachowicz, J.J., J.R. Terwin, R.B. Whitlatch, and R.W. Osman. 2002b. Linking climate change and biological invasions: Ocean warming facilitates nonindigenous species invasions. *Proceedings of the National Academy of Sciences of the United States of America* 99 (24):15497-15500.

Stanley, J.G., S.K. Allen, Jr., and H. Hidu. 1981. Polyploidy induced in the American oyster, *Crassostrea virginica,* with cytochalasin B. *Aquaculture* 23:1-10.

Stauber, L.A. 1950. The problem of physiological species with special reference to oysters and oyster drills. *Ecology* 31:109-118.

Steele, E.N. 1957. The *Rise and Decline of the Olympia Oyster.* Falco Publication, Elma, 126p.

Steele, E.N. 1964. *The Immigrant Oyster* (Ostrea gigas) *Now Known as the Pacific Oyster.* Warren's Quick Print, Olympia, Washington, 179p.

Stenzel, H.B. 1971. Oysters. Pp. N953-N1224 in *Treatise on Invertebrate Paleontology,* Part N, Mollusca 6, Vol. 3, Raymond C. Moore, ed. The Geological Society of America and University of Kansas Press, Boulder, Colorado, and Lawrence, Kansas.

Stevenson, J.C., J.I. Marusić, B. Ozretić, A. Marson, G. Cecconii, and M.S. Kearney. 1999. Shallow water and shoreline ecosystems of the Chesapeake Bay compared to the Northern Adriatic Sea: transformation of habitat at the land-sea margin. Pp. 29-80 in *Coastal and Estuarine Studies: Ecosystems at the Land-Sea Margin Drainage Basin to Coastal Sea,* T.C. Malone, A. Malej, L.W. Harding, Jr., N. Smodlaka, and R.E. Turner, eds. American Geophysical Union, Washington, D.C.

Stroud, R.H. 1992. *Stemming the tide of coastal fish habitat loss: proceedings of a symposium on conservation of coastal fish habitat, Baltimore, Maryland, March 7-9, 1991*. Marine Recreational Fisheries, 14. National Coalition for Marine Conservation, Savannah, GA, 258p.

Sunila, I., J. Karolus, and J. Volk. 1999. A new epizootic of *Haplosporidium nelsoni* (MSX), a haplosporidian oyster parasite, in Long Island Sound, Connecticut. *Journal of Shellfish Research* 18(1):169-174.

Sutherland, J.P., and R H. Karlson. 1977. Development and stability of the fouling community at Beaufort, North Carolina. *Ecological Monographs* 47(4):425-446.

Tamburri, M.N., R.K. Zimmer-Faust, and M.L. Tamplin. 1992. Natural sources and properties of chemical inducers mediating settlement of oyster larvae: A re-examination. *Biology Bullettin* 183(2):327-338.

Tarnowski, M. 1999. A Historical Background for Oyster Landings in Maryland 1916-1998. Maryland Department of Natural Resources, Annapolis, Maryland, 34p.

Tegner, M.J., and P.K. Dayton. 1987. El Niño effects on southern California kelp forest communities. Pp. 243-280 in *Advances in Ecological Research* 17, A. Macfadyen and E.D. Ford, eds. Academic Press, San Diego.

Tenore, K.R., L.F.Boyer, R.M. Cal, J. Corral, N. Garcia-Fernandez, E. Gonzalez-Gurriaran, R.B. Hanson, J. Inglesias, M. Krom, E. Lopez-Jamar, J. McClain, M. M. Pamatmat, A. Perez, D.C. Rhoads, G. de Santiago, J. Tietjen, J. Westrich, and H.L. Windom. 1982. Coastal upwelling in the Rias Bajas, NW [northwest] Spain: contrasting the benthic regimes of the Rias de Arosa and de Muros [*Mytilus edulis*]. *Journal of Marine Research* 40(3):701-772.

Terpstra, L.K. 1998. There goes the neighborhood: The potential private party liability of the international shipping industry for exotic marine species introduction via ballast water in England. *Transnational Lawyer* 11(1):277-309.

Thibaut, T., A. Meinesz, P. Amande, S. Charrier, K. De Angelis, S. Ierardi, L. Mangialajo, J. Melnick, and V. Vidal. 2001. *Elysia subornata* (Mollusca) a potential control agent of the alga *Caulerpa taxifolia* (Chlorophyta) in the Mediterranean Sea. *Journal of the Marine Biology Association* (U.K.) 81:497-504.

Thompson, J.A. 2001. Introduction of *Crassostrea ariakensis* (formerly *rivularis*) to Chesapeake Bay: The Solution to Restoring an Oyster Fishery and Water Quality in the Bay? U.S. Fish and Wildlife Service, Chesapeake Bay Field Office, Annapolis, Maryland, 31p.

Thompson, R.J., R.I.E. Newell, V.S. Kennedy, and R. Mann. 1996. Reproductive Processes and Early Development. Pp. 335-370 in *The eastern oyster*: Crassostrea virginica, V.S Kennedy, R.I.E. Newell, and A.F. Eble. University of Maryland, Maryland Sea Grant College Program, College Park, Maryland.

Tilman, D. 1997. Community invasibility, recruitment limitation, and grassland biodiversity. *Ecology* 78(1):81-92.

Trussell, G.C., and L.D. Smith. 2000. Induced defenses in response to an invading crab predator: An explanation of historical and geographic phenotypic change. *Proceedings of the National Academy of Sciences of the United States of America* 97(5):2123-2127.

Tschang, S., and L. Tse-kong. 1956. A study on Chinese oysters. *Acta Zoologica Sinica* 8:65-93.

Turner, H.J., Jr. 1949. Growth of the North Pacific oyster in New England. Pp. 31 in *Report on Investigations of Methods of Improving the Shellfish Resources of Massachusetts*, H.J. Turner, ed. Division of Marine Fisheries, Department of Conservation, Boston, Massachusetts.

Turner, H.J., Jr. 1950. Growth of the North Pacific oyster, *Ostrea gigas* in Massachusetts. Pp. 8-10 in *Third Report on Investigations of Methods of Improving the Shellfish Resources of Massachusetts*, H.J. Turner, ed. Division of Marine Fisheries, Department of Conservation, Boston, Massachusetts.

U.S. Army Corps of Engineers. 2003. *Decision Document Amendment. Oyster Recovery Project Phase III, Great Wicomico River, VA.* U.S. Army Corps of Engineers, Norfolk District, Norfolk, Virginia.

U.S. Atlantic States Marine Fisheries Commission. 1989. *A Procedural Plan to Control Interjurisdictional Transfers and Introductions of Shellfish.* Fisheries Management Report No. 13. U.S. Atlantic States Marine Fisheries Commission, Washington, D.C., 58p.

U.S. Atlantic States Marine Fisheries Commission. 2002. *Proceedings of the Workshop on the Introduction of Asian Oysters* (Crassostrea ariakensis) *to the Chesapeake Bay.* Special Report No. 74. U.S. Atlantic States Marine Fisheries Commission, Washington, D.C., [Online]. Available: http://www.asmfc.org/PUB/Special%20Reports/SR%2074%20 Asian%20Oyster%20Workshop.pdf [October 20, 2003].

U.S. Bureau of Commercial Fisheries, National Marine Fisheries Service, and U.S. Fish and Wildlife Service, eds. 1960. *Fisheries of the United States.* Control Number 73640272 Washington, D.C.

U.S. Congress, Office of Technology Assessment. 1993. *Harmful Non-Indigenous Species in the United States,* OTA-F-565. U.S. Government Printing Office, Washington, D.C., 391p.

U.S. Department of Commerce, National Oceanic and Atmospheric Administration. 1994. *Ocean and Coastal Resource Management.* [Online]. Available: http://coastalmanagement.noaa.gov/ [October 22, 2003].

U.S. Department of Commerce. 2002. *Molluscan Shellfish Research and Management: Charting a Course for the Future.* Preliminary Proceedings from the Workshop held in Charleston, South Carolina. National Oceanic and Atmospheric Administration, Silver Spring, Maryland.

U.S. Department of Labor, Bureau of Labor Statistics (BLS). 2003. Consumer Price Index. Available at <http://www.bls.gov/cpi/>. Accessed August 9, 2003.

U.S. Environmental Protection Agency (U.S. EPA). 2002. *Mid-Atlantic Integrated Assessment 1997-98 Summary Report.* EPA/620/R-02/003. U.S. Environmental Protection Agency, Atlantic Ecology Division, Narragansett, Rhode Island, 115p.

Ulanowicz, R.E., and J.H. Tuttle. 1992. The trophic consequences of oyster stock rehabilitation in Chesapeake Bay. *Estuaries* 15(3):298-306.

Veitch, F.P., and H. Hidu. 1971. Gregarious setting in the American oyster *Crassostrea Virginica* part 1 properties of a partially purified setting factor. *Chesapeake Science* 12(3):173-178.

Villalba, A., M.J. Carballa, C. López, A. Cabada, L. Corral, and C. Azevedo. 1999. Branchial rickettsia-like infection associated with clam *Venerupis rhomboides* mortality. *Diseases of Aquatic Organisms* 36:(1)53-60.

Virginia Department of Environmental Quality. 2001. *Federal Consistency Information.* [Online]. Available: http://www.deq.state.va.us/coastal/documents/consistency.pdf [October 21, 2003].

Virginia General Assembly. 2002. *House Joint Resolution No.164.* Amendment in the Nature of a Substitution. H.J.R. 164. 2002 Session.

Virginia Institute of Marine Science. 1998. *Oyster Diseases of the Chesapeake Bay.* [Online] Available: http://www.vims.edu/newsmedia/images/oyster-diseases-CB.pdf [May 17, 2003].

Virginia Marine Resources Commission. 2003a. *Virginia Marine Resources Commission Landings Bulletins. Plans and Statistics Department of the Fisheries Management Division.* [Online]. Available: http://www.mrc.state.va.us/page1f8.htm [October 1, 2003].

Virginia Marine Resources Commission. 2003b. *Oyster Ground Production.* [Online]. Available: http://www.mrc.state.va.us/page1h1.htm [October 1, 2003].

Volety, A.K., F.O. Perkins, R. Mann, and P.R. Hershberg. 2000. Progression of diseases caused by the oyster parasites, *Perkinsus marinus* and *Haplosporidium nelsoni*, in *Crassostrea virginica* on constructed intertidal reefs. *Journal of Shellfish Research* 19(1):341-347.

Vrijenhoek, R.C., S.E. Ford, and H.H. Haskin. 1990. Maintenance of heterozygosity during selective breeding of oysters for resistance to MSX disease. *Journal of Heredity* 81(6):418-423.

Walker, K.D., R.B. Rettig, and R. Hilborn. 1983. Analysis of multiple objectives in Oregon coho salmon policy [Strategies with respect to allocation and production of wild and hatchery coho salmon]. *Canadian Journal of Fisheries & Aquatic Sciences* 40(5):580-587.

Wang, R., Z. Wang, and J. Zhang. 1993. *Marine Molluscan Culture.* [In Chinese] Qingdao Ocean University Press, Qingdao, China.

Waselkov, G.A. 1982. *Shellfish gathering and shell midden archaeology*. Ph.D. Dissertation, University of North Carolina at Chapel Hill.

Weiner, R.M., M. Walch, M.P. Labare, D.B Bonar, and R.R. Colwell. 1989. Effect of biofilm of the marine bacterium *Alteromonas colwelliana* (LST) on set of the oysters *Crassostrea gigas* (Thunberg) and *C. virginicia* (Gmelin). *Journal of Shellfish Research* 8:117-123.

Welch, W.R. 1963. The European oyster, *Ostrea edulis*, in Maine. *Proceedings of the National Shellfisheries Association* 54:7-23.

Wells, H.W. 1961. The Fauna of Oyster Beds, with Special Reference to the Salinity Factor. *Ecological Monographs* 31(3):239-266.

Wesson, J., R. Mann, and M. Luckenbach. 1999. Oyster restoration efforts in Virginia. Pp. 117-130 in *Oyster Reef Habitat Restoration: A Synopsis and Synthesis of Approaches*, M.W. Luckenbach, R. Mann, and J.A. Wesson, eds. Virginia Institute of Marine Science, Gloucester Point.

Wharton, J. 1957. *The Bounty of the Chesapeake Bay: Fishing in Colonial Virginia.* University Press of Virginia, Charlottesville.

Whyte, S.K., R.J. Cawthorn, and S.E. McGladdery. 1994. QPX (Quahaug Parasite X), a pathogen of northern quahaug *Mercenaria mercenaria* from the Gulf of St. Lawrence, Canada. *Diseases of Aquatic Organisms* 19(2):129-136.

Wildish, D., and D. Kristmanson. 1997. Benthic Suspension Feeders and Flow. Cambridge University Press, United Kingdom, 409p.

Williamson, M.H. 1996. *Biological Invasions*, 1st ed. Chapman & Hall, New York, New York, 244p.

Williamson, M.H., and A. Fitter. 1996. The characters of successful invaders. *Biological Conservation* 78(1-2):163-170.

Winslow, F. 1881. Deterioration of American oyster beds. *Popular Science Monthly*, New York, 20:20-43, 145-156.

Winslow, F. 1882. *Report on the Oyster Beds of the James River, Virginia and of Tangier and Pocomoke Sounds, Maryland and Virginia.* Appendix II. U.S. Coast and Geodetic Survey for 1881. U.S. Coast and Geodetic Survey, [now the National Oceanic and Atmospheric Administration], Washington, D.C.

Wu, X.Z., and J.P. Pan. 2000. An intracellular prokaryotic microorganism associated with lesions in the oyster, *Crassostrea ariakensis* Gould. *Journal of Fish Diseases* 23(6):409-414.

Yao, Q. 1988. Paleogeographical significance of discovery of paleoforest remains and oyster reef in Shenhu Bay, Fujian. *Journal of Oceanography in Taiwan Strait* 7(3):299-302.

Yonge, C.M., and T.E. Thompson. 1976. Living Marine Molluscs. Collins Publishing Group, London, 288p.

Zabaleta, A.I., and Barber B.J. 1996. Prevalence, intensity, and detection of *Bonamia ostreae* in *Ostrea edulis* L. in the Damariscotta River area, Maine. *Journal of Shellfish Research* 15(2):395-400.

Zhang, X., and Z. Lou. 1956. Oyster. [Chinese]. *Bulletin of Biology* 2:27-32.

Zhang, X., Z. Qi, and Y. Xie. 1959. Moeurs de s'alimenter chez l'*Ostrea rivularis* Gould. [Chinese with abstract in French]. *Oceanologia et Limnologica Sincica* 2:163-179.

Zhang, Y., B.L. Munday, and J. Handlinger. 1995. Mass mortality of flat oysters (*Ostrea rivularis*) associated with a bloom of *Prorocentrum* sp. in the port of Zhanjiang, South China. *Bulletin of the European Association of Fish Pathologists* 15:(2)61-63.

Zhou, M., and S.K. Allen, Jr. 2003. A review of published work on *Crassostrea ariakensis*. *Journal of Shellfish Research* 22:1-20.

Zimmer-Faust, R.K., and M.N. Tamburri. 1994. Chemical identity and ecological implications of a waterborne, larval settlement cue. *Limnology & Oceanography* 39(5):1075-1087.

Zurburg, W., A. Smaal, M. Héral, and N. Dankers. 1994a. In situ estimations of uptake and release of material by bivalve filter feeders in the Bay of Marennes-Oleron (France). Pp. 239-242 in *Changes in Fluxes in Estuaries*, K.R. Dyer and R.J. Orth, eds. Olsen and Olsen, Denmark.

Zurburg, W., A. Smaal, M. Héral, and N. Dankers. 1994b. Seston dynamics and bivalve feeding in the Bay of *Marennes Oleron* (France). *Netherlands Journal of Aquatic Ecology* 28(3-4):459-466.

Appendixes

A

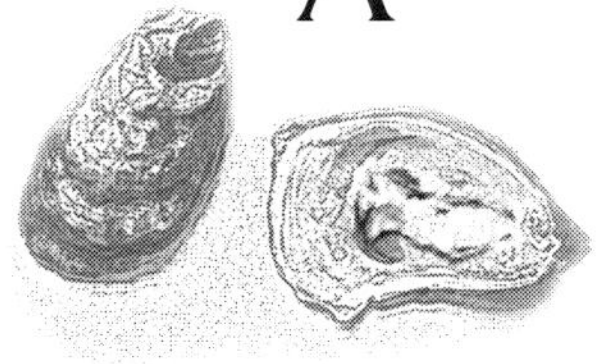

Committee and Staff Biographies

COMMITTEE MEMBERS

James Anderson *(Co-Chair)* is a Professor in the Department of Environmental and Natural Resource Economics at the University of Rhode Island. Dr. Anderson earned his Ph.D. in agricultural and resource economics in 1983 from the University of California, Davis. His research is in the area of fisheries and aquaculture economics. He has worked on bioeconomic interactions between aquaculture and the common property salmon fishery and, currently, the relationship between the seafood market and environmental policies and regulation. Dr. Anderson is the editor of the journal *Marine Resource Economics*. He served on a previous NRC Committee on the Assessment of Technology and Opportunities for Marine Aquaculture.

Dennis Hedgecock *(Co-Chair)* is a Professor at the Bodega Marine Laboratory of the University of California, Davis. He earned his Ph.D. in 1974 from University of California, Davis, in the field of genetics. Dr. Hedgecock is a leading scientist in oyster aquaculture and genetics, though his research covers broadly the population, quantitative, evolutionary, and conservation genetics of marine fish and shellfish. His current research subjects are Pacific oysters and Pacific salmon. He is a former member of the U.S. GLOBEC Scientific Steering Committee.

Mark Berrigan is the Chief of the Bureau of Aquaculture Development, Division of Aquaculture, Florida Department of Agriculture and Consumer Services. He earned his M.S. in marine biology and his B.S. in biology from the University of West Florida. He has broad experience in aquaculture, shellfisheries and multi-dimensional resource management. He is responsible for comprehensive aquacultural development and shellfish resource management in Florida, including the oyster restoration program, aquaculture certification program, and the sovereignty submerged land aquaculture leasing program.

Keith Criddle is head of the Department of Economics at Utah State University. Dr. Criddle earned his Ph.D. in agricultural economics from the University of California, Davis in 1989. His research focuses on the intersection between the natural sciences and economics. He is interested in the management of living resources, particularly fisheries. Dr. Criddle is an associate editor for *Marine Resource Economics*. He has served on the North Pacific Fishery Management Council's Scientific and Statistical Committee since 1993 and was a member of the NRC Committee to Review Individual Fishing Quotas.

William Dewey works as the Division Manager of project development and public affairs at Taylor Shellfish Company. Mr. Dewey earned his B.S. in shellfish biology and fisheries management at the University of Washington, School of Fisheries in 1981. He recently graduated from the Washington Agriculture and Forestry Education Foundation 2-year leadership training program (Class XIX). During his twenty years as a shellfish farmer, Mr. Dewey has taken an active role in the industry, working on water quality issues, health, and farming regulations. He has also served on the Board of Directors of People for Puget Sound and was appointed to the Puget Sound Council in 1996 by Washington Governor Lowry. Mr. Dewey is currently the President of the Pacific Shellfish Institute and serves on the Board of Directors of the National Aquaculture Association and the Interstate Shellfish Sanitation Conference.

Susan Ford is Research Professor Emerita in the Department of Marine and Coastal Sciences at Rutgers University. Dr. Ford earned her Ph.D. from Duke University in zoology in 1984. She is a pathobiologist with experience in the field of shellfish diseases. Her field of interest are invertebrate pathology/parasitology, genetics and mechanisms of resistance to pathogens at the Haskin Shellfish Research Laboratory. She specializes in the oyster diseases caused by *Haplosporidium nelsoni* (MSX) and *Perkinsus marinus* (Dermo). Her research has encompassed projects at the molecular, whole animal, and population level, including numerical modeling of the host-parasite relationship.

Philippe Goulletquer is the Director of the Laboratory of Shellfish Genetics and Pathology at l'institut francais de recherche pour l'exploitation de la mer (IFREMER). Dr. Goulletquer received his Ph.D. from the University of Western Brittany, France in 1989. His dissertation research was in the field of aquaculture and fisheries. His current research is on aquaculture of *Crassoaatrea gigas* and ecosystem management in France. Previously, Dr. Goulletquer worked at France's National Laboratory for Mariculture Ecosystems and as an Assistant Research Scientist at the Chesapeake Biological Laboratory of the University of Maryland. He has served on the ICES Working Group on Introduction and Transfers of Marine Organisms (ITMO) and the Mariculture Committee. Dr. Goulletquer was Co-Chair of the Ad Hoc Technical expert group on the Mariculture impacts on biodiversity for the UN Convention on Biological Diversity (CBD).

Richard Hildreth is Professor and Co-Director at the Ocean and Coastal Law Center, School of Law, University of Oregon. Dr. Hildreth received his B.S.E in 1965 and his J.D. in 1968, both from the University of Michigan. He specializes in ocean and coastal law, property, international environmental law, land-use law, and water resources law. Prior to teaching, he practiced business law with Steinhart & Falconer in San Francisco. He is a co-author of the law school textbook *Coastal and Ocean Law*. He is on the editorial advisory board of the journal *Ocean Development & International Law*. Dr. Hildreth has served on a previous NRC Committee on the Coastal Ocean.

Michael Paolisso is an Associate Professor at the University of Maryland in the Department of Anthropology. He earned his Ph.D. in anthropology from the University of California, Los Angeles in 1985. Dr. Paolisso's research includes applied anthropology, environment and pollution, environmental anthropology, and economic anthropology. He has studied the anthropology of rural Maryland and the cultural responses to the toxic alga *Pfiesteria piscicida*. In addition, Dr. Paolisso is currently the principal investigator of a 3-year study funded by the National Science Foundation and the U.S. Environmental Protection Agency (EPA) to study cultural models of pollution and environment.

Nancy Targett is a Professor of Marine Biology and Biochemistry at the Graduate College of Marine Studies at the University of Delaware. Dr. Targett earned her Ph.D. in oceanography in 1979 from the University of Maine. Her expertise is in biological oceanography and her research focuses on marine chemical ecology and organismal interactions mediated by naturally occurring metabolites, including: plant/herbivore interactions, predator/prey interactions, detoxification of allelochemicals, chemoattraction, and biofouling. She is an associate editor for the *Journal* *of Chemical Ecology* and an Aldo Leopold Leadership Program Fellow. From 1994 to 2000, she held an appointment to the Mid-Atlantic Fisheries Management Council where she chaired several of their committees. Dr. Targett is currently a member of the Ocean Studies Board and chaired the NRC Committee on Marine Biotechnology.

Robert Whitlatch is a Professor of Marine Sciences at the University of Connecticut. Dr. Whitlatch received his Ph.D. from the University of Chicago in 1976 in the field of evolutionary biology. Dr. Whitlatch is a benthic ecologist interested in animal-sediment relationships, trophic dynamics of deposit-feeding invertebrates, life history analysis, shellfish ecology, the ecology of invasive species, and community ecology. He has worked extensively on both oyster reef biology and on the ecology of nonnative species in coastal New England. He is a member of the editorial boards for the *Journal of Sea Research* and *Journal of Marine Research* and has served on numerous peer review panels for the National Science Foundation, U.S. Environmental Protection Agency, and National Oceanic and Atmospheric Administration.

NATIONAL RESEARCH COUNCIL STAFF

Susan Roberts is a Senior Program Officer for the National Research Council's Ocean Studies Board. Dr. Roberts received her Ph.D. in marine biology from the Scripps Institution of Oceanography. She has directed a number of studies including *Decline of the Steller Sea Lion in Alaskan Waters* (2003), *Effects of Trawling & Dredging on Seafloor Habitat* (2002), *Marine Protected Areas: Tools for Sustaining Ocean Ecosystems* (2001), *Bridging Boundaries Through Regional Marine Research* (2000), and *From Monsoons to Microbes: Understanding the Ocean's Role in Human Health* (1999). Dr. Roberts' research interests include marine microbiology, fish physiology, and marine biotechnology.

Kim Waddell is a Senior Program Officer in the National Research Council's Board on Agriculture and Natural Resources. Dr. Waddell received a B.A. in environmental studies from the University of California, Santa Cruz and his Ph.D. in Biology from the University of South Carolina. He has directed the NRC studies *The Future Role of Pesticides in U.S. Agriculture* (2000), *Professional Societies and Ecologically Based Pest Management* (2000), *Environmental Effects of Transgenic Plants: The Scope and Adequacy of Regulation* (2002), and *Animal Biotechnology: Science-based Concerns* (2002). His current projects include two new studies "Biological Confinement of Genetically Engineered Organisms" and "California Research Priorities for Pierce's Disease." Dr. Waddell's research interests include insect ecology and evolution, agricultural biotechnology, and sustainable agriculture.

Denise Greene is a Senior Project Assistant at the Ocean Studies Board and has 9 years of experience working for the National Research Council and National Academies. Mrs. Greene has been involved with studies on Marine Biotechnology, Environmental Information for Naval Warfare, and Defining Best Available Science for Fisheries Management.

Sarah Capote is a Project Assistant with the National Research Council's Ocean Studies Board. She earned her B.A. in history from the University of Wisconsin, Madison.

B

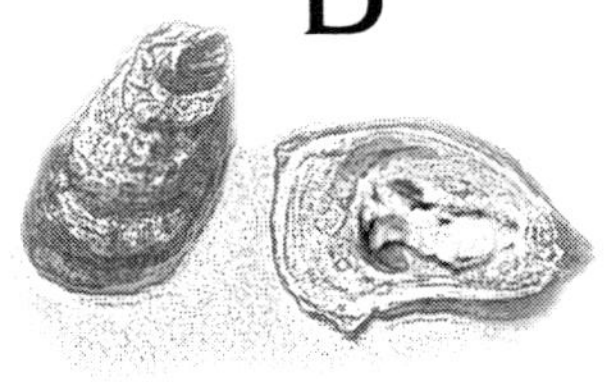

Report to the Virginia Marine Resources Commission

February 21, 2003

Mr. William A. Pruitt, Commissioner
Virginia Marine Resources Commission
2600 Washington Avenue, Third Floor
Newport News, VA 23607

Dear Mr. Pruitt:

On behalf of the National Research Council's Committee on Non-Native Oysters in the Chesapeake Bay, we are writing to express this committee's views about the pending 2003 proposal from the Virginia Seafood Council (VSC) to use *Crassostrea ariakensis* in a field trial. The National Research Council was asked by several state and federal agencies (see attachment B for the study's statement of task and list of sponsoring organizations) to undertake a review of the potential benefits and impacts of introducing this oyster into the Chesapeake Bay. Although the committee's final report will discuss many different aspects of risk assessment for using a non-native species, it will not be completed until June of this year. The committee decided to send this letter because of the importance and time sensitive nature of the pending decision before the Commission. This letter represents the consensus views of the committee and has been formally reviewed and approved by the National Research Council.

The Virginia Seafood Council has submitted a proposal "Economic analysis and pilot-scale field trials of triploid *C. ariakensis* aquaculture" to Virginia's Marine Resources Commission for the 2003 growing season. This proposal is designed as an industry trial with 10 participants and approximately 100,000 animals per site. Four different growing methods would be employed: bags in clam cages, bags on bottom, rack and bag, and floating raft. The animals would be harvested when they reach market size, estimated at 9-18 months. This proposal was originally submitted for 2002 and then revised in response to comments from the Virginia Institute of Marine Science and the Chesapeake Bay Program Living Resources Subcommittee - *C. ariakensis* Ad Hoc Review Panel. The major changes in the new proposal are as follows:

- Genetic (mated tetraploid by diploid) triploids will be used instead of "induced" (chemical) ones,
- The number of field trial participants has been reduced to 10, each with a larger number of animals,
- An economic feasibility analysis is stated as the principal goal, and
- A project manager will be hired specifically to oversee all aspects of the trial.

The purpose of this letter is to identify, based on this committee's work to date, important risks associated with the field trial as proposed. Major sources of risk or concern that are specific to the current Virginia Seafood Council proposal include:

- The process of generating mated triploids is not 100% effective, hence a small number of reproductive diploid oysters will be deployed with the triploids. In the 2000 year class of mated triploids, 3 out 3396 oysters examined were diploid (S. K. Allen, Jr., Virginia Institute of Marine Science; Response to Questions by *C. ariakensis* Ad Hoc Panel 2/3/03). If this frequency of occurrence (about 0.09%) were characteristic of populations of triploids produced by mating tetraploids and diploids, each field site under the 2003 VSC proposal would contain approximately 90 diploids per 100,000 oysters. If these diploids are allowed to become sexually mature and if they are in sufficient proximity to each other, there is a risk that a diploid population of non-native oysters could become established in the Chesapeake Bay. The probability that the reproductive diploids may be in close enough proximity to fertilize successfully has not been quantified, but should be determined for each grow out method.

- Reversion of triploids to diploids increases as the oysters get older, requiring more clearly defined accountability for the inventory to ensure that all oysters are removed by eighteen months. The triploid oysters may undergo gonadal maturation during the proposed trial. Currently, there are no provisions in the proposal for assessing maturation during the length of the trial. The risk of introducing a reproductive population of oysters could be lowered by harvesting animals before they have a chance to produce gametes. In the 2000 year class mentioned above, 25 mosaic animals (partial reversion of triploids to diploids) have been identified to date. In 6 of these mosaics, a small fraction of diploid cells were found in gonadal tissue, but none contained haploid gametes (S. K. Allen, Jr., Virginia Institute of Marine Science; Response to Questions by *C. ariakensis* Ad Hoc Panel 2/3/03). With the large number of oysters proposed for use in this trial a larger number of oysters will be expected to revert to diploid over time, increasing the risk that reproductive non-native oysters could be released into the Chesapeake Bay;
- If diploid *C. ariakensis* are found in the wild in the future, it will not be possible to determine whether or not they originated from this field trial. Genotyping of the broodstock would make it possible to determine whether or not the oysters from this field trial were responsible for introducing diploid *C. ariakensis* into the Chesapeake Bay or neighboring state waters;
- The causes of a significant mortality event may not be identified because regular monitoring for disease is not required. The proposal does not identify resources or responsibility for follow up investigation of a disease event. Furthermore, the position of project manager is contingent on outside funding, posing a risk that the trial will proceed without a responsible party to ensure implementation and coordination of monitoring, data collection, and data management. Both the stated goals of the field trial and safeguards meant to reduce the risk of accidental release of *C. ariakensis* would be compromised without a program manager to ensure enforcement.

A more comprehensive discussion of risks associated with the introduction of a non-native oyster will be provided in the committee's final report, including the potential ecological and economic risks and benefits. These types of risks and benefits have been raised in previous reports (e.g. Chesapeake Bay Program (2002). *Report of the Ad-hoc Panel On the Industry Trials of Triploid Non-Indigenous Oyster Species in Waters of the Chesapeake Bay Basin*, Annapolis, Maryland; Thompson, Julie A. (2001) *Introduction of*

Crassostrea ariakensis to Chesapeake Bay: The solution to Restoring an Oyster Fishery and Water Quality in the Bay? U.S. Fish and Wildlife Service, Chesapeake Bay Field Office, Annapolis MD; and Hallerman, E., Leffler, M., Mills, S., and Allen, S. (2001). *Aquaculture of Triploid Crassotrea ariakensis in Chesapeake Bay: A Symposium Report.* Maryland and Virginia Sea Grant, College Park, MD).

The committee is also concerned that the proposed field trial might be considered "a first time introduction" of *C. ariakensis* as stipulated in the 1993 Chesapeake Bay Program Policy for the Introduction of Non-Indigenous Aquatic Species. The committee firmly supports the Chesapeake Bay Program policy on non-native species introduction and the review process implemented by the ad hoc panel. This process enables participation by the major parties likely to be affected by this important decision. Unless this issue is clarified, the 2003 VSC field trial could preclude the Chesapeake Bay Program's review of future proposals to introduce this species, either as non-reproductive triploids in aquaculture or as reproductive diploids in the wild.

At present, there is insufficient scientific information available to thoroughly quantify and evaluate the risks and benefits of introducing this species into Virginia waters. Even less information is available for assessing the potential spread of *C. ariakensis* in the Chesapeake Bay and into the coastal waters of states along the Atlantic seaboard. If the Commission decides to approve a 2003 field trial, the committee strongly recommends amending the proposal to include measures to reduce the risks described above and to require collection of scientific data necessary for assessing the risk of introducing this non-native oyster. For example, more information is needed on the reproductive cycle of *C. ariakensis* in the field, the causes of mortality events, the fidelity and stability of triploid induction, and the growth rates at different locations under various deployment methods. This information would also be valuable for assessing the economic viability of using *C. ariakensis* in aquaculture.

Sincerely,

Jim Anderson, Ph.D.
Dennis Hedgecock, Ph.D.
Co-Chairs
Committee on Non-Native Oysters in the Chesapeake Bay

ATTACHMENT A

Committee on Non-native Oysters in the Chesapeake Bay

Susan Roberts, Study Director
Ocean Studies Board
500 Fifth Street, NW, NA 752
Washington, DC 20001
(202) 334-2714
(202) 334-2885 FAX

Committee Roster

Jim Anderson, *Co-Chair*
Professor of Resource Economics
University of Rhode Island, Kingston

Dennis Hedgecock, *Co-Chair*
Geneticist, Bodega Marine Laboratory
University of California, Davis

Mark Berrigan
Bureau Chief
Bureau of Aquaculture Development
Tallahasee, Florida

Keith Criddle
Department Head and Professor
Department of Economics
Utah State University, Logan

Bill Dewey
Division Manager
Taylor Shellfish Company, Inc.
Shelton, Washington

Susan Ford
Professor of Marine & Coastal Sciences
Haskin Shellfish Research Laboratory
Rutgers University
Port Norris, New Jersey

Philippe Goulletquer
Director, Laboratory of Shellfish
Genetics and Pathology
IFREMER
France

Richard G. Hildreth
Professor of Law
University of Oregon, Eugene

Michael Paolisso
Professor of Anthropology
University of Maryland, College Park

Nancy Targett
Professor and Associate Dean
Graduate College of Marine Studies University of Delaware, Lewes

Robert Whitlatch
Professor of Marine Sciences
Department of Marine Sciences
University of Connecticut, Groton

ATTACHMENT B

Non-native Oysters in the Chesapeake Bay Statement of Task

This study will examine the ecological and socio-economic risks and benefits of open water aquaculture or direct introduction of the non-native oyster, *Crassostrea ariakensis,* in the Chesapeake Bay. The committee will address how *C. ariakensis* might affect the ecology of the Bay, including effects on native species, water quality, habitat, and the spread of human and oyster diseases. Possible effects on recovery of the native oyster, *Crassostrea virginica,* will be considered. The potential range and effects of the introduced oyster will be explored, both within the Bay and in neighboring coastal areas. The study will investigate the adequacy of existing regulatory and institutional frameworks to monitor and oversee these activities.

The committee will assess whether the breadth and quality of existing research, on oysters and on other introduced species, is sufficient to support risk assessments of three management options: 1) no use of nonnative oysters, 2) open water aquaculture of triploid oysters, and 3) introduction of reproductive diploid oysters. Where current knowledge is inadequate, the committee will recommend additional research priorities.

Study Sponsors

The study is sponsored by the U.S. Environmental Protection Agency, National Oceanic and Atmospheric Administration, U.S. Fish and Wildlife Service, Maryland Department of Natural Resources, Virginia Department of Environmental Quality, Virginia Sea Grant, Maryland Sea Grant, Connecticut Sea Grant, and National Fish and Wildlife Foundation.

Acknowledgments

This letter report has been reviewed in draft form by individuals chosen for their diverse perspectives and technical expertise, in accordance with procedures approved by the National Research Council's (NRC) Report Review Committee. The purpose of this independent review is to provide candid and critical comments that will assist the institution in making its published report as sound as possible and to ensure that the report meets institutional standards for objectivity, evidence, and responsiveness to the study charge. The review comments and draft manuscript remain confidential to protect the integrity of the deliberative process. We wish to thank the following individuals for their participation in the review of this report:

Daniel S. Simberloff, University of Tennesee, Knoxville
Preston Pate, North Carolina Division of Fisheries, Morehead City
Chris Langdon, Oregon State University, Corvallis
James T. Carlton, Williams College, Mystic Seaport, Mystic, Connecticut

Although the reviewers listed above provided many constructive comments and suggestions, they were not asked to endorse the conclusions or recommendations nor did they see the final draft of the report before its release. The review of this letter report was overseen by John E. Dowling, Harvard University, Cambridge, Massachusetts. Appointed by the NRC, he was responsible for making certain that an independent examination of this report was carried out in accordance with institutional procedures and that all review comments were carefully considered. Responsibility for the final content of this report rests entirely with the authoring committee and the institution.

C

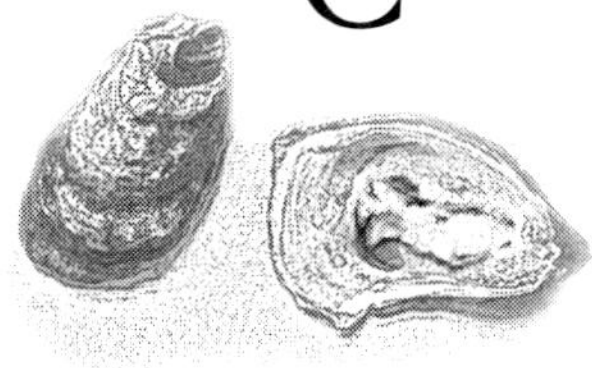

Acronyms

ACOE	Army Corps of Engineers
APHIS	Animal and Plant Health Inspection Service
ASMFC	Atlantic States Marine Fisheries Commission
CBP	Chesapeake Bay Program
CDFG	California Department of Fish and Game
CLO	Chlamydiales-like organism
CMP	Coastal Management Program
COMP	Comprehensive Oyster Management Plan
CWA	Clean Water Act
CZMA	Coastal Zone Management Act
DNR	Department of Natural Resources (Maryland)
EFH	Essential Fish Habitat
EIS	Environmental Impact Statement
EPA	Environmental Protection Agency
ESA	Endangered Species Act
FAO	Food and Agriculture Organization of the United Nations
ICES	International Council for the Exploration of the Sea
IFREMER	Institut français de recherche pour l'exploitation de la mer (French Research Institute for Exploitation of the Sea)

MD DNR	Maryland Department of Natural Resources
NEPA	National Environmental Policy Act
NMFS	National Marine Fisheries Service
NOAA	National Oceanic and Atmospheric Administration
NPDES	National Pollution Discharge Elimination System
NRC	National Research Council
OIE	Office International des Epizooties
PRPB	Puerto Rico's Planning Board
PCR	Polymerase chain reaction
PVC	Polyvinyl chloride
QPX	Quahog parasite X
RLO	Rickettsiales-like organism
SAV	Submerged aquatic vegetation
SPS	Agreement on the Application of Sanitary and Phytosanitary Measures
UNCLOS	United Nations Convention on the Law of the Sea
US FWS	US Fish and Wildlife Service
VEPCO	Virginia Electric and Power Company
VIMS	Virginia Institute of Marine Science
VMRC	Virginia Marine Resources Commission
VSC	Virginia Seafood Council
WTO	World Trade Organization

D

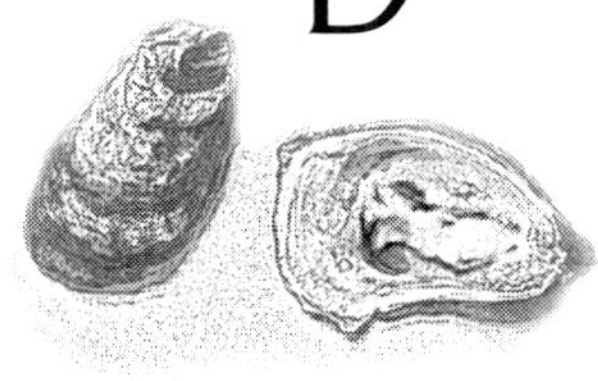

State Law Documents*

U.S. Army Corps of Engineers. 2002. State Program Regional Permit. CENAO-TS-G 97-RP-19. U.S. Army Corps of Engineers, Norfolk, VA. [Online]. Available at: http://www.nao.usace.army.mil/Regulatory/rp-19.htm. [2003, June 19]

Virginia Marine Resources Commission (VMRC). 1998. General Permit #3. 4 VAC 20-336-10 to VAC 20-336-50. Virginia Marine Resources Commission, Newport News. [Online]. Available at: http://www.mrc.state.va.us/garden.htm [2003, June 29]

Virginia Marine Resources Commission (VMRC). 2002. "Pertaining to Importation of Fish, Shellfish or Crustacea," Chapter 4VAC 20-745-10 et seq. Virginia Marine Resources Commission, Newport News. [Online]. Available at: http://www.mrc.state.va.us/fr754.htm [2003, June 29]

Commonwealth of Virginia, Office of the Governor. 2002. Executive Order 23 (2002) Continuation of the Virginia Coastal Resources Management Program Information. Commonwealth of Virginia, Office of the Governor, Richmond.

*Documents are found on the CD-Rom attached to the inside back cover.

Maryland Department of Natural Resources. 2002. Maryland's Coastal Program. Maryland Department of Natural Resources, Annapolis. [Online]. Available at http://www.dnr.state.md.us/bay/czm/coastal_ecosystems.html [2003, July 28].

Virginia General Assembly, House. 1994. House Joint Resolution 450. H.J.R. 450. Requesting the Virginia Institute of Marine Science to Develop a Strategic Plan for Molluscan Shellfish Research and Begin the Process of Seeking Necessary Approvals for In-Water Testing of Nonnative Oyster Species.

Virginia General Assembly, House. 2002. House Joint Resolution 164. H.J.R. 164. Proclaiming Support for the Revitalization of the Virginia Oyster Industry.

E

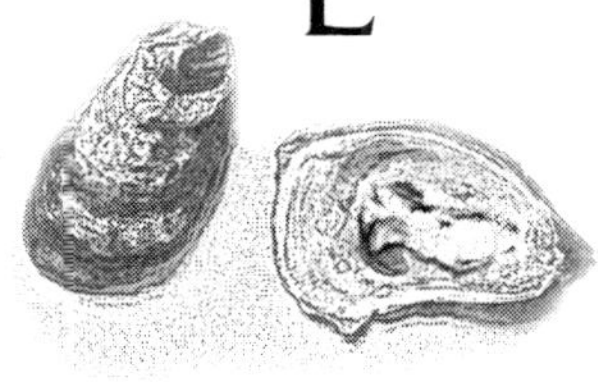

Federal Law Documents*

1999 Presidential Executive Order 13112 on Invasive Species.

18 U.S.C.S. § 42. 2003. 50 Code of Federal Regulations Part 16 (regarding a finding under the Lacey Act).

U.S. Fish and Wildlife Service, Lacey Act Evaluation Criteria, revised March 23, 2001.

U.S. Fish and Wildlife Service Regulations on Injurious Wildlife, 50 C.F.R. Part 16, Injurious Wildlife (Importation or Shipment of Injurious Wildlife, Permits, Additional Exemptions).

Federal Register Vol. 67, No. 144. 2002. Proposed rule (Department of Interior, Fish and Wildlife Service) to amend 50 CFR Part 16.13 to add snakeheads to the list of injurious species.

Clean Water Act Section 117—Chesapeake Bay (33 U.S.C. § 1267).

Chesapeake Bay Program Materials (Chesapeake Bay Program, Annapolis, MD):

*Documents are found on the CD-Rom attached to the inside back cover.

- Chesapeake Bay Program. 1983. Chesapeake Bay Agreement.
- Chesapeake Bay Program. 1987. Chesapeake Bay Agreement.
- Chesapeake Bay Program. 1992. Chesapeake Bay Agreement: 1992 Amendments.
- Chesapeake Bay Program. 1993. Chesapeake Bay Policy for the Introduction of Non-Indigenous Aquatic Species.
- Chesapeake Bay Program. 2000. Chesapeake 2000.

U.S. Army Corps of Engineers, Nationwide Permit 4-67, *Federal Register* 2020, 2063, 2064, and 2078 (January 15, 2002) — Provisions Relating to Shellfish Beds and Shellfish Seeding Activities.

U.S. Army Corps of Engineers Joint Federal/State Public Notice, May 2, 2002.Virginia Seafood Council permit application to introduce 1 million triploid Suminoe oysters.

F

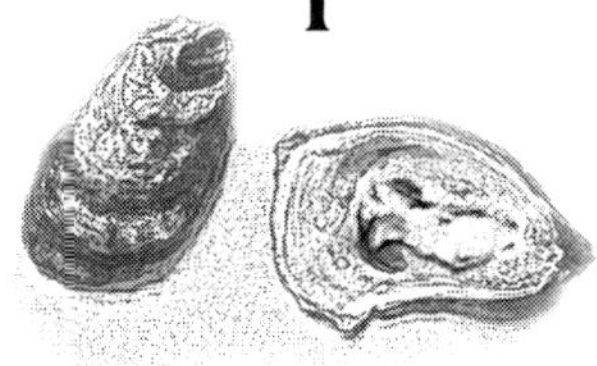

International Law Documents*

International Council for the Exploration of the Sea (ICES). 1964. *Convention for the International Council for the Exploration of the Sea* held in Copenhagen, Denmark, September 12, 1964. 24 U.S.T. 1080; T.I.A.S. 7628. ICES, Copenhagen, Denmark.

International Council for the Exploration of the Sea. 1970. *Protocol to the Convention for the International Council for the Exploration of the Sea* held in Copenhagen, Denmark, August 13, 1970. 27 U.S.T. 1022; T.I.A.S. 8238. ICES, Copenhagen, Denmark. [Online]. Available at: http://sedac.ciesin.org/pidb/texts/exploration.of.the.sea.1964.html. [2003, June 25].

International Council for the Exploration of the Sea. 1994. *ICES Code of Practice on the Introductions and Transfers of Marine Organisms 1994*. ICES, Copenhagen, Denmark.

United Nations, Food and Agriculture Organization. 1995. Article 9 – Aquaculture Development. Pp. 23-26 in *Code of Conduct for Responsible Fisheries*. United Nations, Food and Agriculture Organization, Rome, Italy.

*Documents are found on the CD-Rom attached to the inside back cover.

United Nations. 1982. *Convention of the Law of the Sea* held in Montego Bay, Jamaica. Part XII, Section 1, Art.196. [Online]. Available at: www.un.org/Depts/los/convention_agreements/texts/unclos/closindx.htm. [2003, August 14].

United Nations. 1992. *Convention on Biological Diversity* held in Rio de Janeiro, Brazil on June 5, 1992. Alien species: guiding principles for the prevention, introduction and mitigation of impacts; Articles 5-9.

G

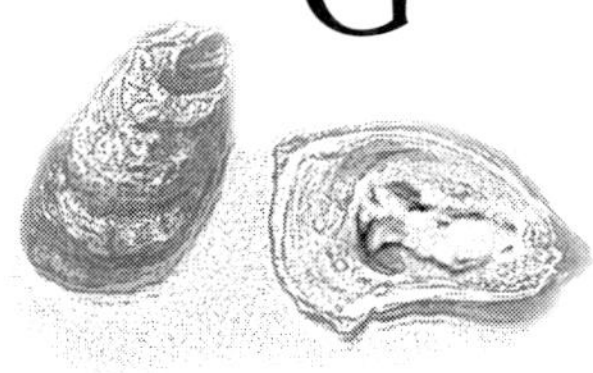

Chesapeake Bay Program Reports*

Chesapeake Bay Program. 2001. *Chesapeake Bay Program Federal Agencies Committee: Recommendations on Suminoe Oyster* (Crassostrea ariakensis) *Aquaculture in Chesapeake Bay*. Chesapeake Bay Program, Annapolis, MD.

Chesapeake Bay Program. 2002. *Report of the Ad-hoc Panel on the Industry Trials of Triploid Non-Indigenous Oyster Species in Waters of the Chesapeake Bay Basin*. Chesapeake Bay Program, Annapolis, MD.

Chesapeake Bay Program. 2003. *Report of the Chesapeake Bay Program Ad-hoc Panel on the Virginia Seafood Council Trials of Genetic Triploid Non-Indigenous Oyster Species in Waters of the Chesapeake Bay Basin*. Chesapeake Bay Program, Annapolis, MD.

---

*Documents are found on the CD-Rom attached to the inside back cover.

H

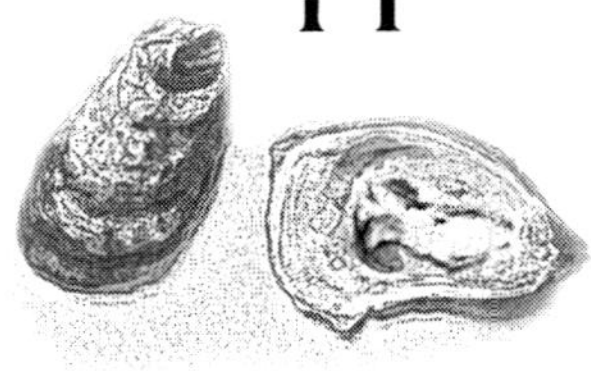

Committee Meeting Agendas

MEETING 1

The National Academies
500 5th Street NW
Conference Room 203
Washington, DC
September 4-5, 2002

WEDNESDAY, SEPTEMBER 4, 2002

OPEN SESSION

11:00 a.m.	**Welcome and Introductions** – Jim Anderson, *Co-Chair*, Dennis Hedgecock, *Co-Chair*, and Susan Roberts, *Study Director*
11:15 a.m.	**Overview of Chesapeake Bay Program Reviews of Non-Native Oyster Proposals** – Fred Kern, *National Oceanic and Atmospheric Administration*
11:45 a.m.	**Discussion**
12:00 p.m.	**Lunch**
1:00 p.m.	**Sponsor Presentation: Environmental Protection Agency** – Mike Fritz

1:15 p.m.	**Sponsor Presentation: National Oceanic and Atmospheric Administration** – Lowell Bahner
1:30 p.m.	**Sponsor Presentation: U.S. Fish and Wildlife Service** – Julie Thompson
1:45 p.m.	**Discussion**
2:15 p.m.	**Break**
2:30 p.m.	**Sponsor Presentation: Maryland Department of Natural Resources** – Frank Dawson
2:45 p.m.	**Sponsor Presentation: Virginia Coastal Program, at Department of Environmental Quality** – Laura McKay
3:00 p.m.	**Discussion**
3:30 p.m.	**Chesapeake Bay Program/Science and Technical Advisory Committee Proposal for Workshop** – Carl Hershner, *Virginia Institute of Marine Science*
3:45 p.m.	**Sponsor Presentation: Maryland Sea Grant** – Frederika Moser
4:15 p.m.	**Discussion**
5:00 p.m.	**Adjourn for day**

THURSDAY, SEPTEMBER 5, 2002

OPEN SESSION

8:30 a.m.	**Breakfast**
8:45 a.m.	**Welcome and Introductions** – Jim Anderson, *Co-Chair*, Dennis Hedgecock, *Co-Chair*, and Susan Roberts, *Study Director*
9:00 a.m.	**Summary of Research on *C. ariakensis*** – Stan Allen, *Virginia Institute of Marine Science*
10:30 a.m.	**Break**
10:45 a.m.	**Discussion**
11:30	**Open session adjourns**

MEETING 2: WORKSHOP

Holiday Inn Select
2801 Plank Road
Fredericksburg, VA 22401
October 7-8, 2002

MONDAY, OCTOBER 7, 2002

OPEN SESSION

8:00 a.m.	**Breakfast**
8:30 a.m.	**Welcome and Introductions** – Jim Anderson, *Co-Chair*, Dennis Hedgecock, *Co-Chair*, and Susan Roberts, *Study Director*
8:45 a.m.	**History of the oyster fishery in the Chesapeake Bay** – Vic Kennedy, *University of Maryland*
9:15 a.m.	**Effects of oysters on water quality** – Roger Newell, *University of Maryland*
9:45 a.m.	**What is the potential for establishment? What are the potential ecological consequences?** – Mark Luckenbach, *Virginia Institute of Marine Science*
10:15 a.m.	**Discussion**
10:30 a.m.	**Break**
10:45 a.m.	**What is the potential for spread? What would be the rate of spread?** – Eileen Hofmann, *Old Dominion University*
11:15 a.m.	**What is the potential for introducing a new disease (human or oyster)? How do you detect viruses?** – Gene Burreson, *Virginia Institute of Marine Science*
11:45 a.m.	**Discussion**
12:00 p.m.	**Lunch**
1:00 p.m.	**Progress with the selective breeding program for *C. virginica*** – Pat Gaffney, *University of Delaware*
1:30 p.m.	***Discussion*** – Stan Allen, *Virginia Institute of Marine Science,* will lead discussion on use of disease-resistant native oysters in restoration
2:00 p.m.	**What is the evidence that restoration has or hasn't worked in MD?** – Ken Paynter, *University of Maryland.*
2:20 p.m.	**What is the evidence that restoration has or hasn't worked in VA** – Roger Mann, *Virginia Institute of Marine Science*
2:40 p.m.	**Discussion**
3:00 p.m.	**Break**
3:15 p.m.	**Brief presentations (15 min) and Panel Discussion: What is the perspective of other states on this introduction?** – Karen Rivara, *East Coast Shellfish Growers Association*
5:00 p.m.	**Workshop adjourns for the day**

TUESDAY, OCTOBER 8, 2002

OPEN SESSION

8:00 a.m.	**Breakfast**
8:30 a.m.	**Welcome and introduction** — Jim Anderson, *Co-Chair*, Dennis Hedgecock, *Co-Chair*, and Susan Roberts, *Study Director*
9:00 a.m.	**Economics** – Doug Lipton, *University of Maryland, Maryland Sea Grant Extension Program*
9:30 a.m.	**Risk analysis modeling**—Eric Hallerman, *College of Natural Resources, Virginia Polytechnic Institute and State University*
10:00 a.m.	**Discussion**
10:30 a.m.	**Break**
10:40 a.m.	**Bill Goldsborough,** *Chesapeake Bay Foundation*
11:00 a.m.	**Larry Simns**, *Maryland Waterman's Association*
11:20 a.m.	**Casey Todd**, *Metompkin Bay Oyster*
11:40 a.m.	**Discussion**
12:00 p.m.	**Lunch**
1:00 p.m.	**George Washington**, *Virginia Watermen's Association*
1:40 p.m.	**Tom Kellum**, *W.E. Kellum Seafood*
2:00 p.m.	**Discussion**
2:30 p.m.	**Break**
3:00 p.m.	**Roundtable on Regulatory Framework** (10 min summaries): Bob Hume, *U.S. Army Corps of Engineers*, Jim Andreasen, *Environmental Protection Agency*, Chris Judy, *Maryland Department of Natural Resources*, Jack Travelstead, *Virginia Marine Resources Commission*, Julie Thompson, *U.S. Fish and Wildlife Service*
4:30 p.m.	**Discussion**
5:00 p.m.	**Public comment session**
5:30 p.m.	**Open session adjourns**

I

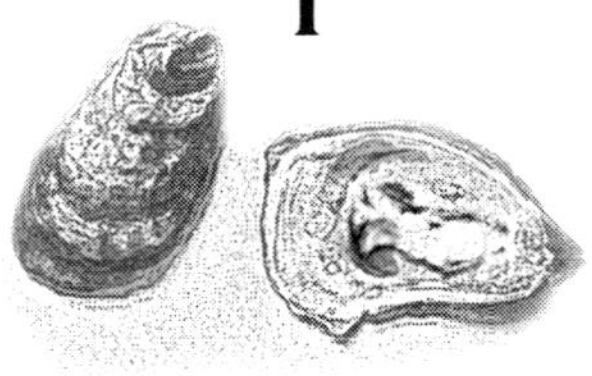

Letters Requesting a Study on Non-Native Oysters in the Chesapeake Bay

Baker, W., Chesapeake Bay Foundation. Letter requesting a study on nonnative oysters in the Chesapeake Bay, 9 January 2002.

Esher, D., U.S. Environmental Protection Agency. Letter requesting the study on nonnative oysters in the Chesapeake Bay, 9 January 2002.

Mikulski, B., U.S. Senate Committee on Appropriations, Subcommittee on VA, HUD, and Independent Agencies. Letter requesting assistance in funding for the study on nonnative oysters in the Chesapeake Bay, 22 January 2002.

Swanson, A.P., Chesapeake Bay Commission. Letter and resolution requesting the study on nonnative oysters in the Chesapeake Bay, 17 January 2002.

CHESAPEAKE BAY FOUNDATION

Environmental Education
Resource Protection and Restoration

January 9, 2002

Ms. Morgan Gopnik, Director
Ocean Studies Board
National Academy of Sciences
2001 Wisconsin Avenue, NW
Washington, DC 20007

Dear Ms. Gopnik:

On behalf of the Chesapeake Bay Foundation, I want to formally request that the Ocean Studies Board undertake at its earliest convenience a study of the oyster *Crassostrea ariakensis* and its proposed introduction into Chesapeake Bay. Recent trials with this non-native species have suggested it would thrive in Chesapeake Bay waters, and while industry support for its use is strong and growing, little is known about the ecological risks that an introduction would carry. Most institutions, agencies and oyster interests in the Chesapeake area agree that a responsible resolution of this issue should start with a thorough and independent technical review. The National Academy of Sciences is best suited to carry out this work.

Oyster restoration is considered one of the most important aspects of the broad effort to save the Chesapeake Bay. Substantial public support exists for various programs to rebuild reef habitat and restock the native oyster, *Crassostrea virginica*. However, *C. virginica* experiences high disease mortality in much of the Bay and will require many years of stocking and selective pressure to develop disease resistance before large reef populations are realized. Recent research by the Virginia Institute of Marine Science suggests that *C. ariakensis* grows faster and survives disease better than the native oyster. As a result, many in the oyster industry advocate using this non-native oyster to support either an aquaculture industry based on sterile triploids or a public fishery based on naturalized populations of reproductive animals released into the Bay. The potential economic benefits are great for this depressed industry, and pressure on public officials to move in this direction is building. Indeed, the prospect of a hardier oyster as a tool for Bay restoration is tantalizing.

Philip Merrill Environmental Center
6 Herndon Avenue, Annapolis, Maryland 21403 • 410-268-8816, fax 410-268-6687

The introduction of a non-native species is not a trivial act, and it carries potential repercussions for the entire east coast. At the very least, public officials must have reasonable assurance that adverse impacts are unlikely or are far outweighed by potential benefits. Unfortunately, the current state of knowledge about this oyster is insufficient to support this conclusion. It is not known, for example, whether *C. ariakensis* would out-compete and eventually eliminate the native oyster, *C. virginica*. Even using accepted protocols for introduction, it cannot be said at this time what the risk would be that an exotic virus would be introduced that could have disastrous affects on other species such as the blue crab, currently our most valuable fishery.

The need for an independent technical study, therefore, is to describe the state of our knowledge of *C. ariakensis*, identify key gaps for which research would be required, and assess the risks inherent in utilizing this species to support Chesapeake Bay fisheries. Under the circumstances of impressive initial results with this oyster and resultant political pressure for introduction, the need is urgent. Accordingly, we ask you to give the earliest possible consideration to a study by the Ocean Studies Board to evaluate these issues.

With great appreciation for your time, I am,

Sincerely,

William C. Baker
President

Many thanks for your consideration!

Bill

Feb-15-02 10:43am From-NRCTASKFORCE 2023342493 T-391 P.02/03 F-986

UNITED STATES ENVIRONMENTAL PROTECTION AGENCY
REGION III
Chesapeake Bay Program Office
410 SEVERN AVENUE
ANNAPOLIS, MARYLAND 21403

FEB 4 2002

JAN 9 2002

Dr. Bruce Alberts
President, National Academy of Sciences
2101 Constitution Avenue, NW
Washington, D.C. 20418

Dear Dr. Alberts:

As Acting Director of the U.S. Environmental Protection Agency's Chesapeake Bay Program Office, I am requesting assistance from the National Academy of Sciences on issues related to the possible establishment of non-native oyster aquaculture in the waters of the Bay. Currently, the species most discussed for such introduction is the Suminoe oyster, or *Crassostrea ariakensis*.

Oyster management in Chesapeake Bay recently has received considerable attention from established research institutions, state and Federal agencies and the public in the Bay region. In the *Chesapeake 2000* agreement, the Chesapeake Bay Program partners (the states of Virginia, Pennsylvania, and Maryland, the District of Columbia, the Chesapeake Bay Commission, and the U.S. Environmental Protection Agency representing the Federal government) committed to "... a ten-fold increase in native oysters in the Chesapeake Bay ..." by 2010. The partners are developing a strategic plan to reach that objective. One of the significant considerations in that plan must be the high impact of disease mortality on native oyster stocks, as well as the appropriate management of harvest mortality.

Because oyster disease is so prevalent in the Bay, the Virginia General Assembly asked the Virginia Institute of Marine Sciences to evaluate the possible role of disease resistant non-native oysters in the revitalization of the oyster industry. Now, based on market-oriented tests, it appears that the non-native Suminoe oyster (*Crassostrea ariakensis*) could resist the diseases decimating the native oyster, grow faster, and fare well in the market. Indeed, the Virginia Seafood Council is developing a proposal to initiate commercial aquaculture production of *C. ariakensis* in the Bay within the next couple of years, starting with non-reproductive triploid organisms.

Anticipating a proposal to initiate *C. ariakensis* aquaculture in the Bay, the Chesapeake Bay Program's Federal Agencies Committee and others have raised concern over the lack of scientific knowledge which would be necessary for any agency to make an informed decision on such a proposal. For example, there appears to have been little or no study of the organism's

Printed on Recycled Paper

ecology, including disease hosting, interspecific interactions, potential for hybridization, reef-building capacity, etc. Also, although we are forming an ad hoc review panel comprised of representatives of the Chesapeake Bay Program partners, no independent assessment of the risks posed by the initiation of non-native oyster aquaculture in the waters of the Bay has occurred. At a minimum, we need to have the following broad questions answered:

1. How effective are institutional, technological, and infrastructure controls to prevent the accidental release of reproductive oysters from aquaculture facilities?

2. If there is an escapement of reproductive *C. ariakensis* into the Bay, how will it compete with the native *C. virginica?*

3. If *C. ariakensis* were to escape and reproduce, how will that affect the natural ecosystem and habitat of the Chesapeake Bay?

4. What is the potential of the Suminoe oyster serving as a reservoir for MSX and Dermo or as a vector for new oyster diseases?

Therefore, recognizing the regional significance of this issue (our actions could affect the ecology of estuaries up and down the U.S. Atlantic Coast), the dire need for guidance in identifying research that would be essential to support an informed decision, and our interest in obtaining an independent evaluation of research results, risk assessment needs, and oyster management options, I am requesting your consideration of a National Academy of Sciences evaluation of these issues. If you are agreeable to considering such an evaluation, I suggest that we could convene a small meeting of interested parties within the next few weeks.

I look forward to hearing from you at your earliest convenience. You may contact me at (410) 267-5709 or Mike Fritz at (410) 267-5721.

Sincerely,

Diana Esher

Diana Esher
Acting Director

01/22/2002 TUE 16:35 FAX ☑002

United States Senate
COMMITTEE ON APPROPRIATIONS
WASHINGTON, DC 20510–6025
www.senate.gov/~appropriations

January 22, 2002

The Honorable Christie Whitman
Administrator
U.S. Environmental Protection Agency
1200 Pennsylvania Avenue, NW
Washington, DC 20460

Vice Admiral Conrad C. Lautenbacher, Jr.
Under Secretary and Administrator
National Oceanic and Atmospheric Administration
U.S. Department of Commerce
14th Street and Constitution Avenue, NW
Washington, DC 20230

Dear Administrator Whitman and Admiral Lautenbacher:

I am writing to request that the Environmental Protection Agency (EPA) and the National Oceanic and Atmospheric Administration (NOAA) help fund a study by the National Academies of Sciences (NAS) on the introduction of non-native species oysters into the Chesapeake Bay.

As you know, oyster restoration is one of the most important aspects of efforts to save the Chesapeake Bay. Oysters are critical to both restoring the Bay's ecosystem and ensuring the economic survival of the Bay's commercial fisheries. That is why I have consistently worked to provide federal resources, including EPA and NOAA funding, for programs to rebuild reef habitat and restock the native oyster, *Crassostrea virginica*. These efforts have met with great success.

Now, recent research suggests that a non-native species, *Crassostrea ariakensis*, may grow faster and survive disease better than the native oyster. Supporters of saving the Bay, including the oyster industry, the watermen, and environmental organizations, agree that we need further information about the potential benefits and impacts of introducing this species into the Bay.

To that end, I request that EPA and NOAA help fund a study by the NAS to further examine these issues. The study is expected to cost approximately $200,000 and be complete within nine months. Representatives of the oyster industry, the watermen, and environmental organizations are currently working with NAS to identify sources of funding for this study, including state and federal resources, and I urge EPA and NOAA to participate in funding this study.

Thank you for your attention to this request. I look forward to working with you on this important issue.

Sincerely,

Barbara A. Mikulski

Barbara A. Mikulski
Chairman, Subcommittee on VA, HUD
and Independent Agencies

Chesapeake Bay Commission

A tri-state legislative assembly

January 16, 2002

Ms. Morgan Gopnik, Director
Ocean Studies Board
National Academy of Sciences
2001 Wisconsin Avenue, NW
Washington, DC 20007

Dear Ms. Gopnik:

I am writing on behalf of the Chesapeake Bay Commission to request that the Ocean Studies Board undertake a study of the oyster *Crassostrea ariakensis* and its proposed introduction into Chesapeake Bay. To amplify the gravity with which we make this request, I am enclosing a resolution adopted by the Commission at its January meeting.

In case you are not familiar with our organization, the Commission is a 21-member tri-state legislative commission created in 1980 to advise the members of the General Assemblies of Maryland, Virginia and Pennsylvania on matters of Baywide concern. The members, mostly legislators from the three states, are responsible for coordinating Bay-related policy across state lines and developing shared solutions.

Oyster restoration is a centerpiece of the Bay restoration effort. Efforts to restore our native species are in full swing, yet despite our reef rebuilding and restocking efforts, the high disease mortality in much of the Bay has hampered progress to restore the native oyster, *Crassostrea virginica*.

Several years ago, the Virginia General Assembly requested the Virginia Institute of Marine Sciences (VIMS) to analyze the viability of growing a non-native oyster in Chesapeake waters. Their tests of *C. ariakensis* revealed that the species appears to thrive in the saltier waters of the Bay, grows quickly, compares favorably in taste to the native oyster, and is highly resistant to the oyster diseases common in our region.

While industry support for its use is fervent and mounting, little is known about the ecological risks that are posed by the introduction of *C. ariakensis* in the Chesapeake Bay. There is near universal agreement among Bay-region leaders that a thorough and independent technical review of this proposed introduction is needed. We believe that the National Academy of Sciences is best suited to carry out this work.

60 West Street, Suite 200 ▪ Annapolis, Maryland 21401 ▪ 410-263-3420 ▪ Fax 410-263-9338

Introducing a non-native species is not a decision to be taken lightly. An independent technical study is needed to describe the state of our knowledge, identify research priorities, and assess the risks inherent in utilizing this species to support Chesapeake Bay fisheries.

The Commission, therefore, requests that the Oceans Studies Board make *C. ariakensis* an immediate priority for study. The Chesapeake Bay Program has liberated $50,000 towards the funding of such a study. Should you agree, we will do everything possible to identify additional sources of money.

Sincerely,

Ann Pesiri Swanson
Executive Director

APS:pc
Enclosure

Chesapeake Bay Commission

RESOLUTION

A legislative commission serving Maryland, Pennsylvania and Virginia.

Resolution Supporting a National Academy of Sciences Evaluation of the Benefits and Risks of *Crassostrea ariakensis* in the Chesapeake Bay

Adopted January 4, 2002

WHEREAS, the decline of native populations of the Eastern oyster, *Crassostrea virginica*, in the Chesapeake Bay and other Atlantic coast estuaries has led the Commonwealth of Virginia, as well as scientific and fisheries industry organizations, to examine the possible role of disease-resistant, non-native oysters in the revitalization of the oyster industry, and

WHEREAS, trials in Virginia using sterile, triploid individuals of an Asian oyster species, *Crassostrea ariakensis*, have shown that it is more resistant to the diseases decimating the native oyster, grows faster, and could fare well in the market, and

WHEREAS, proposals to initiate commercial aquaculture production of *C. ariakensis* in the Bay, using non-reproductive triploid organisms, are forthcoming, and

WHEREAS, possible risks associated with unintentional or intentional establishment of reproductive populations of *C. ariakensis* have been raised, and

WHEREAS, concern has been expressed among many that current scientific knowledge is inadequate to effectively evaluate proposals for the initiation of non-native oyster aquaculture, and

WHEREAS, the history of undesirable consequences of many introductions of marine organisms necessitates a thorough assessment of the impacts (both beneficial and detrimental) of reproducing populations of *C. ariakensis*, and

WHEREAS, the Commission recognizes the regional and national significance of this issue, and the sense of urgency in addressing issues surrounding *C. ariakensis* aquaculture and introduction due to the economic, environmental, and social factors involved;

60 West Street, Suite 200 Annapolis, Maryland 21401 (301) 263-3420

NOW, THEREFORE, BE IT RESOLVED, that the Chesapeake Bay Commission requests that the National Academy of Sciences conduct a study to review current research, evaluate risks and benefits associated with the use of non-native oysters in the Bay, and prioritize additional research needed before a responsible decision on C. ariakensis can be made.

BE IT FURTHER RESOLVED, that this evaluation should, at minimum, address the following questions:

1. What are the probable consequences, if any, of introductions of populations of *C. ariakensis*, either intentional or unintentional, in Atlantic coastal environments?
2. What are the potential risks, if any, of establishment of reproducing populations?
3. What procedures could be taken to minimize those risks?
4. What is the most important research required to reduce the uncertainties in these predictions?

BE IT FINALLY RESOLVED, that the Chesapeake Bay Commission requests that the Academy conduct this study in a time-sensitive manner, if possible within one year.

J

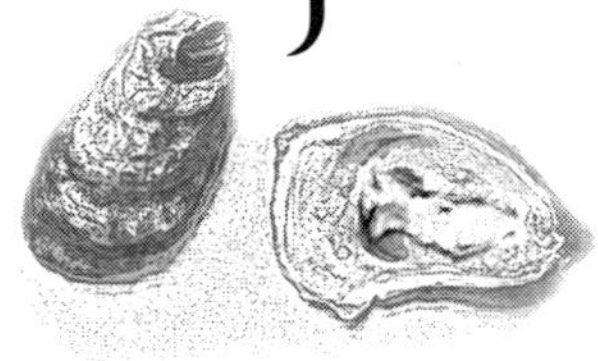

Glossary

Amenity benefits/services – utility (satisfaction, pleasure) derived from sportfishing and other recreation activities or from other activities based on the environment such as experiencing or contemplating scenic beauty, wildlife, or a healthy ecosystem.

Broodstock – adult animals that are spawned to provide larvae for hatchery production.

Clean list – a regulatory approach that prohibits introduction of a nonnative species unless it is included on a list of approved (i.e. not harmful) species.

Cultchless – a technique for growing oysters individually, usually for the half-shell market. The oyster larvae are induced to settle on loose material such as ground shell rather than whole shells.

Diploid – refers to animals whose cells contain two sets of chromosomes, the normal genetic state for oysters.

Dirty list – a regulatory approach that prohibits introduction of species identified as unacceptable and allows introduction of unlisted species.

Dockside value – dollar amount received by watermen at the dock.

Epizootic – a transient disease event in an animal population.

Eutrophication – nutrient enrichment of water bodies, generally referring to elevated levels of nitrogen and phosphorus.

Host – organism in which a parasite or other infectious agent lives.

Infection – presence of a parasite in a host, with or without the development of disease.

Invasive species – a nonindigenous organism that spreads from the site of introduction, becomes abundant, and may displace native species.

Mosaic – animal containing both diploid and triploid cells.

Nominal value – value in current dollars.

Nonindigenous – species found outside its natural geographical range. Also referred to as alien, nonnative, or exotic.

Pathogen – disease-producing organism.

Real price – nominal price adjusted for inflation.

Reversion – production of normal diploid cells in an otherwise triploid animal.

Rogue introduction – a non-sanctioned, direct release of diploid reproductive oysters.

Sed quis custodiet ipsos custodes? – Latin phrase translated as: "Who is to guard the guards themselves?"

Seed – a young oyster, especially one suitable for transplanting to another bed.

Skipjacks – sail-powered wooden vessels native to the Chesapeake Bay that are used for commercial dredging of oysters.

Sociocultural – learned knowledge, values and behaviors that are shared among members of a group, community or region.

Spat –juvenile oysters from the time of settlement through the first year of growth.

Spatfall – the settlement of juvenile oysters onto a substrate.

Triploid – refers to animals whose cells contain three sets of chromosomes rather than the normal two sets. This condition inhibits the ability of the animal to make viable eggs or sperm, reducing fertility to a small percentage of the reproductive capacity of a normal diploid animal.

Zoosanitary –clean and healthy conditions for animal husbandry.

K

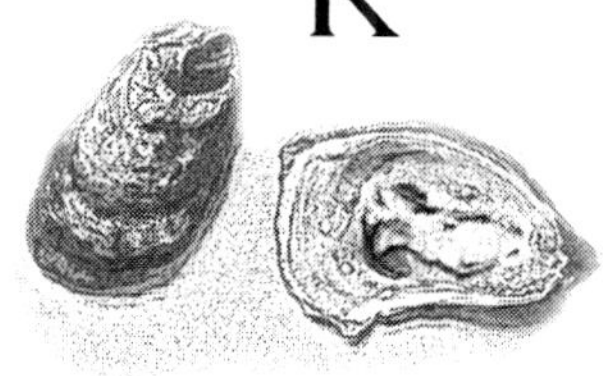

Scientific and Common Names

Scientific Name	Common Name
OYSTERS	
Crassostrea angulata	Portuguese oyster
Crassostrea ariakensis (=rivularis, discoidea, paulucciae)	Suminoe oyster
Crassostrea gigas	Pacific or Japanese oyster
Crassostrea rhizophorae	Mangrove oyster
Crassostrea sikamea	Kumamoto oyster
Crassostrea virginica	Eastern oyster
Ostrea angasi	Australian flat oyster
Ostrea denselamellosa	Asian Milin (meaning densely scaled) oyster
Ostrea edulis	European flat oyster
Ostrea puelchana	Argentinean flat oyster
Ostreola conchaphila	Olympia oyster
Saccostrea glomerata (=commercialis)	Sydney rock oyster
Tiostrea chilensis (=lutaria)	New Zealand flat oyster
OTHER SHELLFISH	
Ceratostoma inornatum	Japanese oyster drill
Corbicula fluminea	Asian clam
Crepidula fornicata	slipper shell
Cyclope neritea	snail

Dreissena polymorpha	zebra mussel
Macoma balthica	Baltic macoma clam
Macoma spp	macoma clam
Mercenaria mercenaria	Atlantic hardshell clam or quahog
Modiolus americanus	American horse mussel
Mulinia lateralis	dwarf surf clam
Mya arenaria	softshell clam
Mytilus edulis	blue mussel
Mytilus galloprovincialis	Mediterranean mussel
Palaemonetes pugio	grass shrimp
Potamocorbula amurensis	Amur River or brackish-water corbula
Rangia cuneata	Atlantic rangia clam
Ruditapes decussatus	grooved carpet shell clam
Ruditapes (=Venerupes) philippinarum	Manila clam
Tagelus plebeius	stout razor clam
Urosalpinx cinerea	Atlantic oyster drill

OYSTER PARASITES AND PATHOGENS

Bonamia ostreae	oyster parasite
Haplosporidium armoricanum	oyster parasite
Haplosporidium costale	agent of SSO disease
Haplosporidium nelsoni	agent of MSX disease
Marteilia maurini	oyster parasite
Marteilia refringens	agent of Aber disease
Marteilia sydneyi	agent of QX disease
Mikrocytos mackini	oyster parasite
Mikrocytos roughleyi	agent of Australian Winter disease
Mytilicola intestinalis	European copepod
Mytilicola orientalis	Japanese parasitic copepod
Nodardia crassostreae	oyster bacterium
Ostracoblabe implexa	oyster shell fungus
Perkinsus marinus	agent of Dermo disease
Polydora websteri	shell-boring polychaete
Quahog Parasite X	agent of QPX disease
Vibrio splendidus	marine bacterium
Vibrio tapetis	agent of Brown Ring Disease

FLATWORMS

Pseudostylochus ostreophagus	Japanese flatworm
Stylochus ellipticus	oyster leech (flatworm)

PLANTS

Caulerpa sp.	Mediterranean seaweed
Chaetoceros calcitrans	diatom
Eichornia crassipes	water hyacinth
Sargassum muticum	Japanese seaweed
Spartina alterniflora	smooth cord grass
Thalassia testudinum	North American seagrass
Zostera japonica	Japanese eelgrass
Zostera marina	eelgrass

FINFISH

Chasmodes bosquianus	striped blenny
Gobiesox strumosus	skilletfish
Gobiosoma bosc	naked goby
Hypsoblennius hentz	feather blenny
Morone saxatilis	striped bass
Opsanus tau	oyster toadfish